AF616755

Studies in Biotechnology Series No. 6

Plant Biotechnology and Biodiversity Conservation

U. Kumar

A. K. Sharma

2005

Agrobios (India)

Dr. U. Kumar

A. K. Sharma
Scientist (Agronomy)
Central Arid Zone Research Institute
Jodhpur

Published by:

Agrobios (India)
Agro House, Behind Nasrani Cinema
Chopasani Road, Jodhpur 342 002
Phone: 91-0291-2642319, Fax: 2643993
E. mail: agrobios@sify.com

First Published 2001
Reprinted: 2005

ISBN : 81-7754-95-5
Rs. 795.00/ US $ 53.00

Published by: Dr. Updesh Purohit for Agrobios (India), Jodhpur
Laser typeset at: Yashee Computers, Jodhpur
Cover Design by: Reena
Printed at: Bharat Printers (Press), Jodhpur

Preface

Biodiversity means the diversity of life. Usually, the measure of biodiversity is genetic diversity, a measure of how many distinct genetic species there are in an area or population. These are sometimes called genetic resources, because they can be used as the starting point for new products or processes. Biodiversity is used in two contexts: the diversity of natural organisms, and the diversity of farmed crops or animals.

In the larger world, biodiversity concerns the vast range of plants (and animals, although these are considered less important from a biotechnological point of view) currently alive. Many of them may produce something useful to humans: a new drug, a new foodstuff, a new material. If they are allowed to be wiped out (and most species of plants live in the tropics, which are now under substantial threat) that potential will be lost for ever. There is substantial concern that the biodiversity of Third World (i.e. not rich) countries is being exploited by rich ones. The biodiversity in India i.e. forests, grasslands, wetlands mountains, desert and marine ecosystems face many pressures. One of the major causes for the loss of biological diversity in India has been the depletion of vegetative cover in order to expand agriculture.

On the farm, biodiversity is considered to be a good thing. If a country plants only one type of crop (for example), then a pathogen can sweep the entire country's crop from the fields. Thus, not having a country plant just a single type (or cultivar) of a crop is protection against pandemics.

The role of biotechnology in this is double-edged. If biotechnology produces a new, wonder-wheat, then that will be planted in place of all other cultivars, and the wheat-growing world will end up being a monoculture, biodiversity will have been reduced. on the other hand, biotechnological methods are such that if we can transform one cereal with a gene you can transform a lot more, so biotechnology could actually increase biodiversity by increasing the number of crops with desirable genes in them. The green revolution, including biotechnology, has been so successful that farmers are no longer under pressure to grow monocultures of the most productive crops.

Authors

Contents

1

Plant Biodiversity : Principles and Conservation

This universe is the creation of the supreme power meant for the benefit of all His creations. Individual species must, therefore, learn to enjoy its benefits by forming a part of the system in close relation with other species. Let not only one species encroach upon the other's rights.

(*Source*: **Ishopanishad**)

Biodiversity is the natural biological capital of the earth

Anon.

Native biodiversities a source of pride for each country, composing as it does a shining part of the national heritage.

Anon.

Our National Food Security depends on our ability to conserve all out biological wealth.

Dr. M.S. Swaminathan

The theme for the 13th World Food Day this year is *Harvesting Nature's Diversity*. Conserving biodiversity for its sustainable use by the present and future generations is integral to UNCED's Agenda 21 and the International Convention on Biological Diversity signed by more than 150 of the nations that attended the Earth Summit in 1992. Since our food and environmental security depends upon the diversity of life on Earth. The contribution of biodiversity to the welfare of our society and also its role in sustainable agriculture and rural development, food security, environmental management and international trade in commodities. Harvesting Nature's Diversity will also set stage for the 1995 International Technical Conference on Genetic Resources to be convened by FAO in Rome.

The world is inhabited by myriads of life forms, animal and plants. These life forms are of great diversity, living in diverse habitats and possessing diverse qualities, which in themselves make very interesting studies. Moreover, these life forms are vital to human survival as they provide food and materials for shelter, clothing, tools and medicine.

In the present day world, many species of plant and animals are being destroyed at a rapid rate. The classic example is indiscriminate logging which destroys not only trees but also other forms of life, as forests harbour many animals and plants that depend on the environment for survival. In the past we placed great importance on the selection of species, protecting and promoting the varieties with high yields and ignoring other "inferior" varieties. The result is a great loss in biological diversities with unlooked-for disastrous effects. The dangers from the loss of biodiversity, especially the threat to agriculture, are given in details in the document being distributed to you today. This loss is all the more regrettable as we posses nowadays modern technologies which can utilize the different genetic qualities in a variety of ways. Some species of a plant may have high yield but are unappetizing; some are both low yielding and tasteless, but these species are hardy and can grow in saline soil. Through modern biotechnologies, wild diversity can also be incorporated into crops and contribute to world agriculture development. We can now extract chemical substance not known to us before from various plants for use in medicine or industry.

Such being the case, the problem before us today is how to conserve what remains of natural biodiversity for future generations?

Variation is the law of nature. It occurs everywhere and every moment. The variations take place at micro levels at short space and small time period, but these become apparent only over a large space and big a time gap. The variations may be linear or cyclic. The variety and variability of organisms and ecosystems is referred to as biological diversity or biodiversity. Similarly, the biological variations initiate at the micro level (bio-molecular level or genes) and become apparent at species and ecosystem level. The biological variations in nature over time and space form the basis of evolutionary processes. Thus, *biodiversity is the degree of variety in nature and not nature itself.* Similarly, the biological diversity is not the same as biological resource although mutually, they form part of each other. The conservation of biological diversity is distinct but related to biological resources. It is difficult to discretely distinguish the value of biodiversity as a distinct characteristic of biological resources (IUCN *et al.* 1990).

Biological diversity means the variability among living organisms from all sources including inter alia, terrestrial, marine and other aquatic ecosystems and the ecological complexes of which they are part. Biodiversity provides the basic biotic resource that sustains the human race. This includes diversity within species, between species and of ecosystems. Biodiversity is the most significant national asset and constitutes an enduring resource for supporting the continued existence of human societies. Biodiversity is not merely a natural resource, it is an embodiment of cultural diversity and the diverse knowledge traditions of different communities across the world. It includes diversity of forms, right from the molecular unit to the individual organism and then on to the population, community, ecosystems and biospheric levels.

The World Commission on Environment and Development (WCED) constituted by the General Assembly of the United Nations published its report in 1987 which provided a major boost and endorsement to the need for conserving the world's rich biodiversity, particularly that of the tropical areas. Despite conflicting views among nations, a broad consensus was reached after bitter negotiations, and 170 countries signed the Biodiversity Convention, which is now ratified by 104 countries (42 Articles were adopted). One of the prerequisite tasks as expressed by Article 7 of the Convention is the identification and monitoring the components of biological diversity. Article 12 calls for Research and Training and suggest programmes for identification, conservation and sustainable use of biological diversity.

ETHICS OF BIODIVERSITY LOSS

Last, but not the least, is the fact that human race has taken upon itself an arrogant role, as though *Homo sapien* is the only species and other species do not matter. In essence the biodiversity loss is the result of undue human interference and a situation has been created where human race is pitehed against all other life. Non-violence is the highest *Dharma.* This was preached by Buddha and, in recent times, by Mahatma Gandhi and Martin Luther King. In fact, Gandhi has clearly mentioned time and again that *since human being has no power to create life, therefore, he has no right to kill life.* The western philosophers have come to the conclusion that present environment problems (including biodiversity loss) are an outward expression of a crisis in the mind and spirit of human being.

The ethics of biodiversity loss is now a major consideration for saving biodiversity. This is particularly, so because such loss is forever and there is no way to recreate lost biodiversity.

A few things stand out very clearly:

- Realism demands that in the area of biodiversity there will always be tremendous interdependence between gene-rich/technology-poor and gene-poor/technology rich countries. With the global climate changes on the horizon, there are likely to be newer challenges, candidate genes will have to be identified to evolve new cultivars resistant/tolerant to such changes. Such cooperation between North and South must increase if human race has to succeed. No nation can be self-sufficient in biodiversity. There would always be newer demands.
- Realistic national policy on conservation and utilization of biodiversity have to be evolved by a National Biodiversity Conservation Board in the best interest of each country.
- *In situ* and *ex situ* forms of conservation are both important and mutually reinforcing and supportive. It is high time that FAO gets involved in *in situ* conservation particularly through genetic reserves.
- Equally important is to support and build biotech capability in the South by biotech-rich North. This would be to the mutual benefit of all countries.
- Bridges need to be built with NGOs and women's groups regarding on-farm conservation and utilization by involving communities.
- There is considerable interdependence between industrial (gene-poor/technology rich) and developing (gene-rich/technology poor) countries. Biodiversity is a fit case for North-South co-operations – particularly skills, knowledge and trainings should be imparted by North to South as far as biotechnology is concerned.
- The spectrum of biodiversity ranges from DNA to individual, to population of a species, to community to ecosystems and the whole thing is a continuum.
- There has been narrowing of species base: from a large number of species, that early human beings used, to wheat, rice, maize and potato accounts for over 60 per cent of calories and protein intake by the human race. Furthermore, there has also been a narrowing of genetic base in each species resulting in a few cultivars being cultivated on a large scale.

- In order to meet the food needs for escalating population, there is need for conservation and utilization of genetic resources coupled with conservation of soil and water, integrated nutrition and pest management and crop diversification.

- Bridges have to be built between scientific and technical groups working in the area of conservation and utilization of biodiversity and the local communities and women's groups. The latter have considerable local technical knowledge and wisdom that need to be utilized by the farmer.

PLANT GENETIC RESOURCES

The theories of source material and origin of cultivated plants must be fundamental for scientific plant breeding. N. I. Vavilov has often been considered as the father of plant genetic resource activities. He proposed that crop improvement should draw from wide genetic variation. For this he collected plant and their wild relatives from most parts of the world. These provided gene pool from which cultivars could be bred which could be suitably adapted to cultivation. Vavilov revealed that genetic variation in cultivated species was concentrated in certain regions of the world which he termed *Centers of diversity.* He identified a number of these centers which have been the principal area from which breeders have collected raw material for their work. In recent years, however, the radical change which have taken place in modern agriculture have presented a formidable threat to the diverse resources held in the vavilonean centers.

Importance of Genetic Diversity in Crops

Every major crop that is grown in the field has a very narrow genetic base and contributes to their *genetic vulnerability*. The value of local land races of long-established cultivars and the gene which they may hold, often remain unappreciated until new foreign cultivars are introduced into a region. Therefore, the value of local gene pool for the development of regionally adapted cultivar cannot be under estimated.

History of Plant Genetic Resources Conservation

Food and Agricultural Organization (FAO) of the United Nations Organization made first efforts on plant exploration in 1961. FAO placed on a formal footing in 1965, and panel on Forest Gene Resources was established in 1968. At the same time FAO formed its crop Ecology and Genetic Resources Unit. During this early period of activity. FAO conducted international technical conference under the auspices of FAO and the

International Biological Programme (IBP). The real work began with establishment of International Board for Plant Genetic Resources (IBPGR) by Consultative Group on International Agricultural Research (CGIAR). Now IBPGR works in close association with FAO, and its primary function lies in the development and administration of the International plant genetic resources conservation network. The major tasks of the Board was the promotion, collection, conservation and evaluation of plant genetic resources of species of major economic importance and their wild relatives. It has defined over 50 priority crops needing urgent action and priority regions requiring survey. The board made efforts to establish the standards, methods and procedures for exploration, evaluation and for conservation of stocks for both seeds and vegetative material, to be applied particularly in the network of replicated storage centers of gene banks.

The Genetic Variability of Crop Plants

The pattern of genetic diversity seen in crop plants results from the interaction of following important factors:

Gene mutation	Enhance variation in natural populations.
Migration Recombination	Migration can increase variation in populations by adding allied genetic material
Selection	Selection is the process by which undesirable genes or gene combinations are eliminated from populations. Act to reduce variation.
Genetic drift	It Act to reduce variation and covers two important concepts: (A) The random change in genetic equilibrium in plant populations, especially small populations. (B) The founder principle which is an important feature of crop evolution

The reproductive biology of the plants also play key role in plant variation. Variation in outbreeders is greater because of heterozygosity at different loci, but in inbreeders homozygosity reaches a high level. A successful gene combination will reproduce more frequently. So there is some degree of differential fertility in populations between different genotypes.

PLANT GENETIC RESOURCES EXPLORATION

There are various sources from which the entire range of germplasm for any major crop can be assembled. Gene pools may be indigenous, introduced or locally developed. The various resources that need to be conserved may be categorized as follows.

Source (Raw material of the plant breeder)	Arability
Wild relatives	Wild relatives of crops share common ancestors with crops but have remained in the wild as products of nature. They are termed domesticates, because they cannot survive in nature
Land-races and Primitive cultivars	Local crop varieties developed in primitive agricultural systems, rather than being deliberately bred, farmers selected them over many generations.
Obsolete cultivars	Obsolete cultivars, left over from the early days of plant breeding, are now mainly found in germplasm collections.
Advanced breeding line, Mutations and other product of plant breeding programmes	Advanced breeding lines and stocks are plants which have been developed by breeders for use in modern scientific plant breeding. They include cultivars not yet ready for release to farmers.
Modern cultivars	The high-yielding elite cultivars have been developed by scientific plant breeding for modern intensive agriculture.

Genetic Resources Exploration

The exploration and collection of plant genetic resources need sound scientific principles and knowledge. A botanical collector and the plant genetic resources collector are not the same. A botanical collector look for uniformity or trueness to type, while plant genetic resources collector look for the fullest possible recovery of genetic variation of species irrespective of the relative frequency of rarity of ay genes or linked genetic complexes. An optimum sampling strategy should enable the collector to obtain, with 95% certainty, all the alleles occurring in the population at a frequency greater than 5%. In order to achieve this, it is desirable to understand something of the population structure of a species and its eco-geographical distribution.

Sampling Strategies

These are based on an extensive knowledge of the pattern of genetic variability of populations. In nature extensive geographical variation is factors are a major determinant of genetic diversity, and many species display differentiation into ecotypes in response to their habits. All environmental factors including agricultural environment, and agro ecotypes are most clearly distinguished in primitive cultivar. Climatic factors such as maximum and minimum average temperature, precipitation and the seasons of dormancy and growth, light intensity and day length are all reflected in corresponding developmental characteristics (Bennett, 1970).

Geographical variation pattern include, such characteristics as disease resistance morphological features and other conspicuous differences as well as variation in quantitative characters. The relevance of this to plant breeding is obvious, for although varieties or strains may look alike, they may differ greatly in factor useful to the plant breeder, especially in differences in physiological characters. Allard (1970) indicated that variations in populations are due to much genetic variation and heterozygosity. Sampling method must be used to ensure the collection of representative within-population variation as well as that associated with geographical patterns of variation.

Seed Crops

Bennett and Allard indicated that, in most circumstances, sample size should not exceed 50 plants per population and in no circumstances it is desirable to collect more than 100 plants per population. In situation where populations are not so large as in the case of cereal, it may not be possible to collect a large sample. The collection of samples which are too small carries with it the great problem, of genetic drift upon consequent immediate regeneration of the seed. Larger samples enable at least a proportion to be put into long-term storage right away, avoiding the need for immediate regeneration. Sampling is of two types:

Non-Selective	It is applicable to wild vegetation, as used in ecological studies, but could not be utilized in a field of primitive wheat, for instance, without causing much damage.
Selective sampling	In selective sampling phenotypes which are distinguishable, where in terms of a morphological character, or the manifestation of physiological character, such a drought tolerance or disease resistance are collected separately. Only a bulk sample should be collected from each population.

Vegetatively-Propagated Crops

It is applicable for those crops which are not sexually fertile and do not produce seed, therefore, vegetative propagules like rhizomes, corms, bulbs, tubers and roots are collected. It cannot be assumed that the population structure of a vegetatively propagated crop conforms to the same pattern as a seed crop. Each plant provides only a single type e.g., potato. The following are the problems with vegetatively propagated materials:

1. It is bulky and difficult to transport.
2. It is difficult to keep alive.
3. It must be collected at the right stage of maturity or otherwise it will not grow.
4. The plant propagules are produced underground, therefore, no plant is visible at maturity.
5. It is less difficult to collect cultivated material than wild material, because of the better defined population structure within a cultivated field.

Collecting Sites

The plant materials can be collected from farmers fields, kitchen or orchard gardens, market and wild habitats. Most sampling is done from farmer's field. In areas where genetic erosion has been intense and the widespread cultivation of modern cultivars is common, such garden can be an important source of genetic diversity. Sometime a collector may arrive at a particular locality after harvest of the crops, and find it necessary to make collection from a farmer's store. This is often important for vegetatively-propagated crops. The collections are also made from markets, although the amount of information one obtains in market can be quite small compared to that obtained directly from farmers.

Field Documentation

Field documentation is mainly done on the evaluation and characteristic of the crop plant. The identifying details of any collected germplasm have immense importance, if the collection is made form international community. Avoid confusion and duplication of collected germplasm. Other information which needs are:

- **Taxonomic identification of the material**
- **The source of the germplasm**

  Wild habitat

  Farmer's field

  Farm store

Market etc.

- **Local habitat conditions**

  Soil type

  Rainfall

  Temperature, etc..

CONSERVATION

The strategy of conservation depend on the nature of the material and on the objective and scope of the activity. The nature of the material is defined by the length of the life cycle, the mode of reproduction, the size of the individuals and its ecological status, whether wild, weedy or domesticated (Frankel, 1970). The purpose to which the material is put will determine the degree of integrity which is essential or desirable to maintain. Conservation strategy must take into consideration the time dimension; whether it be for short-medium-or long-term storage and where the storage will be located. The genetic make-up of the material and the type of sample collected will reflect the breeding system of the material and how it will be regenerated.

There are two basic approaches to germplasm conservation namely *in situ* and *ex situ* conservation.

In Situ Conservation

This is an approach, which is applied mainly to wild species related to crop plant, to forest and pasture species. It is often recommended that these species should be preserved, maintaining the genetic integrity of their nature state, as communities in stable environments. Most nature conservation programmes are aimed at the level of ecosystems or multi species communities. The application of in situ conservation of landraces is not, therefore, a feasible option, and because of the rapid loss of genetic variation through cultivar replacement, *ex situ* conservation in one form or another is the most practical and safe approach for such material.

Ex Situ Conservation

This type of conservation includes the use of botanic gardens and arboreta on the one hand and genebank on the other. Botanic gardens and arboreta represent the oldest forms of conservation in Europe data back to the 15th and 16th centuries. Little material has survived since then however apart from a few trees, so that most are in fact of later origin. Guldager (1975) suggested four alternative objectives for *ex situ* conservation:

1. To establish and maintain conservation stands characterised as far as possible by the same genotype frequencies as the original populations (provenance)-static conservation of genotypes).
2. To establish and maintain conservation stands characterised as far as possible by the same gene frequencies as the original populations, thus avoiding total loss of any allele static conservation of gene pools.
3. To establish conservation stands in which gene frequencies are allowed to change freely according to natural selective forces evolutionary conservation.
4. To establish conservation stands in which gene frequencies are deliberately changed by man in order to conserve characteristics important for plantation economy in a region and at the same time eliminate undesirable characteristic-selective conservation.

Guldager (1975) further suggested following conservation differences:

Genotype frequency gene frequency conservation	It is truly achievable in vegetatively-propagated material. The aim is to prevent the loss of genetic information.
Static conservation	It takes the form of the storage of seeds or vegetative material in gene banks. It reduces the life processes to a low level. It is safe and the cheapest method.

TYPES OF BIODIVERSITY

The term biodiversity includes three different aspects, which are closely related to each other. Following are the types of biodiversity:

Genetic diversity

It refers to the variation of genes within the species. This constitutes distinct population of the same species or genetic variation within population or varieties within a species.

Species diversity

It refers to the variety of species within a region. Such diversity could be measured on the basis of number of species in a region.

Ecosystem diversity

In an ecosystem, there may exist different landforms, each of which supports different and specific vegetation. Ecosystem diversity is difficult to measure since the boundaries of the communities, which constitute the various sub ecosystems are elusive. Ecosystem diversity could best be understood if one studies the communities in various ecological niches within the given ecosystem; each community is associated with definite species complexes. These complexes are related to composition and structure of biodiversity.

Agro-biodiversity

The agricultural biological diversity more commonly referred to as the agro-biodiversity has been fast emerging as a strong, evolutionary divergent line from the biodiversity, which deals with the life forms at large. It has been specifically recognized to differentiate between concern for ecosystems versus agro-ecosystems, wild forest flora and fauna versus agriculture related plants, reptiles, insects, avian and microbes; *in situ* conservation of wild forms versus on farm conservation of landgraves and traditional/primitive cultivars or *ex situ* conservation of plant genetic resources, etc.

Agro-biodiversity in a traditional farming system is as follows (adopted from Altieri, 1991 and UNDP, 1995):

- Rich in plant and animal species
- A wide diversity of niches in the local environment utilized
- Reuse of organic residues, consuming biomass enabled
- Ecosystem functions, such as pest, weed and disease management enhanced.
- Locally available resources consumed to an advantage.
- Reduction of risk and optimization of resources use.
- Associated with farmers' time tested local knowledge about resources.

BIODIVERSITY AT DIFFERENT LEVELS

Global Level

It is estimated that there exists 5-30 million species of living forms on our earth and of these, only 1.5 million have been identified and include 3,00,000 species of green plants and fungi, 8,00,000 species of insects,

40,000 species of vertebrates and 3,60,000 species of microorganisms. Recently it has been estimated that the number of insects alone may be as high as 10 million, but many believe it to be around 5 million.

The tropical forests are regarded as the riches in biodiversity. According to the opinion of the scientists more than half of the species on the earth live in moist tropical forests, which is only 7% of the total land surface. Insects (80%) and primates (90%) make up most of the species.

Table 1.1: Estimated number of species worldwide

Taxonomic group	No. of species
Bacteria	3600
Blue green algae	1700
Fungi	46983
Bryophytes	17000
Gymnosperms	750
Angiosperms	250000

The species diversity in tropics is high as:

- In tropics, as the conditions for evolution were optimum and for extinction fewer.
- In tropics, species diversity was conserved over geological time. Due to low rates of extinction prevailing there; and
- Biological diversity is the result of interaction between climate, organisms, topography, parent soil materials, time and heredity.

However, these explanations need experimental observations and confirmation.

Country Level

India is located in south Asia, between latitudes 6° and 38° N and longitudes 69° and 97° E. The Indian landmass extending over a total geographical area of about 3029 million hectares, is bounded by Himalayas in the north, the bay of Bengal in the east, the Arabian sea in the west, and Indian Ocean in the South. The Wide variety in physical features and climatic situation have resulted in a diversity of ecological habitats. This richness in biodiversity is due to immense variety of climatic and altitudinal conditions coupled with varied ecological habitats. The Indian region having a vast geographical area is quite rich in biodiversity with a sizable percentage of endemic flora and fauna. These vary from the humid tropical Western Ghats to the hot desert of

Rajasthan, from the cold desert of Ladakh and the icy mountain of Himalayas to the warm costs of peninsular India.

Table 1.2: Number of recorded biota in India

	Taxon	No. of Species
FLORA	Bacteria	850
	Algae	2500
	Fungi	23000
	Lichens	1600
	Bryophyta	2700
	Pteridophyta	1022
	Gymnosperms	64
	Angiosperms	17000
	TOTAL	**48736**

(based on available data)

An estimate shows that about 200 species of Pteridophyta and about 4950 species of Angiosperms are endemic in India. In India, about 1,15,000 species of plants and animals have been identified and described. For example, the following crops arose in the country and spread throughout the world: rice, sugarcane, asiatic vignas, jute, mango, citrus, banana, several species of millets, several cucurbits, some ornamental orchids, several medicinal and aromatics. Infact, the country has been recognized as one of the world's top 12 megadiversity nations. This region is also a secondary centre of diversity for grain amaranthus, maize, red pepper, soybean, potatoes and rubber plant.

In flora, the country can boast of 45,000 species which accounts for 15 per cent of the known world plants. Of the 15,000 species of flowering plants, 35 per cent are endemic and located in 26 endemic centres. Among the monocotyledons, out of 588 genera occurring in the country, 22 are strictly endemic.

The North Eastern region boast of being unique treasure house of orchids in the country. The important Indian orchids are *Paphiopedilum fairieyanum, Cymbidium alóiflium, Aerides crispum,* etc.

Information regarding other flora and fauna are patchy. Hundreds of new species may be present in the country awaiting discovery. The Western Ghats in Peninsular India, which extend in the southern states, are a treasure house of species diversity and has about 5,000 species. It is estimated that almost one-third of the animal varieties found in India have taken refuge in Western Ghats of Kerala alone.

The Indian Gene Centre is among the twelve megadiversity regions of the world. More than 20 crop species were domesticated here. It is known to have more than 49,000 species of plants 18,000 species of higher plants, including major and minor crop (166) and their wild relatives (326). Around 1,000 wild edible plant species are widely exploited by native tribes. These include 145 species of roots and tubers, 521 of leafy vegetables/greens, 101 of buds and flowers, 647 of fruits and 118 of seeds and nuts. In addition, nearly 9,500 plant species of ethno-botanical uses have been reported from the country of which around 7,500 are the ethno-medicinal importance and 3,900 are multipurpose, edible species.

Wild Plant Wealth

More than 75 per cent of Indian diversity (in the number of species in genera such as *Dendrobium*, *Bulbophyllum*, *Liparis*, *Coelogyne*, *Paphiopedilum*, *Vitis*, *Citrus*, *Musa*, *Rhododendron*, *Hedychium*, *Elaeacarpus*, and *Elaeagnus* occurs in this region. Over 30 species of legumes and 45 of grasses mainly occur in the temperate belt of the northeast Himalayan region. The legume flora is represented by species of *Astragalus*, *Caragana*, *Medicago*, *Melilotus*, *Parochetus*, *Trifolium*, *Trigonella* and *Vicia*. Among grasses, a large diversity is noticed in genera such as *Agrostis*, *Alopecurus*, *Bromus*, *Calamogrostis*, *Dactylis*, *Festuca*, *Glyceria*, *Lolium*, *Muehlenbergia*, *Phleum*, *Poa*, *Stipa* and *Trisetum*.

Table 1.3: Endangered/potentially endangered wild useful taxa of Askot Wildlife Sanctuary of Kumaun

Taxa	Local name	Elevation range(m)	Uses
Aconitum heterophylum	Atis	3500-4500	M
Angelica glauca	Chipi/ Gandraini	3000-4000	M, Ed
Ephedra gerardiana	Somwalli	3000-4500	M
Megacarpaea polyandra	Rukhi	3300-4200	M, Ed
Mahonia borealis	Bhains / Kirmor	1800-2500	M, Ed, Fu
Pleurospermum angelicoides	Chorak	2000-3000	M, Ed
Quercus lanuginosa	Riank	1500-2200	Fd, Fu, Hb
Rheum australe	Dolu	3000-4000	M,
Rhododendron	Poksin	3000-4000	M, Fu

Taxa	Local name	Elevation range(m)	Uses
anthopogon			
Taxus baccata	Huner	2200-3300	M, Fu, Hb
Thalictrum pauciflorum	Mameri	3000-4500	M

(M - Medicinal, Ed-Edible, Fd-Fodder, Fu-Fuel, Hb-House building, Ct-Cultivated tools, Re - Religious).

Local inhabitants gather a wide range of plants from wild habitats, which provide edible tubers, green leafy vegetables, edible fruits and nuts etc. The more promising of these have been domesticated and/or protected under homestead management.

ECOSYSTEM DIVERSITY IN INDIA

These ecosystems harbour and sustain the immense biodiversity. India is one of the 12 megabiodiversity centres in the world. This attributed to the immense variety in physiography and climatic situations resulting in diversity of ecological habitats ranging from tropical subtropical, temperate alpine to desert.

1. Desert
2. Forests
3. Grasslands
4. Wetlands
5. Mangroves

Desert Ecosystem

According to the world bio-geography classification, India represents 2 major realms and 3 basic biomes including 12 bio-geographical regions. However the Wildlife Institute of India has suggested that India be divided into following ten bio-geographic regions.

1. Trans Himalayan
2. Himalayan
3. Indian desert
4. Semi-arid
5. Western Ghats
6. Deccan Peninsula
7. Eastern Himalayan

8. Gangetic Plains
9. North East India
10. Islands and Coasts

Desert ecosystem is characterized by low precipitation, arid lands, which are largely barren except far sparse or seasonal vegetation cover. Species in this habitat are adapted to `an extremely harsh, water-scarce environment. Covering 2% of the landmass, deserts in India re classified into three distinct types:

1. The sandy thar desert of Western Rajasthan and adjoining areas of other states.
2. The salt desert of Kutch in Gujarat.
3. The high altitude cold desert of Jammu and Kashmir, Leh Laddakh and Himachal Pradesh.

The sandy Thar desert in India covers about 2,78,330 sq. km. of which 1,96,150 sq. K,. is in Rajasthan, 62,180 sq. km. in Gujarat and 20,000 sq. km. in Punjab and Haryana. The sparse vegetation is mainly composed of *Prosopis cineraria, Capparis decidua, Ziziphus nummularia, Acacia nilotica, Clotropis procera* and *Prosopis juliflora.*

The salt desert of Rann of Kutch is spread over 9000 sq. km. It is characterized by a typical salt marsh, salt bush plants community of halophytes.

Extending over the North Himalayan ranges, the cold desert is characterized by extremely low temperature going down below-45ø C and low rain fall ranging from 500-800 mm annually. The vegetation is sparse alpine steppe and mostly herbaceous and shrubby. Some common species are *Salix daphnoides, Myricaria elegan, Cicer microphyllum, Polygonum affine* and *Potentilla bifurea* etc.

Arid Zone

Indian arid track commonly known as the Great Indian desert is infect, Eastern part of the Thar desert. Indian arid zone lies between 24° to 35.5° N latitude and 70° 78 east longitude cover an area of approximately 317000 sq km out of which nearly 196150 sq km, is occupied by hot desert located in part of Rajasthan, Haryana, Punjab and Gujarat.

Rajasthan is situated in the North-west part of India and lies between 23° 3 to 30° 12 North latitude and 69° 3 to 78° 17 East longitude. It covers an area of approximately 342239 sq km (Census, 1981). Representing nearly 11 per cent of the total area of Indian sub-continent. About 61 per cent of hot arid zone of the country lies in western Rajasthan. It spreads over 11 out of 31 districts namely Jaisalmer, Barmer, Jodhpur, Bikaner, Sriganganagar, Churu, Nagour, Sikar, Jhunjhunu, Pali, and Jalore. Out of all these 11 districts Bikaner needs a special mention. It is situated in North-west part of the State between 27° 11 to 29° 3 North latitude and 71° 54 to 76° 12 east longitude. Having an average altitude of 228 M above sea level. The total area covered by this district is 27336.2 sq km.

Climatic features

Arid ecosystems constitute a significant part of the earth's dry climates. The limits of the arid lands, however, defy precise definition as they encompass the true deserts, the extremely arid and less arid regions. The best way to classify is to demarcate them as those dry regions where precipitation is scarce and meets less than one-third of the annual water need (PE) of the area. These regions represent the extreme dry climates with harsh environmental conditions. The natural vegetation in these regions is mainly dominated by xerophytes, rangeland grasses and short lived annuals having a good degree of drought resistance, to survive under the persistent environmental constraint viz.; moisture availability.

The Indian arid ecosystem is characterised by low annual precipitation varying from slightly above 500 mm to less than 100 mm, with high coefficient of variability (30 to more than 70 per cent) and erratic intra-seasonal distribution leading to protracted droughts during crop growing season. Apart from the low precipitation, these regions in general experience extreme variations in diurnal and seasonal temperatures, high wind regime especially during the summer periods, coupled with high evaporative demand of the atmosphere. Skies tend to be clear in these regions almost through out the year.

Even during the main rainy season, bright sunshine hours usually amount to 50 per cent or more of the possible. Peak radiation in summer can reach as high as 275 Wm^{-2} and on average over the year upto 200 Wm^{-2} in many of the arid areas making these as regions of highest solar insolation on the globe.

The floristics of desert vegetation

Indian desert has 682 species belonging to 352 genera and 87 families of flowering plants. Of these, 9 families, 37 genera and 63 species are introduced. Poaceae and Leguminosae are the largest families amongst monocotyledons and dicotyledons, respectively. Incidently, Poaceae is the largest family of 57 genera and 111 species. Phytogeographically, 37 per cent of the botanical species represent African elements, 20.6 per cent Oriental elements, 14 per cent species being tropical and 10.3 per cent cosmopolitan. Nearly 9.4 per cent species are endemic to this region (Bhandari, 1995). Beside, a large number of species show polymorphism, which is more pronounced in herbaceous and graminaceous plants. Diversity in *Cenchrus, Indigofera, Eleusine, Dactyloctenium, Cymbopogon, Heteropogon, Lasiurus, Euphorbia, Tribulus, Cucumis* and *Tephrosia* is highly staggering and striking.

Many of these species are economically important plants. It has been documented that 40 species belonging to 21 families yield leaves that can be used as vegetables; 27 species coming from 10 families yield edible fruits; 8 species yield fibre; 3, crude rope; 8, oils and 7, gums or resins. Beside, many of these are sources of medicines.

Vegetation community Types

These species are organised into communities depending upon climatic (mainly rainfall), edaphic (Soil texture, depth) and biotic factors. However these can all be grouped into six types: mixed xeromorphic thorn forest, mixed xeromorphic woodland, mixed xeromorphic riverine forest, lithophytic scrub desert, psammophytic scrub desert and halophytic scrub desert. These are described below:

Mixed Xeromorphic Thorn Forest	*Abutilon indicum, Acacia leucophloea, Acacia senegal, Anogeissus pendula, Azadirachta indica, Barleria acanthoides, Barleria prionites, Bauhinia racemosa, Boswellia serrata, Capparis decidua, Cassia auriculata, Commiphora wightii, Cordia gharaf, Dichrostachys cinerea, Dipteracanthus patulus, Euphorbia caducifolia, Grewia tenax, Grewia villosa, Maytenus emarginatus, Mimosa hamata, Moringa cocanensis, Prosopis cineraria, Salvadora oleoides, Sarcostemma acidum, Securinega leucopyrus, Ziziphus nummularia.*
The ground flora in the low rainfall zone	*Achyranthus aspera, Aristida funiculata, Blepharis sindica, Boerhavia diffusa, Bothriochloa pertusa, Brachiaria ramosa, Cymbopogon jwarancusa, Eleusine compressa, Eleusine hirtigluma, Enneapogon brachystachys, Eremopogon foveolatus, Hackelochloa granularis, Heteropogon contortus, Indigofera cordifolia, Indigofera tinctoria, Lepidagathis trinervis, Melanocenchris jacquemontii, Oropetium thomaeum, Pupalia lappacea, Sehima nervosum, Tephrosia petrosa, Tephrosia purpurea, Tragus biflorus, Tridex*

procumbens.

Mixed Xeromorphic Woodlands

Acacia jacquemontii, Aerva persica, Balanites aegyptiaca, Calotropis procera, Capparis decidua, Capparis decidua, Capparis dicidua, Celosia argentea, Cenchrus ciliaris, Cenchrus setigerus, Convolvulus microphyllus, Crotalaria burhia, Crotalaria burhia, Dactyloctenium sindicum, Desmostachya bipinnata, Eleusine compressa, Heliotropium subulatum, Indigofera oblongifolia, Prosopis cineraria, Prosopis cineraria, Pulicaria wightiana, Salvadora oleoides, Salvadora oleoides, Tephrosia purpurea, Ziziphus nummularia.

Mixed Xeromorphic Riverine Thorn Forest

Acacia cupressiformis, Acacia nilotica, Acacia nilotica, Ailanthus excelsa, Albizzia lebbeck, Aristida adscensionis, Boerhavia diffusa, Cenchrus biflorus, Cenchrus ciliaris, Cenchrus setigerus, Chloris virgata, Crotalaria burhia, Cynodon dactylon, Cyperus arenarius, Dactyloctenium aegyptium, Desmostachya bipinnata, Digera muricata, Digitaria adscendens, Ficus bengalensis, Ficus religiosa, Heliotropium subulatum, Indigofera cordifolia, Moringa oleifera, Prosopis cineraria, Pulicaria wightiana, Salvadora persica, Salvadora oleoides, Solanum surratense, Tamarindus indica, Tamarix auriculata, Tecomella undulata, Tephrosia purpurea, Vernonia cinerascens, Volutarella divaricata, Xanthium strumarium, Ziziphus mauritiana.

Lithophytic Scrub Desert

Acacia senegal, Aristida hirtigluma, Boerhavia diffusa, Boerhavia elegans, Bonamia latifolia, Capparis decidua, Cleome brachycarpa, Cleome papillosa, Crotalaria burhia, Dactyloctenium sindicum, Eleusine compressa, Eragrostis sp., Indigofera cordifolia, Leptadenia pyrotechnica, Mollugo cerviana, Oropetium thomaeum, Orygia decumbens, Prosopis cineraria, Sarcostemma acidum, Tribulus terrestris, Ziziphus nummularia.

Psammophytic Scrub Desert

Acacia jacquemontii, Acacia senegal, Aerva persica, Aerva pseudotomentosa, Aristida funiculata, Calligonum polygonoides, Calotropis procera, Cenchrus biflorus, Citrullus colocynthis, Clerodendron phlomoides, Crotalaria burhia, Cyperus laevigatus, Dipterygium glaucum, Haloxylon salicornicum, Leptadenia pyrotechnica, Maytenus emarginatus, Panicum turgidum, Sericostema pauciflorum.

Halophytic Scrub Desert

Aeluropus lagopoides, Chloris virgata, Cressa cretica, Cyperus rotundus, Dactyloctenium aegyptium, Desmostachya bipinnata, Dichanthium annulatum,

Echinochloa colonum, Eleusine compressa, Euphorbia granulata, Fagonia cretica, Haloxylon recurvum, Peganum harmala, Portulaca oleracea, Salsola baryosma, Schoenfeldia gracilis, Sporobolus helvolus, Sporobolus marginatus, Sporobolus marginatus, Suaeda fruticosa, Trianthema portulacustrum, Zygophyllum simplex.

Dominance diversity trends in Indian arid ecosystem

Dominance diversity relations undergo change upon degradation. Woody perennials in the older alluvial plains in the Luni basin had low diversity and low equitability on sites having degraded grass-covers such as Oropetium-Eragrostis or Dactyloctenium-Eleusine type. Intermediate to high diversity and equitability characterised the sites having optimum cover of Dichanthium-Desmostachya. The degraded nature of former group of sites was confirmed by the geometric dominance diversity curves whereas those sites in the latter group exhibit lognormal dominance diversity curves representing no degradation.

Change in vegetation quality and palatability

Thus, grazing lands have not only declined in area and yield, but the quality of the forage has also declined. The majority of annual species that get hold after perennial species have died, are quite often less palatable and become unpalatable as they mature to bear seeds with awns. Further, the terrain becomes barren after December by which time all annuals also die. Thus, this makes land vulnerable to wind erosion in summer.

Ecologically, the species diversity upon disturbance may first increase, followed by a sharp decline later on. There are chances that species may even become endangered due to excessive use, as in the case of *Commiphora wightii*. With loss or changes in the form of diversity, large scale changes in food webs, energy flow and biogeochemical cycles are inevitable, but still to be studied. This pushes the fragile arid ecosystem toward more a point of disaster than stability.

The vegetation cover of arid tract of Rajasthan may primarily be categorized into two major parts: indigenous crop system and native trees, shrubs and herbs; and the second one is exotic varieties and races of various crops and plants introduced or migrated and florushing well due to changing weather conditions.

Cotton, gawar, bajra, rice, sugarcane, moth, til, and jawar are important *kharif* crops of the region where as gram, wheat, sarson, jow, find a similar status as *rabi* crops.

Due to completion of Rajasthan Canal Project (renamed Indira Gandhi Nahar Project – (IGNP) phase I and II, north-west Rajasthan is witnessing a total face lift. The whole scenario relating to climate and floristic has taken a tern from 1991 to 1998.

Increasing irrigation facilities due to net-work of main canal, distributaries and lift canals has opened up doors for new crops. *Kharif* production since 1961-62 has gone up from 0.068 ha to 370.27 ha in 1996-97. So is the case with *rabi* which has shoot up from 13.76 ha in 1961-62 to 471.73 ha in 1996-97. Collective increase in *kharif* and *rabi* comes to be 1.444 ha in 1961-62 to 842.00 ha in 1996-97. This leads to introduction of newer crops like rice, sugarcane, cotton, and wheat and subsequent introduction of new races of these crops in the area.

Interaction between indigenous crop varieties along with their wild relatives is a must with introduced varieties and races.

As such their arises a natural need to conserve and protect indigenous races which were more suited to this area along with good outcome of interactions of these races with the introduced one.

Same has happened with native trees, shrubs, and herb varieties. Due to completion of IGNP Phase I and II and phase III work under progress some evil natural effects have start showing their presence such as water lodging, salinization, vertical water movement, etc.

In the Third Meeting of the working group on "Environmental Aspects" sponsored by NWDA National Water Developmental Agency on 17th Dec., 1997 concluded that people are facing environmental problems like:

1. Massive displacement of people due to the submergence of area by water rise (water lodging).
2. New diseases on agricultural crops.
3. Unwanted weeds have started occupying exter.·ive area.
4. Important medicinal plants of the desert are rapidly disappearing.

Threat caused due to these problems, to biodiversity of the area has worried the participants and to conserve it steps has already been taken by creating protective zones preserving natural habitat. Therefore, conservation of biodiversity work has already been initiated in situ at ecosystem level.

This however, does not seems to be the long term solution of the problem. It is necessary to conserve them at genetic level. This will be done by collecting pollen of various plants species and varieties from different habitat and their subsequent cryopreservation.

The rich florestic composition which form the backbone of socio-economic status of this arid track not only need preservation and protection it too should be monitored in a way that the existing flora get rich and richer in the coming days.

Forest Ecosystems

The forests cover of the country in about at 639,600 sq. km according to the Forest Survey of India Assessment (1995). This represent 19.46% of India's total geographical area India is endowed within diverse forest types ranging from the tropical wet evergreen forests in north-east to the tropical thorn forests in the central and western India. The forests of the country provide several essential services to the mankind. Forests are the source of a number of food items, fuel-wood, fodder and timber. Other economic uses indeed providing raw material for forest based industries. Some of the minor forest produce including gum, resin, honey, etc. forests perform important ecological functions such as maintaining delicate ecological and hydro-biological balances, conserving soil, controlling floods, drought and pollution. Forests provide habitats for innumerable plants, animals and microorganisms. Forests are the source of recreation and religious inspirations.

Grasslands

Grasslands, which are also called as steppes, prairies, pampas and savannas in various parts of the world, are vegetation types with predominance of grass and grass like species. In India the total area under grasslands is about 3.9% or 12 millions hectares. Grasslands in the country also exhibit as diversity ranging from semi-arid pastures in Daccan Peninsula, Humid Semi waterlogged grass land of Tarai belt, rolling shola grassland on the hilltops of Western Ghats and the high attitude alpine pasture of Himalayas. Five distinct types of grasslands have been recognised in India.

- *Sehima - Dichanthium* type
- *Dichanthium - Cenchrus - Lasiurus* type
- *Phragmites - Saccharum - Inperata* type
- *Themeda - Arundinella* type
- Temperate alpine type

The significance of about 10,000-odd known grass species in the world is enormous in leading vegetation cover to about 17% of the earth land surface, besides providing a number of cereals and fodder to mankind. About 15% of world grasses are represented in India (Vates et. al., 1999).

The significance of tropical grasslands has been well realized for their highest productive capacity. Grassland ecosystems are also expected to play crucial role in the global biospheric responses to climatic change and in augmenting vegetation in forests facing plunder. The W. W. F.'s Biodiversity Conservation Prioritisation (BCP) report on diwindling biodiversity of grasslands is a timely reminder on the extent and ferocity of environmental deterioration that has degraded nearly half of the earth's vegetative surface (Daily 1995; Vates et. al. 1999). It is estimated that the Indian grasslands harbour about 1256 species belonging to 245 genera.

Wetlands

Wetlands are transitional zones that occupy intermediate position between dry land and open water. These ecosystems are dominated by the influence of water. They encompass diverse and heterogeneous habitats ranging from rivers, flood plains and rainfed lakes to swamps, estuaries and salt marshes.

India by virtue of its extensive stretch and varied terrain and climate, supports a rich diversity of inland and coastal wetland habitats. It is estimated that India has about 4.1 million ha. The significance of tropical grasslands has been well realized for their highest productive capacity. Grassland ecosystems are also expected to play crucial role in the global biospheric responses to climatic change and in augmenting vegetation in forests facing plunder. The W. W. F.'s Biodiversity Conservation Prioritisation (BCP) report on diwindling biodiversity of grasslands is a timely reminder on the extent and ferocity of environmental deterioration that has degraded nearly half of the earth's vegetative surface (Daily 1995; Vates et. al. 1999). It is estimated that the Indian grasslands harbour about 1256 species belonging to 245 genera.

Mangroves

Mangroves constitute an important economic resource providing fodder, fuel, wood, tannin, edible fished, hides, honey, wax, various chemicals and medicines. Mangroves play important role in stabilizing shorelines and protecting them from cyclones. Mangrove also harbours a variety of plants and animals.

Mangroves are salt-tolerant ecosystems in tropical and sub-tropical regions. These ecosystems are largely characterized by assemblage of unrelated tree genera, that share the common ability to grow in saline and dual zones.

India harbours some of the best mangroves swamps in the world, located in the alluvial deltas of Ganga, Mahanadi, Godavari, Krishna and Cauvery rivers and on the Andaman and Nicobar group of Islands. The total area covered by mangroves in India is about 6,700 sq. Km. amounting to the 7% of the world mangrove. The largest stretch of mangroves in the country lies in the Sunderbans in West Bengal covering an area of about 4,200 sq km. The predominant mangrove species are *Avicennia officinalis. Exoecaria agallocha, Heritiera fomes, Bruguiera parviflora, Ceriops decandra, Rhizophora mucronata* and *Xylocarpus granatum.*

SPECIES DIVERSITY

Biogeographically, India is situated at the tri-junction of three realms namely Afro-tropical, Indo Malayan and Paleoarctic realms, and therefore has characteristic element form s in each of them. This assemblage of three distinct realms makes the country rich and unique in biological diversity. Based on the available data, India ranks tenth in the world and fourth in Asia in plant diversity and ranks tenth in the world and fourth in Asia in plant diversity ad ranks tenth in the number of mammalian species and eleventh in the number of endemic species of higher vertebrates in the world.

Status of Survey

At present, 1.75 million species have been recorded so far in the world (Global Biodiversity Assessment, 1995). India's contribution to this record stands at 7% Surveys conducted so far have inventoried over 49,000 species of plants and 81,000 species of animals. As until now, only 70% of the area have been surveyed, it is estimated that the flora and fauna already identified are only a part of what actually occur in India. The list is being constantly added to especially in the case of lower plants and invertebrate animals. Survey and inventorisation of India's biodiversity is still far from competing, especially for the lower groups of plants and invertebrate animals.

As noted earlier 49,000 species of flowering and non-flowering plants representing about 12% of the recorded world's flora, have already been identified. Significant diversity has been recorded in pteridophytes with 1022 species and Orchidaceae with 1082 species. Comparative statement of recorded number of plants species in India and World is given in table (2.1).

A total of 81,251 animal species have been recorded in India which represents 6.67% of the faunal species recorded in the world; of these vast majority are insects with over 60,000 species. The vertebrate fauna is also diverse and varied.

Endemic species

Endemism of Indian biodiversity is significant. About 4,900 species of flowering plants are 33% of the recorded floras are endemic to the country. Thee are distributed over 141 genera belonging to 47 families. These are concentrated in the floristically rich areas of North East India, the Western Ghats, North West Himalayas and the Andaman and Nicobar Islands.

The Western Ghats and the Eastern Himalayas are reported to have 1,600 and 3,500 endemic species of flowering plants, respectively. These constitute two of 18 hot spots identified in the world. It is estimated that 62% of the known amphibian species are endemic to India of which a majority occur in Western occur in Western Ghats.

Endemic species are the plants, which are limited in their distribution i.e. they are restricted to a small area and are not found elsewhere in the world. It may be due to:

1. Poor adaptability of a species in a wide range of ecological
2. Presence of some geographical barrier, e.g. Sea, Mountains etc.
3. Failure of dispersal of reproductive organs (propagules, seeds, runners etc.).
4. The species might have been comparatively young and not have enough time to spread.

Some examples of endemics include *Metasequoia* living gymnosperm endemic in China, *Sequoia* (red wood tree) endemic in coastal valleys of California, USA, *Primulla* and *Potentilla*, at high altitudes of Himalayas, *Ginkgo biloba* endemic in Japan and China.

Cultivated Plants/Agro biodiversity

The agricultural biological diversity is a strong, evolutionary divergent line from the biodiversity, which deals with the life forms at large. Several multifaceted, scientifically established, threats to agro-biodiversity have emerged, particularly in many technologically advanced countries, in which agriculture plays key role in national economy and the **Gross Domestic Product** (GDP), the more important and relevant issues are:

1. Safe
2. Their management under a sustained system
3. Providing for further opportunities for agro-evolutionary processes.
4. The efficient, judicious and equitable use of this diversity.
5. To attain progressive increases in yielding ability and wider adaptability of improved varieties by appropriate use of conserved germplasm.

Two contrasting dimensions of agro-biodiversity may be clearly viewed in the present context, namely:

1. Its occurrence in the traditional ago-ecosystem in which cultivated plant diversity is maintained, more or less in equilibrium and
2. Its manipulation in the intensive agro cropping systems, which are externally guided for their varietal components and packages of cultivation practices.

The traditional agro-ecosystems' maintain and protect biological diversity for the co-existing cultivated plant species and associated biota including the micro-flora, insects, reptiles, amphibians and avian. Beside in such systems, the abiotic and environmental components have been sustainably maintaining and influencing opportunities for micro-evolutionarily processes to occur within and across the naturally adapted and evolutionarily fittest diversity.

India is an acknowledge centre of crop diversity. It is considered to be the homeland of 167 important cultivated plants species and 320 species of their wild relatives. India is considered to be the centre of origin or 30,000 - 50,000 varieties of rice, pigeon pea, mango, turmeric, ginger, Pepper, banana, bitter gourd, okra, coconut, cardamom, jack -fruit sugarcane, bamboo, taro, indigo, sun hemp, amaranthus, gooseberries, etc. The gene bank of National Bureau of Plant Genetic Resources (NBPGR) has a collection of over 1,44,000 varieties. The details of the active germplasm holding and base collection of NBPGR are given below:

Table 1.4: Active Germplasm Holding and base Collection at NBPGR.

Crop Groups	Active germplasm holding	Base collection
Cereals	12,086	43,409
Pulses	38,695	22,269
Millets Minor Millets	10,349	14,488
Oil Seeds	19,808	14,278

Crop Groups	Active germplasm holding	Base collection
Vegetables	12,146	5,681
Medicinal and Aromatic Plants	870	942
Pseudocereals	4,739	736
Tuber	2,053	-
Crops/Spices	4,060	-
Horticultural/Ornamentals	2,212	-
Fibre Crops	-	3,212
Released Crop Varieties	-	904
Reference Samples	-	53,161
Grand Total	**1,07,018**	**1,59,080**

Genetic Diversity

Genetic diversity is defined as variations in the genetic composition of individual within or among species. India being one of the 12 megabiodiversity countries possesses rich genetic diversity. Studies in genetic diversity are not as yet very widespread and sharply focused. However, studies in genetic diversity of wild crop relatives and domesticated animals have been carried out.

There are several hundred species of wild crop relatives distributed all over the country A major centre for wild relatives is North-Eastern hills and Tamil Nadu hills are rich in wild relatives of millets, wild relatives of wheat and barley have been located in the Western and North Eastern Himalayas.

Table 1.5: Wild relatives of crops in India

Crops	Number of wild related
Millets	51
Fruits	104
Spices & Condiments	27
Vegetables & pulses	55
Fibre Crops	24
Oil - seed, tea, coffee, tabacco, sugarcane	12
Medicinal plants	3,000

Table 1.5 has the statement of wild relatives of crops recorded so far.

ENDANAGERED PLANT SPECIES

About 427 endangered plant species have been listed by the Botanical Survey of India in its publications on the floristic synthesis.enumerated for Red Data book. This contributes to about 20 per cent of India's total floristic wealth of higher plants. Examples of endangered species occurring in the North - Eastern region including the Eastern Himalayas are as under:

Eastern Himalayas	*Acer laevigatum, A. molle, Anglica nubigena, Bunium nothum, Carum villosum, Pimpinella wallichii (Northeastern region also), Panax pseudogineseng, Calamus inermis, Phoenix rupicola (Northeastern region also), Lactuca cooperi, Berberis affinis, Engelhardtia wallichiana, Coptis teeta, Aquilaria agallocha (Northerastern region also), Boehmeria tirapensis.*
Northeastern region	*Heracleum burmanicum, Peucedanum sikkimensis, Pimpinella evoluta, P. flaccida, Trachelospermum auritum, Ilex embeloides, I. khasiana, I. venulosa, Amorphophallus bulbifer, A. sylvaticus, Dioscorea laurifolia, D. arbiculata, Hopea shingkeng Musa velutina, Hedychium aurantiacum, H. gracilimum, H. dekianum, H. gratum, H. greenii, H. hookeri, H. marginatum. H. rubrum.*

The array of biological resources including their genetic resources are renewable in nature or in similar *ex-situ* conservation with their proper management can support human needs indefinitely. Thus, it is appropriate to treat these as the fundamental sources for sustainable development. Their perpetuation in the existing forms and to the extent feasible on existing basis is therefore, essential to suitable explore factors from such a big reservoir for tomorrow's need.

The available evidences, however, indicate that human activities are eroding the prevalent biological resources and greatly reducing the biodiversity of the planet. Estimating the precise rate of loss, or even the current status of the species, is stupendous task because no monitoring system, whatsoever, systematic, is likely to match the huge diversity in its existing form. Thus, much of the basic information remains lacking, especially on the species rich tropics and the hotspots of agro-biodiversity. At least 10 per cent of the India's wild flora and a large fraction of its wild fauna are threatened, with many on the verge of extinction. This may not be surprising, while considering the fact that in the past few decades India has lost at least 50 per cent of its forests, polluted over 70 per cent of its water bodies, built or cultivated on much of its grasslands, and degraded many coastal areas. Further, to bio-prospecting this habitat, ill effects of destruction, hunting, over exploration, pesticide pollution, excessive botanical and zoological collection and a host of other activities have together taken a heavy toll. The endemic species present to date, as enlisted in the box below, need a systematic management attention.

SPECIES DIVERSITY AND ECOSYSTEM STABILITY

Diversity of genes within species increases its ability to adapt to adverse environmental conditions. When these varieties or populations of these species are destroyed, the genetic diversity within the species is diminished. In many cases, habitat destruction has narrowed the genetic variability of species lowering the ability to adapt to changed environmental conditions. The greater the variability of the species, the more is the ecosystem stability. It has been considered to be related to the cycling and recycling of nutrients, which in turn, increases the efficiency of the resource use in the ecosystem.

The survival and well being of the present day human population depends on several substances obtained from plants and animals. The biodiversity in wild and domestic form is the source for most of humanity's food, medicine, clothing, housing, cultural diversity, intellectual and spiritual inspiration. It is believe that one-fourth of the known global diversity is in serious risk of extinction and calls for an integrated approach for conserving global biodiversity. Establishment of nature reserves or biospheres with lot of biophysical variability, maintenance of corridors with different nature for the possible migration of the species in response to climate change, etc. are the immediate steps to be taken for conserving the very precious biological diversity. Global network for gene banks, microbiological resources centres, and marine parks are also important. At the same time conservation must be coupled with socio-economic development, especially in countries where population pressure threatens the national biotic resources.

HARVESTING NATURE'S DIVERSITY

From the beginning of humanity's efforts to harness nature and the environment some 12,000 years ago, rural peoples have engaged in a process of domesticating nature's resources by planting crops and raising animals for food, medicine, and other basic needs. Modern agriculture, fisheries and forestry is based on this long process of diversification and adaptation of beneficial plants and animals to meet a wide range of environmental conditions and varying human needs.

Yet, nature's legacy to use of diverse genetic resources is at risk. Genetic erosion – the reduction of diversity within and between a spices – is global threat to agriculture.

This biological heritage which spans generations is threatened by the recent rapid pace of change, undesirable side effects of industrialization and continuing expansion of the world's population.

Today, our ecological resources are being exploited at rates which exceed their sustainable yield. Human transformation of natural habitats whether for subsistence or commercial purposes, poses the largest threat: land lost to highways and the urban sprawl, overgrazing of pasture land, draining of wetlands, repetitive slash and burn cultivation in forests, overzealous logging, unsustainable fuel wood collection, indiscriminate and excessive use of fertilizers and pesticides, over-shifting, air and water pollution – all severely damage our natural resources.

Every year, 5-7 million hectares of cultivated land are degraded. Between 1980 and 1990, tropical forests were destroyed at an annual rate of 15.4 million hectares – and with it, all the biological diversity these forests contain. Since the beginning of this century, about 75 per cent of the genetic diversity among agricultural crops has been lost. Hundreds of tree species are endangered in whole or in significant parts of their gene pools. Almost 20 per cent of livestock breeds in developing countries are in peril. In aquatic environments, many species are endangered by pollution, diversion of waterways, environment degradation and the release of exotic species.

If genetic diversity is lost, neither human beings nor nature can select to adapt to meet changing needs in agriculture and even society at large. The vulnerability of crop varieties and animal breeds would increase to their limits of survival. Selection can only be done where biodiversity exists.

However, there are some grounds for hope. Global food production has improved since 1992, which enhances the prospects for food security. It is encouraging to note that world cereal stocks are now above the minimum level considered necessary for world food security.

By the year 2025, food production will have to expand by an estimated 60 per cent to meet increased demand, mostly in developing countries. Developing agricultural crops and animal breeds through genetic improvement could play a significant role to increase food production. Sustainable agriculture must be institutionalized if we are to conserve and use our productive base and provide refuge to the important gene pools of wild plants, trees and animal species, which are adapted to local climates and topographies and resistant to pests and diseases.

BIODIVERSITY, BIOPRODUCTIVITY AND BIOTECHNOLOGY

The most profound influence on the biodiversity has been that of biotechnology particularly because of the transfer of genes across the taxonomic/phylogenetic barrier. This area is indeed knowledge-intensive and needs considerable infrastructure and leads to value-addition of the products. Increasing emphasis on biotechnology should not mean that science-based conservation and rights of the local people have to be ignored. Both aspects are important.

Figure 1.1 indicates relationship between biodiversity and bio-productivity. Low productivity and low diversity is characteristic of harsh habitats. The pre-green revolution agriculture has high diversity and low productivity. From here the world moved to green revolution agriculture which has high productivity but low diversity. While it has paid rich dividends in making food-deficient countries self-reliant and sufficient in food, there is no doubt that it has not been environmentally-friendly. For this purpose, countries have to move to the right hand top square where a combination of high diversity and high productivity would have to be achieved. This would need apart from change in agronomic practices, added inputs of genetics and plant breeding and biotechnology.

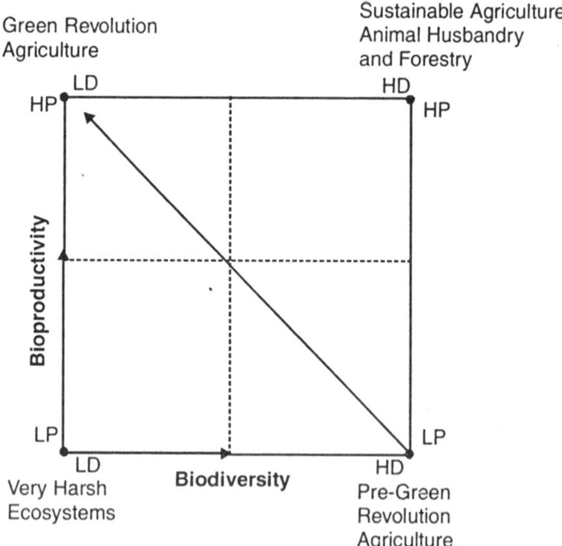

Fig. 1.1: A relation between biodiversity and bioproductivity.

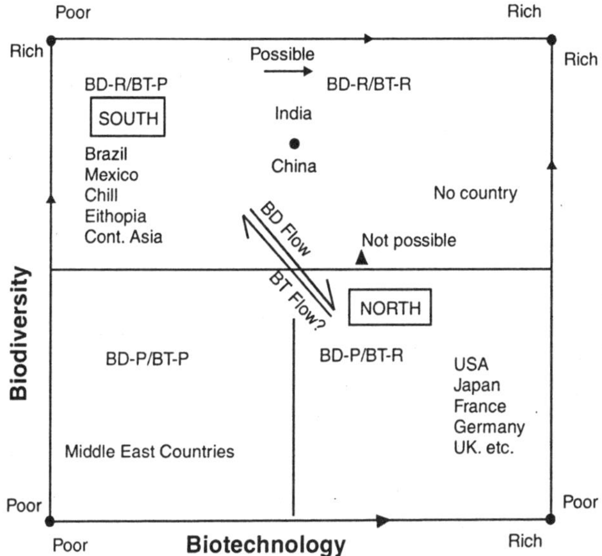

Fig. 1.2: A relation between biotechnology and biodiversity

The relationship between biodiversity and biotechnology is given in Figure 1.2. Based on the extent and nature of biodiversity and biotechnology available, the nations can fall in four groups; those that are:

poor in both biodiversity and biotechnology capabilities (example: Middle East Countries)

1. Poor in biodiversity but rich in biotechnology (most industrial countries of the North).
2. Rich in biodiversity but poor in biotechnology capability (most countries in south).
3. Rich in both biodiversity and biotechnology (no country).

The flow of biodiversity has been from biodiversity-rich and biotechnology-poor South, to the countries in the North. The flow of biotechnology from latter to former has not been commensurate with the former. The two groups of countries, though essentially interdependent, have to reorganize their policy on biodiversity, biotechnology and bioproductivity, so as to complement and supplement one another.

The countries have to look into the intellectual property rights and breeder's and farmer's rights in an objective manner so that the interests of both developing and industrial countries are kept in view. There is an urgent need for capability-building in gene-rich countries which can be a major plans of future North-South cooperation.

AIR POLLUTION AND PLANT DIVERSITY

Air pollution can influence plant species in diverse way and thus affect ecosystem at various levels of organization. In the past few decades, dieback and decline of forests in the U.S. and Europe have been linked with air pollution levels in those areas. Early information on the effècts of air pollution on plant diversity came from studies conducted near large point sources of pollution where increased mortality, declining vigor, and compositional changes in communities were reported downwind of individual operations. But the application of abatement technology, strict air quality regulations, and the shift to tall stacks has altered the nature of air quality and prevailing concentrations of air pollutants in ambient air. Sulfur dioxide, HF, and heavy metals, the causes of severe damage during the past, are now of limited importance. The major focus at present has shifted to the impacts of regional elevated concentration of oxidants, especially O_3, and acid rain. Thus, the recent trend of change in air quality has increased the complexity of the studies assessing the effect of air pollution on plant diversity.

Loss in diversity has many unfavorable implications related to ecosystem functions such as energy flow and biogeochemical cycles. By virtue of pertubing energy fixation at producer level, air pollution has the potential to partub the energy flow in natural and managed ecosystems. Human activities are critical to the relationship between biological diversity and ecological processes. The current problems of air pollution and global climate change have clearly emphasized the need to understand the biodiversity response to these anthropogenic stress factors. Living things exist with specific ecological processes, and an unfavorable alteration in any of these has serious implications for the existence of each other. The values placed on biodiversity are strongly linked to the human influences on it and their underlying social and economic driving forces. This area requires a serious research effort throughout the world in view of the increased loss of biodiversity during the last few decades. Long-term monitoring programs are required at selected areas of long-term or short-term pollution history, and data on structural and functional attributes of biotic components need to be collected periodically to evaluate changes in diversity, resiliency, and productivity.

The prevention of changes ecosystem functioning and biodiversity that will affect the composition of the atmosphere is likely to be one of the most important policy issues of the future. Developing countries require a special effort in this direction, as the rate of biodiversity loss linked with other activities is already several magnitudes greater than developed countries. A global effort is warranted for such problems.

Responses of Air Pollution on Plants

The plant response to air pollutions depends upon the chemical toxicity and exposure pattern of the pollutant and the sensitivity level of the species. Plant species differ in their perception and response to air pollutants due to differential expression under influence of genetic and/or environmental factors leading to constitutive and inducible differences among the plants within and between the populations to the same air pollutant (Taylor et al., 1992, Mansfield et al., 1984). Pollutions may influence plant growth and development through multiple pathways and mechanisms over widely varying timescales. Effects may be direct due to deposition on plant surfaces, resulting in impaired physiology and metabolism of foliar surfaces, or indirect, as a consequence of altered nutrients or toxic metals availability or due to altered susceptibility to other stresses. Air pollutions in combination may induce additive, antagonistic, or snyergistic impacts on plants depending upon the specific pollutions, their concentration and exposure duration, and environmental conditions. Lower-level chronic exposures are generally more representative of today's regional-scale air pollution situations.

Variability in species response has been suggested to be governed by a sequence of events, which either reduces penetration of pollutant in leaf interior (pollutant avoidance) (Mansfield et al., 1984), and/or enhances the ability of plant tissue to withstand the pollutant and its products after reaching the target (pollutant tolerance) (Hippeli and Mechanisms, 1996). These sequences of events initially depend upon constitutive resource availability; therefore regulation of gene expression and acceleration of secondary metabolite production may modify the response pattern (Barnes et al., 1999).

Air pollutions enter the plant body through stomatal pores in the leaves. Darrall (1989) has reviewed the stomatal response of plants to a variety of air pollutions as of partial stomatal closure, wider opening, or no response depending upon pollutant concentration and exposure duration, environment mental conditions, and plant species and their cultivars. Some plants are known to exhibit intrinsically higher stomatal conductance and tend to be more susceptible to damage than others (Reich, 1987). Implications of stomatal responses to air pollutant also depend on responses of other critical physiological processes such as carbohydrate assimilation and patitioning (Wolenden et al., 1992). Variability in inter and intraspecific response to air pollutants with respect to photosynthate partitioning may alter the nature of competition within a plant community (Grime et al., 1986, Mansfield et al., 1988). After entering into the leaf interior, air pollutants may be metabolized sequestered, accumulated, or excreted. Detoxification of pollutants and their products as the cell surface and interior has been widely documented in the literature (Kangasjärvi et al., 1994, Polle et al., 1993, Iqbal et al., 1996).

Impact of Air Pollution on Plant Diversity at Different Levels of Organization

Biological diversity, or in shortened form biodiversity, means the variability among living organisms inhabiting the Earth. In a broader perspective, biodiversity is ecological diversity referring to the number of species in a given area, the ecological roles of the species, the changes in composition across a region, and their grouping together with processes and interactions within and between systems (Heywood et and Baste, 1995).

Three attributes of biodiversity are suggested: composition (number of alleles or species); structure (physical arrangement of biotic components and biomass distribution); and function (natural processes of energy flow and nutrient cycling) (Noss, 1990). It is well established that the stresses from regionally transported gaseous pollutants and acidic deposition can cause structural and functional changes in plant communities (Lauenroth et al., 1985, Kozlowski et al., 1985), leading to subtle degradation of ecosystems (Rosenberg et al., 1979). Air pollutants can affect various levels of ecosystem organization, starting from an individual leaf to an entire ecosystem (McClenahen, 1978, McLaughlin, 1985). Although the species-level diversity is the most-studied element of biological diversity in relation to atmospheric pollutants, genetic- and community-level diversities are also very important.

Genetic Diversity

This element of biodiversity is the foundation of all other aspects of diversity. Loss of plant species of small population sizes can lead to reduced genetic diversity, lowered fitness, and increased extinction risk (Schnoewals et al., 1983). Intrinsic variations within populations serve as the immediate source of genetic diversity. Taylor and Murdy (1975) demonstrated plant-to-plant variation of foliar injury in the herbaceous annual *Geranium carolinianum* L. due to SO_2 exposure. Even after a 30-year period, the populations continued to exhibit a high degree of variability. Bell et al.,

(1991) have also shown variations across the taxa for most physiological and growth parameters in response to air pollution. Evolution of SO_2 resistance in grassland species in industrialized regions of the U.K. has been reported (Bell et al., 1991). The ecological and biological issues affecting the genetic variations are physiological mechanisms of resistance, speed and frequency with which adaptation rises, and the role of microevolution (Tylor et al., 1992). (Roose et al., 1982) have shown that microevolution may have negative physiological costs resulting in the plants being less competitive under conditions of other stresses. Since plants follow dissimilar mechanisms of resistance to air pollutants and consequently have their own fitness costs, a shift in competitive interactions may favor the species with minimum cost of resistance. When the cost of resistance is very high and the rate of selection against resistant individuals is low, species replacement takes place (Antonovics et al., 1971).

Dunn (1959) had suggested long ago that ambient levels of O_3 in the Los Angeles basin were high enough to drive the selection of resistant genotypes of *Lupinus bicolor*. Recent studies on *Populus tremuloides, Trifolium repens,* and *Plantago major* have provided variations in genetic diversity among populations due to evolution of resistance to ambient O_3 under field conditions (Reiling et. al., 1992, Berrang et al., 1991, Macnair, 1993). Inherent resistance of populations to O_3 is correlated with the O_3 level at the source of the plant material. Direct involvement of O_3 in this genetic diversity other than any environmental factors has been proved as the percentage of resistant individuals increased in populations exposed to elevated levels of O_3 over long periods of time (Heagle et al., 1991, Davison and Reiling, 1995).

Genetic diversity has profound implications on community structure. Reduction in population size and restriction of exchange of alleles among populations and to air pollution effects will increase the chances of spatial isolation of the species. It has been hypothesized that the evolution of resistance due to air pollution contributes to the loss of genetic diversity within plant species (Bergmann et al., 1989, Karnosky et al., 1991). Development of resistant populations of grasses and herbs growing in the vicinity of coal-fired plants has been reported (Singh et al., 1994). A reduction in genetic variability of populations and the evolution of resistance and adaptation to gaseous air pollutants have been observed by Ayzaloo and Bell (1981) and Coleman and Mooney (1990). Genetic diversity can be affected at the community level by mortality or reduced growth and reproduction of sensitive species which are subsequently replaced by more-resistant species. Parsons and Pitelka 1991) were of the view that significant losses of genetic diversity may only occur due to population extinction or genetic drift at restricted areas experiencing intense pollution episodes

(Taylor and Murdy, 1975, Murdy, 1979, Heggestad and Heck, 1971). The chronic level of pollution in recent years should not actually be a cause of great concern for genetic diversity.

Community Diversity

Plant community structure depends on various environmental factors. Air quality is an important factor which has changed substantially due to rapid industrialization and urbanization in the recent past. Air pollutants have not been normal constituents of the environment in which plants have evolved over the centuries. Thus, nature has not provided a continuous selection pressure to isolate genes for resistance. However, genetic modification and recombination may provide morphological and physiological system the exclude and tolerate excesses of the pollutants (Reinert et al., 1982).

Secondary Effects of Air Pollution on Plant Diversity Loss

Competitive relationships between plant species alter because of increased vitality or dieback of sensitive plant species (Narayan et al., 1994, Singh et al., 1994, Miller et al., 1983, Keller et al., 1988, Barrang et al., 1989). Intraspecific competition was suggested to be a major factor determining relative abundance of sensitive and tolerant aspen clones in clean and polluted areas. Direct evidence of air pollution effects on competition comes from the study of (Steubing and Fangmeier, 1987), who exposed the under story of a beech forest in artificial chambers at an SO_2 concentration simulating the ambient peak level. Forb species showed reduction in leaf area and reproductive capacity, while grass species and the vine *Hedera helix* proved to be resistant. In clover fescue plots, (Heagle et al., 1989) observed resistance of clover species leading to good performance compared with fescue. Bennett and Runeckles (1977) found that competitive ability of ryegrass increased compared with clover de to increased tillering by ryegrass under O_3 treatment. (Armentano and Bennett 1992) have clearly suggested that intraspecific competition is more sensitive to pollution than interspecific, but pollution can alter species competitive abilities. The effect is initially visible in clonal substitution followed by species.

Forest systems, while resilient to stresses, may be strongly controlled by a balance in competitive potential of a few key species. Responses of individual species and the developmental stage of the ecosystem at the time of stress are key determinants of competition response. Changes in relative growth rate between species is suggested to be a useful indicator of subtle changes in competitive potential (McLaughlin, 1985) Shifts in the relative interspecific competitive potential of yellow poplar and white oak in mixed deciduous forest stands in eastern Tennessee have been reported (Doyle, 1983). Competition can both compensate for and amplify the effects of air pollution.

Plant diversity loss is implicated in altered host-parasite relationships (Treshow et al., 1975), plant-pollination relationships (Carlson and Dewey, 1971), plant-pathogen relationships (Manion, 1981), etc. A high degree of association between attach by bark beetles and prior oxidant damage was noted for ponderosa pine in Southern California (Stark et al., 1968). The alterations of resource allocation to various organs due to air pollution may influence the ability of the plant to resist diseases (Manion, 1981, Skelly and Miller). The damage to coniferous forests in Norway by an insect, *Exteleia dodecellar*, was linked to the acid rain problem (Huttunen). A similar case of insects damage to spruce was reported in middle Europe (Edmunds and Alstad, 1982). Alterations in honeybee (*Apis mellifera* L.) populations due to SO_2 and fluoride pollutants have been linked with detrimental effect on fruit yield (Carlson and Dewey, 1971). Reduction in vigor of mycorrhiza lobolly pine due to acid rain may have adverse effect on nutrient and water uptake and growth potential (McMool et al., 1988).

The change in species composition of a natural ecosystem can affect the nutrient cycling adversely by changing the nutrient reserves in various parts of soil profile (Materna, 1984). The rate of decomposition also differ among plant species. Erosion of soils due to tree mortality in forested area may accelerate. The hydrological function of forests is another area of conern related to plant diversity loss.

AZOTOBACTER AND BIODIVERSITY

Azotobacter is a free-living aerobic bacterium dominantly found in Indian soils. It is a non-symbiotic heterotrophic bacterium capable of fixing on an average 20 Kg N/ha/year. Besides, this it also produces growth-promoting substances and is shown to be antagonistic to pathogens. *Azotobacter* are found in the soil and rhizoshpere of many plants and their populations ranges from negligible to 104 gl of soil depending upon the organic mater, pH, temperature, microbial interactions etc. *Azotobacter chroococcum* is the most prevalent species found but other species described include *A. agilis, A. vinelandii, A. beijrinckii, A. insiginis, A. macrocytogenes* and *A. paspali.*

A large fraction of microbial population can dissolve insoluble inorganic phosphorus occurring in soil. Amongst the fungi, the most efficient phosphate solublizers were found in the *Aspergillus* (*A. awarmori*) and the *Penicillium* (*P. digitatum*) group. Other genera include *Curvularia* and

Trichoderma. Among the bacteria, the most efficient isolates were from *Pseudomounas* (*P. strioata, P. rattonis*) and *Bacillus* (*B. subtilis*) group. A few species of cyanobacteria have also been reported to dissolve Mussorie rock phosphate. Microorganisms solublizing phosphates are usually isolated from the rhizosphere portion and their ability to dissolve PO_4 are checked on the medium containing phosphorus. The efficient cultures are selected after a series of tests with growth in specific media containing insoluble phosphate source.

Our agriculture system depends on microbial activities and there appears to be a tremendous potential for making use of microorganisms in increasing crops production. There exists an immense diversity among microorganisms and even within a group of microorganisms. Conservation of this diversity is essentially needed to get continuous benefit of microbes. The maintenance of this potent battery of prototype and new microorganism is the very basis of impressive investment in long term basic and applied microbiological research. The importance of conservation is growing steadily as the number of species preserved is continuously increasing by isolation and genetic manipulation but there is also a need for awareness of the importance of proper preservation of micro-organisms because important microbes preserved improperly may be lost forever or face genetic instability. A few attributes should be kept in mind while preserving the microbial cultures.

There has been little effort to develop type culture collections for microorganisms in India. At research institutions, as part of the research programmes, there exists a small collection of strains, which might have been isolate during the course of research or obtained from some other collections. While such small collections have no doubt, localized importance, they cannot serve the large national interest and the authenticity of the cultures obtained from such personal collections cannot be established or documented. Such an activity should be organized at the national level so that the whole scientific community has access to it. The deposit of personal cultures at an established culture collection centre or gene bank carries its own importance because these will be preserved in the best possible way(s) and be well characterized, authentic and recognized. Thus, the future of microorganisms is in conserving them and picking the best possible strains as per requirements. There is a good collection of bacteria established at IMTECH, Chandigarh and collection of blue-green algae, Fungi and *Rhizobium* at IARI, but there is a need to establish a repository of important micro organisms at National level to conserve these diverse microbial genetic resources.

2

Ways to Conserve Plant Biodiversity

WHY CONSERVATION?

Indian region is a treasure house of wild genetic resource. Wild species and relatives of crop plants contain valuable genes that are of immense genetic value in crop improvement programmes. The important wild related species and types in various crop groups, prevailing under different phytogeographic zones of the country needs particular attention in the agro-biodiversity management system for a sustainable use to help maintain food, nutritional and agricultural economic security.

It is assumed that these genetic resources shall play a positive and unique role in the development of new cultivars, including restructuring the existing ones that are handicapped due to poor expression or lack of one or the other attribute. The most important characters derived from the wild relatives have been the resistances to biotic and abiotic stresses. Such resistance has been mostly observed to be invariably simply inherited, easily transferable and clearly expressed. Ever since the successful translocation by Sears, of resistance factors in wheat against yellow stripe rust (*Puccinia striiformis*), this approach has been followed in several crops against different diseases, using various techniques and degrees of manipulation. Nevertheless, for certain crops with unique biological cycle and narrow genetic variability in their cultivated types, such as potato and groundnut, breeders and more likely to turn increasingly to the wild relatives of the crops for useful characters.

Conservation of biodiversity through a network of protected areas including National Parks, Sanctuaries, Biosphere Reserves, Marine Reserves, Genc Banks, Wetlands, Coral Reefs etc.

1. Conservation of microorganisms, which help in reclamation of wastelands and revival of biological potential of land.
2. Protection of domesticated plant and animal species in order to conserve indigenous genetic diversity.
3. Maintenance of corridors between different nature reserves for the possible migration of species in response to climate, or any other disturbing factor.
4. Rehabilitation of rural poor/tribals displaced due to creation of protected areas.
5. Protection and sustainable use of genetic resources/germplasm through appropriate laws and practices.
6. By establishing database at various levels to document support for protecting traditional skills and knowledge for conservation.
7. Multiplication and breeding of threatened species through modern techniques of tissue culture and biotechnology.
8. Discouragement of monoculture introduction.
9. Control of over-exploitation through cities and other agencies.
10. Restriction on introduction of exotic species without adequate investigations.
11. To establish conservation parks for the rare, endemic document local resources and support for threatened species.

All these efforts need a change in the attitude of bureaucrats in order to solve the problem, develop harmonious association of the forest departments and forest communities towards being partners in conserving biodiversity, adopting new management approaches to the whole problem, understanding the multi-product benefits of biodiversity, through incorporating the latest in technical advancement in research and management aspects of forests.

The conservation of biodiversity is linked with the maintenance of ecological stability and productivity. These attributes are important in sustained development and stable national economy (Mc Neely 1988). The concept of biodiversity is very vast and seated at a *microlevel*. In holistic sense, it can be linked with most of the biological processes in nature. Its conservation involves a number of parameters such as number of species, their population dynamics, distribution, habitat, structure, microhabitats, physical environment, climate, present management and past history. Therefore, the efforts for conservation of biodiversity should be in tune with the processes and its occurrence in space and time, from micro level to *megalevel.*

Biodiversity is the variety and variations occurring in nature, which has sustained the harmonious existence of life on earth. The components of this diversity are so interdependent that any change in the system leads to a major imbalance and threatens the normal ecological cycle.

Acknowledging the need for conservation, the concern for conservation of biodiversity at global level figured for the first time. In the discussions at the UN Conference on the Human Environment held in Stockholm in 1972. The UNEP identified conservation as a priority area in 197. It was only towards 1980's that systematic and concentrated efforts to look at biodiversity conservation profile at international level started with constitution of an AD hoc working group of experts on biological diversity by UNEP in 1987. Eventually an expert group was constituted by UNEP which started it's work in 1989.

Culmination in Conversion on Biological Diversity (CBD) at the UN Conference on Environment and Development (UNCED) held at Reo de Janeiro in June 1992. This convention entered into force on 29th December, 1993. At present, 166 countries are parties to the convention.

This international treaty is a historic treaty in that it not only reflects the commitment of global community for conservation and sustainable used of biodiversity but also visualises sharing of benefits arising out of utilisation of genetic resources with the countries of origin.

Objectives

According to this, the main objectives of the convention are:

1. The conservation of biological diversity.
2. The sustainable use of components of biodiversity.
3. The fair and equitable sharing of benefits arising out of the utilisation genetic resources.

The objective of equitable sharing of benefits has to be further seen in the reaffirmation of sovereign rights of states over their own biodiversity which entitles a country to equity and fairness in sharing of benefits arising out of the utilisation of their genetic resources. The 23 preamble paragraphs of the convention recognise and reaffirm.

1. The intrinsic value of biological diversity.
2. The sovereign rights of states over their biological resources;

3. The fundamental requirement of *in situ* conservation of ecosystem and natural habitats.
4. The supporting role of *ex situ* measure.
5. The vital role of local communities and women in the conservation and sustainable use of biological diversity.
6. The desirability of sharing equitably the benefits arising from the use of traditional knowledge, skits, innovations and practices.
7. The importance of and the need to promote regional and global co-operation for conservation, and
8. The requirement of substantial investments to conserve biological diversity.

Implications

The significant implications of the provisions of the convention and the main obligations imposed on the contracting parties are summarized below:

1. The parties are required to take measure for *in situ* conservation of biological diversity, promote rehabilitation and restoration of degraded ecosystem, and ensure protection of threatened species. (Article 8)
2. The parties are obliged to respect. Preserve and maintain knowledge, innovation and practices (8).
3. The parties are also to adopt measures for *ex situ* conservation of components of biological diversity, for complementing the *in situ* efforts (Article 9).
4. The parties are to facilitate access to genetic resources on mutually agreed upon terms with prior informed consent of the country providing these resources (Article 15).
5. The recipient country is to share in a fair and equitable way the results of the research and development and the benefits arising from the commercial and other utilisation of genetic resources with the party providing such resources. The convention calls for transfer of relevant technologies, including biotechnology on fair and most favourable terms from the developed to developing nations which provide genetic resources used for the development of these technologies (Article 16).
6. It also calls on the private sector to facilitate access to and transfer of such technologies developed by them (Article 16.4).
7. The contracting parties are to cooperate in this regard to ensure that patents and other intellectually property lights are supportive of and do not run counter to the objectives of the convention (Article 16.5).

8. Recombinant DNA technology (or genetic engineering) are increasingly gaining ground and are being seen as potentially useful for various sectors by experts. Recognizing that introduction of these biotechnologies requires utmost care and caution based on precautionary principal (outlines in Agenda 21) particularly because recall of an introduced organism in nature would well be possible, the convention commits the parties to consider an international protocol for safe transfer, handling and use of any living modified organism resulting from biotechnology (Article 19.3)
9. The parties are also to take measure for facilitation access on a fair and equitable basis and on mutually agreed upon terms to the results and benefits arising from biotechnologies based upon genetic resources provide/transferred (Article 19.2).
10. The developed country parties are committed to contribute to a fund to enable developing country parties to meet the "agreed full incremental cost' for implementing the provisions for their convention (Article 20.2)
11. The financial mechanism is to operate within a democratic and transparent system of governance and "function under the authority" of the conference of the parties. (Article 21)

Important points to note about convention are the facilitation of access by developing countries is linked with equitable sharing of benefits; thus making it a two-way process. Technology flows are also found on the principle of equity with the developing countries providing resources and traditional technologies, and developed countries sharing and transferring technologies including biotechnologies and providing financial resources to help developing countries to meet their commitments and realize benefits.

From the foregoing discussion, it is amply clear that the convention is the first global, comprehensive agreement to address all aspects of biological diversity, genetic resources, species and ecosystems which would have revolutionary and far reaching implications. Being based on considerations of equity and shared responsibility, the convention envisages reciprocity of arrangements between developed and developing countries, thereby promoting a renewed partnership between them.

The main implementation measures for the convention are to be through national strategies, plan or programmes, to be developed in accordance with each country's situation and capabilities. Although in its text, the convention commits the contracting parties to take substantive action in many areas as under.

Action Plans

1. Development of National plans, strategies or programmes for conservation and sustainable use of biodiversity and integrating these into relevant sectoral or cross-sectoral plants, programmes and policies.
2. Inventorisation and monitoring of components of biodiversity and of processes adversely impacting it; developing and strengthening of *in situ* mechanisms for biodiversity conservation both within and outside protected areas.
3. Development of *ex situ* measures for biodiversity conservation, as a complement to *in situ* approaches.
4. Restoration of degraded ecosystem and recovery of endangered species.
5. Adopting measures to avoid and minimize adverse impacts on biodiversity.
6. Protecting and encouraging customary use of biological resources that are compatible with conservation or sustainable use requirements.
7. Adopting economically and socially sound measures that act as incentives for conservation and sustainable use of components of biodiversity.
8. Promoting and encouraging research contributing towards achieving the objectives of the convention.
9. Developing educational and public awareness programmes with respect to conservation and sustainable use of biodiversity.
10. Facilitating access to genetic resources on mutually agreed terms and prior informed consent, and taking measures for fair and equitable sharing of benefits arising from utilisation of the resources thus transferred;
11. Facilitating access to and transfer of technology, including biotechnology to developing countries under fair and most favourable terms.
12. Facilitating exchange of information relevant to biodiversity.
13. Promoting scientific and technical co-operation with other parties.
14. Consideration of a protocol for safe transfer, handling and use of living modified organisms resulting from biotechnology and
15. Providing new and additional financial resources by the developed country parties to enable the developing country parties to meet the agreed full incremental costs for implementing the provisions of this convention.

CONSERVATION FOR SUSTAINABLE USE: A HOLISTIC APPROACH

In view of the large scale exploitation of useful diversity of various plant species from forest and open areas, particularly for medicinal and other economic plants, and the prevalence of fragile ecosystems in many parts of the country and also the existence of diversity in several useful co-existing biological species, it becomes important to conserve these plant species/co-existing species and systems either by way of domestication and cultivation or by other *ex situ* or *In situ* conservation methods for their sustainable use. The biological factors that determine the conservation approaches and the most common methods used for germplasm conservation of different PGR categories, were clearly defined viz., ecosystems/ agro-ecosystems, specific habitats, gene pools, special genetic stocks, etc. A balanced application of these, *ex situ/* on farm/ *In situ/* in vitro conservation approaches and technologies, would be needed so as to suitably register location specificity, case-to-case basis, cost-effectiveness, feasibility for retrieval and networking, etc. There is a need to give emphasis on `cultivation of wild forms' (domestication), rather than `collecting from the wild alone' because the former lays more emphasis, over the latter approach, on botanical identity, quantity for supply, genetic improvement, quality control, etc., of the conserved material. Such cultivation (domestication) may have to be initiated in the respective agro-habitats showing micro-climates similar to the niche of wild species.

Conservation strategies must have a holistic approach and also dynamism and vibrance of genetic evolution that should be adopted for sustainable utilization of natural resources linked equitable sharing of benefits accrued from the use of plant-biological diversity of the Indian origin.

A collaborative approach should be followed by the national agencies and regional bodies responsible for policy, planning, research and development for agriculture (including horticulture) and forestry to cover agro-biodiversity and biodiversity (wild species) simultaneously.

Diversity mapping should be done along with inventorization at micro-level for effective management, conservation and use of crop plant species and their relatives.

Priorities drawn for future survey, collection and inventorization of PGR diversity should be based on gaps identified.

Agro-ecological niches should be clearly identified along with the prevalent crops and cropping systems, the custodians of diversity, i.e., indigenous community(ies), their ethnic groups and economic background and other developmental and ecological factors related to the area for consideration of their suitability for on-farm conservation may be demarcated for action-oriented programme.

Sub-regional consortia and regional networking should enable to safeguard the regional interests for genetic diversity. Contiguity of adjoining areas housing similar diversity for particular species gene pool can be availed, to share the benefits of *on farm* conservation/*In situ* conservation across the geo-political barriers.

Ample research back up should be developed, ensuring funds and other resources, at the national gene bank to enable effective scientific monitoring, guidance and future direction to such activities.

Linkages should be strengthened with other governmental departments, state universities, non-governmental organizations and communities and gradually the role of each interested party should be clearly defined for getting better results.

Specially designated genetic reserves, gene sanctuaries and/or genetic gardens should be earmarked for wild species of food value or other economic importance. Conversely, such species should be carefully monitored and inventoried in the existing biosphere reserves. Strategies, approaches and technology, including biotechnology, should be developed/standardized for gradually bringing these wild economic species under cultivation.

TYPES OF CONSERVATION

The two major elements involved are conservation and utilization of biodiversity. Both are most important and have to be pursued with equal emphasis.

Conservation of biodiversity has been possible primarily through *in situ* conservation, i.e. *on site* conservation in the natural habitats. Such conservation is relatively inexpensive and takes care of all the species (plants, animals and microorganisms as an interacting system) in a particular habitat. *In situ* conservation also warrantees the continuation of the ecological processes and organic evolution, thus helping the concerned biota to continuously evolve in response to newer environmental changes.

The conservation of cultivated crops and domesticated animals has been primarily possible through *ex situ* conservation ever since agriculture evolved. Today there are many ways to achieve this: seed, pollen; sperm, ovule/ova, embryo, tissue and even DNA banks. These banks are working for conservation. The material is in the bio-banks available for distribution. The CGIAR system is reported to have over 500,000 individual entries of plant materials of over 140 crop plant species. This wealth is supposed to be held in trust for the benefit of humankind through use of science and technology.

No country, howsoever rich in biodiversity, can be self-sufficient in genetic resources. There are instances where crops introduced form other countries have become major crops in the countries of adoption, e.g. wheat, maize red pepper, tobacco, cashew, rubber, tomato, potato, etc. in India. Therefore, countries are dependent on genetic resources from other countries. It is also important that each country has not only to ensure conservation of its own biodiversity but also seek to enrich genetic diversity of species which have, over a period of time, become naturalized crops and have entered their trade. Thus it is an interdependent effort rather than an independent one where each country each country is concerned with itself.

BIODIVERSITY CONSERVATION THROUGH BIOTECHNOLOGY

Biotechnology is the *third wave* in biological sciences and represents such an interface of basic and applied sciences, where gradual and subtle transformation of science into technology can be witnessed. Biotechnology is also defined as the applications of scientific and engineering principles to the processing of material by biological agents to provide goods and services. United States Congress's Office of Technology Assessment defined biotechnology as any technique that used living organisms to make or modify a product, to improve plants or animals or to develop microorganisms for specific uses. The document focuses on the development and application of modern biotechnology based on new enabling techniques of recombinant-DNA technology, often referred to as genetic engineering.

Since biotechnology involves the use of all life forms for human welfare. Therefore, extinction of wild species and destruction of ecosystems has been a major concern of policy markers and environmentalists. A discussion on biotechnology involving biodiversity is relevant, because biodiversity is being utilized to provide genes from wild species for biotechnological exercises.

Several biotechnological tools are now available to tackle various specific problems and to enhance the potentials of grass covers such as feeding value, propagation and persistence. Basically, in range grasses, several tools of biotechnology, comprising endosperm and anther culture, somaclonal variation, protoplast culture and fusion, transformation, etc, can address breaking the apomictic barrier for recombination for recombination of desired traits. The grass biotechnologists are actively engaged to achieve this goal. Besides, the mapping of genes, isolation, protoplast culture and fusion, transformation, etc. The grass biotechnologists are actively engaged to achieve this goal. Besides, the mapping of genes, isolation, cloning and characterization of important genes/traits in range species as well as in other crops can ultimately improve the value of grass cover through genetic transformation in the pasture species. It is envisaged that by following these approaches anti-quality factors can be suppressed and the digestibility of grasses can be increased. The vast array of gene mapping technologies now allows even tracking the genes for polyphonic traits and the metric traits like seed production; the biomass would also be hopefully improved. The current extensive research on nitrogen fixation in the world's crop plants would certainly go a long way in adding to the fertility and diversity of grass covers. Propagation of specific genotypes of endangered but useful species on a large scale in a short time frame is possible through the modern biotechnological approaches.

Genetic resources are renewable, provided they are well managed. Conservation and management of biodiversity, closely linked to the conservation and wise use of natural ecosystem, is of major importance for sustainable development. The tropical and sub-tropical areas are rich in term of plant genetic resources. India is particularly rich in biodiversity, including the ecosystem, species and genetic diversity, species and genetic diversity, by virtue of its varied climates and physical features.

In view of the unique importance of the grass covers in terms of total agro biodiversity scenario, it is appropriate to the realistic and assume a the following as priority concerns in the future context. The updating of grass cover/grazing resources of India is required to be done on priority. Almost three decades have passed when grass cover of India was documented. The environmental degradation has caused enormous impact not only on the flora and fauna but the area represented activity. Remote sensing and GIS approach have been refined and made more widely available in the country during this period. These can be suitably used for the required updating.

1. Collection, evaluation and exploitation of promising grasses/legumes, which have not yet been tested for their prudential as forage species.
2. Inventory of grazing routes, grazing system and designing sustainable production system for migratory grazers of Himalayas as well as of the Thar Desert.
3. Soil and water conservation may be taken up an integral part of grassland development.

4. Greater thrust is required to be made for selection of native types from stress prone habitats i.e. ravines, riverine, alkaline, acidic, hot desert, cold desert and steep slopes etc, their characterization, conservation and value addition.
5. *In situ* conservation of genetic wealth of above mentioned stress prone areas through botanical gardens can help in preserving several species which are facing extinction are less important today but may be of much value in future.

Ever since primitive man learned the art of farming and realized the economic utility of plants, he started saving selected seeds or vegetative propagules from one season to the next. Infact, he was practicing a type of germplasm conservation and management. The primitive cultivars and their wild relatives constitute a poll of genetic diversity (germplasm or gene bank) which is invaluable for breeding programmes. However, the population and needs of man have steadily increased over time leading to persistent exploitation of natural resources. The pressure thus created led to the way for gradual exclusion of traditionally used genotypes. In this way, some of the valuable gene pools might be lost unless coordinated efforts are made towards the conservation of genetic stocks all over the world.

Realizing the danger of erosion of genetic resources, the objective of providing necessary support to collection, conservation and utilization of the plant genetic resources is concern of the world now.

Germplasm Conservation Using Cell and Tissue Culture Objectives

1. To ensure the maximum level of genetic stability of the conserved material that is consistent with practical storage periods and with the available technology and resources.
2. To maintain accurate and detailed records concerning the biology of the conserved material, and its know history prior to collection. It is important to known the disease status of the material, and its full culture history including media details, the use of growth regulators, frequency of transfer etc. Cultures and derived plants should be characterized at some level of morphology performance and any known genetic characteristics, including analysis of gene products and DNA fingerprinting where this is possible and appropriate. There is a responsibility to disseminate this information as widely as possible.
3. To develop and improve technology at all levels, and to increase the efficiency and utility of in-vitro techniques in the conservation of plant germplasm.

4. To construct and maintain a policy for the in-vitro conservation of those plant genetic resources appropriate to the collection with due regards to Internationals. National and Regional/Local priorities.
5. Conservation of somaclonal and gemetoclonal variations in cultures.
6. Maintenance of recalcitrant seeds.
7. Conservation of cell lines producing medicines.
8. Storage of pollen for enhancing longevity.
9. Conservation of rare germplasm arising through somatic hybridization or other methods of genetic manipulations.
10. Delaying the process of ageing.
11. Storage of meristem culture for micro propagation, micro grafting and production of disease free plants.

Applications of In vitro Conservation

Techniques for in vitro conservation of plant germplasm may be applied in any situation where the basics of tissue culture and cell culture are available, and technologies discussed above can be applied to wild species, the crops of agriculture and horticulture and tress. Methods for minimal growth storage are available to most laboratories, and it becomes a matter for local decisions as to whether attempts at conservation should be made and in some way be formalized. This would mean, a commitment to adequate documentation for conservation purpose and to disseminating information about the collection, however limited, to suitable parties.

Cryogenic storage requires relatively simple equipment and facilities, but these are often costly which limits its application. Liquid nitrogen is a relatively common material that can be obtained in most parts of the world and its availability only occasionally restricts the use of cryopreservation in conservation. As this technique provides dramatically extended storage periods, it is perhaps even more important that care is paid to the quality of documentation and record keeping associated with the conservation processes. Major organisations such as IBPGR and ATCC (American Type Culture collection) are in a position to apply, develop and promote in-vitro technology in the service of genetic conservation. Importantly, they are sufficiently large and international to be a vehicle for collection, collation and dissemination of information. They have a responsibility to improve in-vitro techniques and to be involved in edition whilst such organisations can be seen as an effective world wide model with regard to in-vitro conservation. Other national and regional organisations should be prepared to take on a similar rate within their sphere of influence. There are also obvious applications of this technology in all aspects of research and development, including the work of commercial organisations.

Progress in the development of plant cell and tissue culture techniques for long term germplasm conservation has been quite significant especially in view of the level of activity and small number of workers involved. The techniques of cell culture in particular have been so refined that only a few species are likely to the recalcitrant to known procedures. During the past decade studies on the freeze preservation of cell culture have trended from basic aspect.

The evidences available so far indicate that freeze preservation is the most reliable approach to the long term preservation of cell culture which possesses the biosynthetic capacity for synthesis and accumulation of secondary metabolites (Kartha, 1987).

Accumulation of secondary metabolites requires a certain degree of differentiation, often accompanied by increase in cellular and volume as well as lengthening of generation period. Therefore the identification of appropriate growth stages most conductive to freezing becomes critical in devising an appropriate cryoperservation strategy of cell producing secondary metabolite.

Germplasm conservation using cell, tissue, organ culture has been reported in 60 species including various crops (Withers 1985, Bajaj 1987). From the information available, conservation is possible in both monocotyledonous and dicotyledonous families. Temperate species figure more prominently then tropical species.

In biotechnology laboratories, freeze preservation and storage of plant material can be of enormous value in maintenance of stock cultures and samples that await screening for future used. Almost an indefinite number of replicates from germplasm of large number of plant species can be stored in-vitro using little space in contrast to the thousand of hectares of land that would be required for the same number of plants if maintained *in situ*. Thus, the material that would otherwise be discarded due to resource can certainly be stored without reservations.

Recent Development in *In Vitro* Conservation

The role of the gaseous environment within the culture vessel in growth and development has been considered with view of extending the relatively low-technology approaches available for in-vitro conservation. Volatile compounds in the gaseous culture environment, together with carbon dioxide, will affect growth rates in-vitro (Deproff, 1985, Gould and

Murashige 1985, Beicher et. al. 1987). Other physical factors will be also moderate to the biological effects of the gaseous environment. The most significant of these to date are the total volume of the gas space above the culture medium and the diffusion distance from the culture tissue to the point of gas exchange with the external atmosphere i.e. the seal of the vessel.

In a study using regenerated carrot plantlets range of culture, vessels were constructed with identical gaseous volumes but differing diffusion distances (Baltson et. al., 1988). Biomass measurements at various stages during a 30 day culture cycle showed that growth increase was independent of the gaseous volume above the cultures atleast within the range studies, but was significantly affected by changes in the diffusion distance. Increasing the diffusion distance from 50 to 150 nm, which effectively doubled the period between culture transfers, halved biomass increase. This provides an inexpensive technique that provides some degree of extended storage for in-vitro conservation systems.

It is also possible that the advent of gas permeable culture vessels might assist conservation in-vitro at the level of unfrozen storage (Tanaka et. al., 1988). These containers of various designs are made of relatively inexpensive plastic. Their benefits are for organised plantlet cultures in the main, as well as they allow autotrophic metabolism to develop to a significant extent, reducing the dependence of the culture upon high levels of exogenous carbohydrates, and similarly reducing the need of complex growth regulators regime. This simplification of the culture environment allows for significantly prolonged growth in culture vessel, with basic nutrients being added using, for example, a hypodermic syringe for sterile transfer of basic medium.

Recent progress in the techniques of induction and manipulation of somatic embryogenesis (Ammirate, 1989) may also have considerable implication of prolonged maintenance of tissue-derived in-vitro, with the possibility of a high degree of genetic stability. The technique may be particularly applicable when the material of interest is an isolated cell or protoplast population that can be manipulated into somatic embryogenesis relatively simple and at high frequency. In particular, techniques have been developed that allow for the desiccation of somatic embryos in a way that reflects the physiology and development of the zygotic embryo of the orthodox seed (Ammrito, 1989). The desiccation system relies upon pre-treatment of the somatic embryos with compound such as ABA, IBA and sucrose, added to the culture medium and has been successfully demonstrated for species as diverse as grape (Gray, 1987), soybean (Bendrof and Slawinska, 1988) and wheat (Carman, 1988), whilst not applicable to species whose zygotic embryos are in any case recalcitrant, (Grant, 1986). These developments offer the potential for extended storage of germplasm derived from in-vitro system by storing dried, somatic embryos in a manner similar to that used for conventional seed storage. This might provide for long-term storage without an absolute requirement for the relatively advanced technology associated with liquid nitrogen storage systems. If the degree of drying and its associated cellular adaptations do accurately reflect the characteristics of the orthodox seed it might be possible that the material could be kept at room temperature, or the more conventional at –29° C of the mechanical freezer and yet achieve the level of genetic stability required for effective in-vitro conservation (Roberts et. al. 1981). Further investigation will be required to see if dried somatic embryos can indeed produce the storage characteristics, without genetic alterations of the zygotic orthodox counterparts, coupled with a coating technique used to produce artificial seeds (Redenbaugh et. al. 1988), this system has even greater potential.

The relatively recent development of a field collection system where material in the field is taken directly into a protective by possible temporary in-vitro environment (Yidana et. al., 1987) is a significant enabling technology to be used in conservation. The technique employs various levels of sterilization for explants in the field and incorporates antibiotics into the growth medium. Explants could be collected *in situ* even inmost remote habitats and transported back to the laboratory with the minimum of loss, and without the need for transported back to the laboratory with the minimum of loss, and without the need for transporting whole plants or large pieces of them.

To date the successful cryopreservation options available for in-vitro plants tissues cultures have largely been based upon slow cooling techniques (Kartha 1985, Withers 1987) although faster rates have been beneficial at times (Grout and Henshow, 1978; Kartha, 1987; Withers et. al., 1988). At slow cooling rates there is degree of cellular dehydration following extracellular freezing and then the plunge into liquid nitrogen that typically concludes such slow looking protocols.

To take pure water through the glass transitions temperature to form a glass, as opposed to crystalline ice requires or ultra rapid cooking rates under practical circumstances complete vitrification of any significant volume of water is rarely, it ever obtained (Angell, 1982; Reid et. al., 1985; Taylor, 1987). A working definition of the vivification does allow for some crystalline ice embedded in a glassy matrix where ice crystal size does not exceed 10 nm diameter (Rienle, 1968). Aqueous solutions such, as those within and surrounding cells in suspension, will vitrify more readily than water alone, and at lower cooking rates, as the solute concentration increases solutions undergo vitrification more readily and so cells incubated with very high levels of cryoprotectant solution can e taken through vitrification as an alternative to the more conventional techniques of cryopreservation. The benefit is that following the appropriate pre-treatments required to achieve vitrification, the need for a controlled rate of cooling into liquid nitrogen is removed and simple quenching is likely to be sufficient.

Recently, the techniques of vitrification have been successfully applied to *Asparagus* and *Brassica* cells cultured in-vitro adding another potential valuable option to the methodology for cryogenic storage (Uragami et. al, 1989; Langis et. al., 1989).

In the case of *Brassica campestris* suspension cells were first equilibrated in 1.5 M ethylene glycol solution and then dehydrated in a complex medium estimated as being approximately 19 osmolal. The high concentration of extra-cellular solute well result in water loss from the cell, leaving a high concentration of intra-cellular solute also. In this instance vitrification was achieved by plunging a 0.5 ml plastic straw containing cells and medium directly into liquid nitrogen. To avoid crystallization during thawing a rapid warming rates has been employed, which was close to 600° C per min. in this instance. Significant levels of survival were recorded using tripheny tetrazolium chloride (TTC) assay as indicator, and these results confirmed by callus re-growth.

TISSUE CULTURE APPROACHES

Generally, in vitro maintained plant cells, tissues and organs are preserved under appropriate conditions for long terms storage as *gene banks*.

In-vitro methods are the most useful for conservation vegetatively propagated plant, species with recalcitrant seeds and genetically engineered materials. The potential advantages of in-vitro conservation include:

1. Requirement for little space for preservation of a large number of clonally multiplied plants.
2. Maintenance of the material in an environment, free of pest or pathogens.
3. Protection against dangers of natural environmental hazards.
4. Availability of nucleus stock to propagate a large number of plants rapidly whenever necessary and
5. Minimizing the obstacles generally imposed by quarantine system on the movement of live plants across national boundaries, since they are raised and maintained in aseptic environment. The present chapter deals only with the application of plant cell and tissue culture in germplasm

conservation. Plant tissue culture based biotechnology has opened new avenue for the propagation and improvement of plant species. A wide range of plants has now been regenerated through techniques of tissue culture. Many palm species including *Cocus nucifera*, Oil palm and *Phoenix dactylifera* (Data palm) multiplies mainly through seeds, and there is no efficient method of vegetative propagation.

The development and application of plant tissue culture system of propagation and conservation would therefore have obvious benefits.

1. Many long lived forest trees including both angiosperm and gymnosperm do not produce seed until a certain age and must be propagated vegetatively when it is desired to reproduce similar to the parental genotype.
2. One area of research currently being investigated is the possibility of very rapid clonal propagation of fruit trees by means of tissue culture techniques. These techniques have special relevance to certain fruit tree cultivars, which are either difficult and/or expensive to propagate by conventional means.
3. Somatic embryogenesis for higher frequency plant regeneration and synthetic seed production (Gupta et. al., 1993).
4. Generating genetic variability for better genetic traits (Larkin and Scowcraft, 1981) somatic hybrid production for increased fields.
5. Cryopreservation of valuable gene pool for future use. (Withers, 1985).
6. Production of cell line and recovery of plants for disease resistance, drought resistance, salinity resistance.
7. In-vitro propagation, improvement and conservation of rare/vulnerable/ endangered plant species (Withers, 1986).

In Vitro Conservation

Conservation in-vitro is wholly dependent upon the techniques of plant cell, tissue and organ culture, and is appropriate *in situ*ations where conventional seed storage cannot or is not to be employed.

In particular in-vitro conservation is used for vegetatively propagated material species with recalcitrant seed and material biology. This later material may have been costly and limited in quantity and have characteristics whose stability during repeated mitotic divisions and following meiosis are unknown. It is, therefore prudent to conserve stocks of this material to ensure consistency in subsequent experimentation of production processes, and as definitive source of reference material.

The material stored in vitro may be protoplast, isolated cells grown in suspension or on semi-solid medium, meristem cultures at various stages of development or organised plantlet. It can be assumed that genetic stability within the in-vitro systems increase as the complexity of the cultured material, with completely differentiated plantlets in culture having the least risk of genetic alteration during an in-vitro excursion (Karp, 1989). Consequently, if the storage conditions are to permit some level of growth and metabolic activity then organised plantlets are the conservation material of choice, but where growth is to be completely inhabited as in cryopreservation excised shoot meristems may be the more suitable explains (Ford - Lloyd and Jackson 1986; Withers 1987) where cell or protoplast cultures are to be maintained and there is no choice as to a preferred level of organization then care must be taken to ensure the maximum genetic stability, Preferable by opting for cryopreservation.

At the simplest level conservation in-vitro is achieved by repeated sub-culture continuing the continuing the protection from adverse environmental conditions and pathogens that is inherent in the in-vitro systems. This is expensive both in terms of labour and the basic material of the culture process and whilst least demanding in terms of the level of available technology, has a considerable risk of material risk of material loss due to human error or the failure in-vitro security e.g. the invasion of pathogen during transfer.

Modes of Propagation

There are large number of threatened/rare/endangered plant species on which in-vitro propagation of threatened plants could be achieved by the following methods.

1. Clonal propagation
2. Somatic embryogenesis
3. Organogenesis
4. Callus differentiation

Clonal Propagation

A variety of plants species can be propagated through forced axillary shoot bud proliferation. Thus plants produced through this methods are morphologically and genetically similar to the parental plant. Main benefits of these methods are:

1. Rapid multiplication of superior clones and maintenance of uniformity.
2. Multiplication of disease free plants.

3. Multiplication of sexually derived sterile hybrid.

Clonal propagation is achieved by placing sterilized shoot tip or axillary bud on culture medium that it's sufficient to induce formation of multiple buds.

In-vitro propagation of plant using apical shoot or nodal segments was obtained in *Anogeissus sericea* (Kaur et. al., 1992). Propagation of *Nepenthes khasiana*, a threatened plant species endemic to Meghalaya was obtained using nodal segment from the field grown plants (Rathore and Tandon, 1989) and from seedling raising explants (Rathore et. al., 1991, Latha and Seent, 1994). *Gerbera auranitca*, threatened ornamental species recovered by the culturing axillary shoot buds on nutrient medium (M.S. medium) containing 5 MMBAP, which differentiated into shoots and differentiated shoots include root on IBA (5 to 10 MM) containing medium. A wide range of plants has now been regenerated through this technique.

Somatic Embryogenesis

The process of development of embryogenesis. This was first demonstrated in Carrots (Steward 1959) where bipolar embryos developed from single cell. Further, the tissue derived from these somatic embryos spontaneously developed into embryos again. However, S. E. is influenced by plant extracts, growth regulator and by the physiological state of calli.

In vitro propagation of threatened medicinal plants through somatic embryogenesis have been obtained in *Podophyllum hexandrum* (Arumugam and Bhojwani, 1990), Santalum album (Bapat et. al. 1990).

Mura Costa et. al. (1993) have reported in-vitro plant regeneration in *Ocotea catharinensis*, an endangered Brazilian hard wood forest plant through somatic embryogenesis.

Research work is also being done to produce artificial seeds via this pathway, but a lot of work has to be done in the area of embryo encapsulation, storage and conversion into plants.

Organogenesis

The process of regeneration of plants from the callus called as organogenesis. In *Meconopsis panicutlata*, regeneration was observed through organogenesis (Sulaman, 1994) and same type of mode of regeneration has been observed in *Meconopsis simplicifolia*, an endangered plant species of Himalayan region (Sulaman et. al., 1993). *Rheum embodi* - a threatened plant was regenerated from shoot tip and leaf explant through organogenesis (Lal and Ahuja 1998).

A number of such examples can be sited with have developed the complete plant through regeneration by organogenesis thus helping in conservation of germplasm.

Callus Differentiation

In vitro regeneration has been reported in *Coptis teeta*, an endemic and endangered plant of Arunachal Pradesh through callus differentiation into shoots and subsequent induction (Tandon and Rathore 1992). In *Valeriana wallichii*, a threatened plant was propagated through callus differentiation and recovered complete plantlets (Mathur and Ahuja 1991).

In vitro Storage of Germplasm and Cryopreservation

Assuming that differentiated plantlets are to be stored in-vitro, with all that this implies for genetic stability, and then lowering growth temperature to slow metabolism and development has previously reduced the labour and expense for repeated subculture. The security of the cultures is also improved as the frequency of the physical interventions of subculture is also reduced. Similarly, the use of osmotic stress and the addition of growth retardant to the culture medium have effectively prolonged the interval between culture transfers (Withers 1987, Pritchard et. al., 1986).

The greatest genetic stability is achieved by storage of propagules, capable of regeneration in-vitro liquid nitrogen i.e. **cryopreservation** (Kartha 1985, 1987, Grout and Morris 1987, Withers 1987) where possible these propagules will be excised meristems, capable of direct regeneration into plantlets, but they may, of necessary, by isolated cells or protoplasts.

Table 2.1: In vitro regeneration of threatened/rare/ vulnerable /endangered plant species

Name of Plants	Explant Used	Medium	Mode of Regeneration	Reference
Aconitum carmichaeli	Shoot tips & axillary buds	MS	Clonal multiplication	Hatano et. al. (1987)
Aconitum nepellus	Nodal shoot	MS	Axillary shoot proliferation	Robert cervelli (1981)
Aconitum novebora-acense	Nodal shoot	MS	Axillary shoot proliferation	Robert Cervelli (1981)
Anogeissus sericea	Nodal segment (seedling ex-plant)	MS	Axillary shoot proliferation	Kaur et. al. (1992)

Name of Plants	Explant Used	Medium	Mode of Regeneration	Reference
Caralluma edulis	Shoot segments	MS	Clonal multiplication (shoot tip)	Hatanco et. al. (1987)
Commiphora wightii	shoot segments	MS	Axillary shoot proliferation	Barve and Metha (1993)
Coptis teeta	Hypocotyl segment	MS	Callus culture	Tandon and Rathore (1991)
Gerbera aurantiaca	Axillary bud	MS	Axillary shoot proliferation	Mayer Vanstaden, (1998)
Nepenthes Khasiana	Mature nodal segment	MS	Axillary shoot proliferation	Rathore and Tandon (1989)
Nepenthes Khasiana	Axillary bud	MS or Wood Pt. Medium	Axillary shoot proliferation	Latha and Seeni (1994)
Ocotea catharinensis	Zygotic embryos	MS	Somatic enbryogenesis	Casta et. al. (1993)
Oncidium varicosum	Seeding	Knudson medium	Root tip culture	Kerbeay (1983)
Picrorhiza Kurroa	Axillary bud	MS	Axillary shoot proliferation	Upadhyay et. al. (1989)
Podophyllum hexandrum	Zygotic embryos	MS	Somatic embryogenesis	Arumugan and Bhojwani, (1989)
Rauvolfia serpentina	Low temperature	MS	Storage	Sharma and Chandel (1992)
Rheum embodi	Axillary bud	Liquid	Axillary shoot proliferation	Lal and Ahuja, (1989-93)
Rumex acetosella	Axillary bud	MS	Somatic embryogenesis	Calafic et. al. (1987)
Trichopus zeyanicus	Axillary bud	MS	Axillary shoot proliferation	Krishan et. al. (1995)
Valeriana wallichii	Axillary bud	MS	Callus culture	Mathur and Ahuja (1991)
Wrightia tomentosa	Axillary bud	MS	Axillary shoot proliferation	Purohit et. al. (1994)

Cryopreservation

In this system stability is imposed by ultra low temperature and storage is at, or close to -196° C using liquid Nitrogen (or the vapour immediately above it), as practical and convenient oxygen. At such temperature normal cellular chemical reactions do not occur as energy level are too low to allow sufficient molecular motion to complete the reaction. Water exists either in a crystalline or glassy state under these conditions and such high viscosity (> 1013 poises) that rates of diffusion are insignificant over time spans measured at least as decades. The majority of the chemical changes that might occur in a cell are therefore; effectively prevented and so the cell is stabilized o the maximum extent that is practically possible.

Unfortunately, that is not to say that biological material successfully. cooled to -196° C is in state of complete suspended animation. Certain type of chemical reaction can still occur at these temperatures. Such as the formation of the free radicals and macromolecular damage due to ionizing radiation. The only real threat to genetic stability comes from such reactions, especially those that damage nucleic acids. Any damage that does occur will necessarily be cumulative as enzyme repair mechanism are also totally inhibited at these low temperatures.

While there are as yet few quantitative data on genetic stability at ultra low temperature for higher organisms, studied by Ashwood, Smith and Grant (1977) indicate that to reach a D10 level (Where D10 is the radiation dose resulting in 90% mortality of the population) a frozen cell population would have to be exposed to back ground radiation for some 32,000 years. It is also noteworthy that dimethyl sulphoxide, probably the most commonly used cryoprotectant in vitro preservation and may aid in radiation damage. (Finkle, et. al., 1995).

The potential of conservation system for in-vitro material based upon cyrogenic storage is therefore, clear and the technique has become relatively widely used.

Limitations of *In vitro* Conservation

There are practical limitation to any conservation system that does not provide a maximum genetic resource whilst, at the same time, conserving the potential of the system to produce whole plants. The level of limitation that is acceptable depends upon individual circumstances. However, where resources are not themselves limiting and the pressure of the time allow for significant research and development, the technique of cryopreservation must be the current in vitro conservation method of choice.

Yet, application of drying to somatic embryos, and the construction of artificial seeds, presents and exciting prospect for the future. Once successfully processed, the dried somatic embryos might be effectively stored at -20° C using a simple domestic refrigerator, which is globally available technology. However, the pre-treatments necessary to achieve desiccation without loss of viability are often complex and currently it is only the tissue of somatic embryos that can be treated in this way.

It is likely that the somatic embryo shares much of its genetic expression and developmental pathways with the zygotic embryo. Both have a potential to withstand extremes of desiccation when the zygotic embryo, and that both have a potential to withstand extremes of desiccation when the appropriate sequence of genes are in operation. In a developing orthodox seed (Roberts, 1973) the hydrated, zygotic embryo will respond to chemical stimuli from the parent plant to develop desiccation resistance and thus dry to the low water content characteristic of the see. It appears that a somatic embryo can be triggered artificially in much the same way (Ammirato, 1989). This technique has obvious potential in circumstances where embryogenesis is both possible and desirable, and where the zygotic embryo does not show recalcitrant characteristics (Grout, 1986). However, it is not clear whether the techniques for drying could be feasible for other somatic cell and tissue types, that have patterns of development and differentiation far removed from that of the embryo whilst it is true that leaf callus cell for instance, contain the genes used in the developing embryo to tolerate desiccation during seed formation. It is another matter to be able to exploit this aspect of the genome independently of the rest of the embryo development. (Morris and Grout, 1990).

As cryopreservation has developed as a technology, and surviving material examined more critically, the concepts of optimized procedures and modified culture conditions by support recovery growth following thawing have been considered in more depth.

It is probably an exceptional occurrence when a cryopreserved cell is thawed to physiological temperatures with structure and functions completely unaltered, ready to resume typical activity. More commonly, surviving cells will have varying cells will have varying degree of damage that may will be able to repair, depending upon available reserves and metabolic integrity, and may require specific "recovery" media or past thaw conditions to ensure a reliable return to a known and predictable performance. Those that cannot effect satisfactory repair will die, sometime after thawing.

Different circumstances of cooling and warming may well generate different lesions and consequently differing requirements for repair, and it is not improbable that, even after repair, recover ells might show a range of alteration in both structure and function.

OTHER METHODS

Cold Storage

Germplasm conservation by storing material in cultures at low and no-freezing temperature the ageing of plant material is slow down but not completely stopped as in freeze prevention.

Low Pressure and Low Oxygen Storage

Attempts have been made to development low pressure Storage (LPS) and Low Oxygen Storage (LOS) as feasible techniques for future conservation of cultured plant materials. In LPS, the atmospheric pressure surrounding the tissue cultures is reduced resulting in partial decrease of the pressure created by gases in contact with the plant material. On the other hand, in LOS, the atmospheric pressure (760 mm Hg) is not reduced but the inert gases (Particularly nitrogen) are combined with oxygen to create low oxygen pressure.

These are found useful in both short term and long term storage. There is also belated recognition that nature is infinitely more inventive than chemists and that natural molecules have with stood many tough tests of survival for millions of years. An effort is now onto revive culturing of plant in-vitro. A biotechnology firm in USA. *Phytera*, has for example set up and stored plant cell cultures of 5,000 species in glass vials.

3

The Role of Plant Genetic Resources in Conserving Biodiversity

All forms of life on Earth are parts of a huge, inter-dependent system which involves the physical components and the biological community of life. Human existence depends on a balanced play of forces in the biosphere to derive renewable and lasting resources for sustenance and economic development. Development has to be both people-centered and conservation-based so that progress will be sustained. Safeguarding the biodiversity – the totality of genes, species, and ecosystems in a region – is fundamental to the developmental processes. Diversity provides balance to the human environment. The value of diversity at different levels among living organisms and the ecological complexes is particularly apparent in agriculture, the long and complex food cycle.

LOSS IN BIODIVERSITY AND GENETIC DIVERSITY WITHIN A CROP

When one-sixth of today's human population entered the Industrial Age, at least 200,000 to 300,000 plant species existed, among which 50,000 species were edible and 5,000 were used as human food. Today, wheat, rice and maize supply more than 60% of the calories and proteins that the humans obtain from plants. Twenty-three other food crops among a total of 103 species supply more than 95% of the remaining energy and protein. In wheat and rice today, about 80% and 63%, respectively, of the total production areas are planted to the closely related semi-dwarf (HYVs) that replaced numerous varieties grown in the past. The widespread planting and intensive cultivation of the HYV's lead to genetic vulnerability – a situation in which the uniform crop varieties can be imperiled by serious pest outbreaks and/or damaged by adverse weather. Development projects and human neglect have wiped out many wild relatives which form another rich pool of useful genes. Multiple cropping of rice has displaced many other summer crops, some of which are protein-rich legumes. Meanwhile, numerous plant species, particularly the tropical forest trees, are disappearing at an alarming and accelerated pace. Recent estimates place the annual loss in tropical rain-forest species at 5-10% annually and the denuded areas will continue to expand from the present extent of 17 million hectare per cent.

Nowhere in the world is the loss of biodiversity and intra-specific genetic diversity more serious than what is happening in the Asia Pacific Region because its agriculture has the greatest antiquity and intensification. This Region also has the highest population growth of about 2.1% per annum (China and Japan excluded), partly as a result of sustained growth in food production since the advent of the Green Revolution in wheat and rice.

THE ROLE OF PLANT GENETIC RESOURCES IN SUSTAINING FUTURE DEVELOPMENTS

In implementing sustainable agricultural systems, we need technologies that will raise and support crop and livestock yields without degrading the productivity level of the natural resources base. Dramatic increase in crop yields usually depletes the soil and water resources and is akin to mining. Ecologically-sound farming strategies should ensure that the farmers can afford the necessary inputs from indigenous sources and will meet local needs.

Strategies in Sustainable Agriculture

1. The conservation and development of plant and animal genetic resources (syn. *Germplasm*).
2. Integrated pest management (IPM).
3. Conservation and management of soil and water resources.
4. Crop diversification.

Human food supply has been narrowed to a small number of staples, each of which is subject to the vagaries of whether leading to drought and/or flood, other natural disasters, disease and insect attacks, seasonal limitation on adaptation, changes in production costs and market demands, and, more recently, changing trends in consumer preference for special quality features or nutritional contents. On the other hand, crop diversification reduces the ill effects of monoculture of a single crop: pest epidemics, soil fertility depletion, low water use efficiency, restricted nourishment, and unstable farm income. Hence, genetic conservation should extend beyond the small number of presently grown species that command economic importance.

For the commercially important crop plants, the present extent of genetic conservation is generally inadequate or incomplete. Major cultivars have been collected but often duplicated in different gene banks. Minor cultivars, primitive forms and wild relatives are under represented in most gene banks. Meanwhile, untold numbers of traditional varieties and their wild relatives have vanished from farmer's fields or natural habitats. Little effort has been given to *in situ* conservation in original habitats. For the conserved materials, genetic erosion (loss due to human failure) in gene banks continues at alarming rates as insufficient government support and inefficient management by under-staffed and inadequately trained workers prevailed in recent years, in spite of the recent growth in modern gene banks. A rewarding use of the conserved germplasm requires an adequate supply of viable materials, systematic evaluation of their potential worth, wide dissemination of information about the germplasm, and collaborative efforts, both inter-disciplinary and inter-institutional, in enhancing the use of novel germplasm. Otherwise, exotic materials will be overlooked by plant breeders because of their apparently poor agronomic appearance. With the advent of cellular and molecular techniques, the pool of usable genes and alleles is now extended beyond the crop's own genus. However, the sophisticated recombinant DNA and somatic hybridization techniques will not attain cultivar development and release without the inputs of the applied disciplines which traditionally handle crop improvement.

Mere conservation without the capability to rapid and large scale multiplication and to enable a superior strain to reach a large number of growers will not achieve the goal of conservation. *In vitro* techniques for the mass propagation of an urgently needed stock has yet to be used.

RECENT DEVELOPMENTS IN PLANT GENETIC RESOURCES (PGR) CONSERVATION ACTIVITIES

Plant genetic resources (PGR) conservation is a young and multi-faceted technology. It began systematically in the mid=-1960s, drawing on diverse disciplines and different technologies. During the past three decades, the world has witnessed a flurry of intense activities in the this field, especially the Asia-Pacific Region: field collecting, establishment of national PGR centers and modern gene bank facilities, germplasm exchange and collaborative use, training projects, and heated debates on germplasm ownership and associated rights in FAO sessions. Public awareness on the importance of PGR conservation was greatly enhanced, partly due to the activist-promoted controversies and strife between the North (germplasm-poor nations) and the South (germplasm-rich nations). Eventually, PGR programmes won increased recognition and support.

In terms of numbers, about 2.5 million accessions (samples) are kept by the 100-plus PGR programmes in world, while unduplicated (unique) accession amounted to slightly over one million. About one-third of the holdings are cared for by the International Agricultural Research Centers in the CGIAR system. As an example, the International Rice Research Institute (IRRI) in the Philippines holds 83,000 accessions of rice, the largest collection for a single crop. On the other hand, the true extent of genetic diversity in the existing collections is yet to be studied and revealed. What proportion of the collections can be supplied in viable state for users is also open to question. For the sake of efficiency and security, duplication in different collections should be consolidated while storage at duplicate site (or sites) for a representative collection is also needed.

An encouraging development in recent years is the rapid growth of modern gene banks with refrigerated seed storage facilities. Such gene banks in the developing countries have increased from three in the mid-1970s to 30 in 1991, often with assistance from the industrialized nations. However, the new gene banks often face huge electricity bills, problems in maintaining sophisticated refrigeration equipment, and excessive workload in regenerating aged seed stocks and entering past records into a centralized database. Several national programmes are about to collapse.

The PGR workers represent the most handicapped link in the long and complex chain of operations in genetic conservation. Deficiency in both quantity and quality is a common problem. Improved training, both academic and practical, are needed. The workers also need increased support from the decision-makers and enhanced career incentives so that they will say in this painstaking and ill-glamorous profession.

Greater collaboration at international and inter-institutional levels are imperative to enhance germplasm conservation, evaluation, research and use. Pooling of the meager resources will be increasingly essential in the face of world-wide decline in natural and financial resources. A scheme of collaboration between germplasm-poor, developed nations and germplasm-rich, developing nations has been proposed by this author (Chang, 1993) and is shown below:

Less-Developed Countries Serving as:	Developed Countries Serving as:
Sources of genetic diversity and germplasm	Sources of funding and sophisticated research
Site for seed rejuvenation in home environment	*Ex situ* preservation and distribution sites
Site of *in situ* conservation	Purification and disease elimination of stocks
Field research sites staff	Training centre for gene bank

All sectors of the human society must join hands in PGR conservation. Therefore, community groups, home gardeners, schools teachers and children, and NGOs will add to the overall effectiveness and magnitude of conservation efforts.

REWARDS FROM USING PLANT GENETIC RESOURCES

At this point, one might ask: Do we derive from PGR conservation which is labour-intensive, slow-moving and costly? The answer for the Asia-Pacific Region is a resounding "yes" and "very much so". The HYVs of wheat and rice were based on semi-dwarfing gene or genes in rice were isolated from agro-looking obscure varieties. The large number of pest resistance genes in rice were isolated from agro-economically inferior land-races.

The wild rice, *Oryza nivara,* collected in north-central India has furnished a dominant gene for resistance to the highly destructive grassy stunt virus. The 18 million hectares of hybrid rice in China are solely based on teh cytoplasmic male-sterility (*CMS*) found in a wild rice plant (*O. Sativa f. Spontanea*) on Hainan Island. Through inter-disciplinary evaluation at IRRI and collaborating national research centres, promising sources of tolerance to adverse ecological factors (drought, flood, salinity, soil toxicity, extreme temperatures) have been identified from non-productive land-races and used in breeding. Wild relatives of rice, including some of a different chromosome group or basic number, are being increasingly used in wide crosses. One of the wild species, *O. officinalis,* was collected in the vicinity of Bangkok. Another promising wild relative, *Porteresia coarctata,* thrives in the tidal swamps of Bengal Bay and has high tolerance to salinity.

In spite of the enormous contributions from novel germplasm, it is difficult to assign monetary values to their productive offspring. An earlier study on the rice HYVs ascribed about one-fourth each of the increased production to irrigation, fertilizers and improved varieties in eight Asian countries during 1964-66 to 1974-76 (Herdt and Capule, 1983). A recent assessment of the economic value of rice HYVs used in India indicated a 13.4% increase in yield from 1975 to 1984, of which 5.6% can be attributed to the improved genetic materials. The conclusion was that the rice PGR efforts have yielded handsome returns (National Research Council, 1993).

LOOK AHEAD

To meet the increasing need for food in the populous Asia-Pacific Region, the main options facing the policy-markers are: (i) expansion for cultivation and/or irrigation, (ii) restoration of abandoned or damaged land, and (iii) sustenance and improvement of productivity on existing arable land. The first choice is either slowing down due to rising costs or leading to enormous environmental and social costs. Transmigration on a massive scale has limited application. The second option is both slow and costly. The third approach of raising and/or stabilizing land productivity remains the only workable option for most parts of the Region. However, inputs are needed for either favoured or unfavoured areas. For the unfavoured areas, traditional farming principles may be re-tooled to advantage. In all situations, plant genetic resources of the past or of an improved nature are an essential component.

Table 3.1: Promising Uses of Plant Genetic Resources in upgrading and/or sustaining agricultural production under two levels of inputs

High Input System	Low Input System
Raise crop yields/productivity level	Facilitate farming diversification & stabilize production/farm income
Reinstate genetic diversity among major cultivars of a crop, e.g. supply new *cms* sources	Improve biological control of insects, disease and weeds
Provide durable tolerance to pests and adverse environments	Facilitate crop diversification & deployment to conserve soil & water resources e.g. supply biofertilizers
Raise soil & water use efficiency	
Facilitate crop diversification & increase land productivity	Enrich nutritional sources
Improve nutritional balance among food crops	Reduce cash inputs

For the favoured farming areas, such as irrigated rice lands, stabilizing the crop production by reducing damages by pests and/or adverse weather and lowering production costs should be the prime goals. Pest control through biological control and more durable forms of genetic resistance should be pursued. The yield plateau in wheat and rice HYVs which has persisted since the 1970s should be overcome by combined inputs from plant physiologists, biochemists and genetic engineers. For the less favoured rainfed areas where either an excess or a deficit of water destabilizes yields, the problem involves complexes of quantitative plant characters. Accordingly, the kind of genetic elements we should search for in PGR will go beyond the conventional major genes that were so effective in the past, to novel gene-complexes, cytoplasmic combinations and nuclear-cytoplasmic inter-actions. The genetic pool will be extended beyond a crop's own genus and may even transcend the monocot-dicot boundary. Transgenic cultivars of the future will include genetic materials from bacteria, yeasts and animals.

With the mass migration of farm youth to the urban areas, policy-makers need to provide better incentives to the rural communities. Plant genetic resources and biotechnological innovations are powerful tools in raising and sustaining agricultural productivity, but their effectiveness will depend on the combined concern and inputs of human society, of which the farmers will remain in the forefront. The troubled farming sector surely needs our care and help more than ever before – so that the farmers, along with their plants and livestock, will stay as the hub of the food-producing wheel.

THE EXTENT OF USE OF PLANT GENETIC RESOURCES IN RESEARCH

Recently (June, 2001), N.S. Dudnik, I. Thormann, and T. Hodgkin of International Plant Genetic Resources Institute (Crop Sci. 41:6-10: 2001) indicated that establishment of *ex situ* germplasm collections has been the result of several decades of global efforts of conserve plant biodiversity. The centers of the CGIAR alone maintain over 600000 accessions of plant species, and over six million more are kept in institutions around the world. While these vast collections may hold benefits for society even if they are used only rarely, increasing their use today clearly enhances their value.

The genotypic and phenotypic diversity maintained in the collections can be invaluable for breeding programs but also for basic research into evolution, gene expression, resistance, and other areas. Obstacles to use of Plant Genetic Resources (PGR) have been identified, and include lack of information on material, difficulty in accessing collections, underproduction of seed and the difficulty in moving specific genes into *good genetic backgrounds* as well as identifying obstacles to the use of PGR, it is important to quantify the level of use more accurately and to obtain better information on current patterns of use. They further indicated the extent and frequency of use of literature to estimate the extent and frequency of use of plant genetic resources in applied plant genetic research.

Case Study

They selected four journals for inclusion in the survey: *Crop Science, Euphytica, Plant Breeding*, and *Theoretical and Applied Genetics.* These were chosen as being among the foremost international journals publishing research in agricultural genetics. Moreover, they were felt to provide a comprehensive overview across multiple disciplines and in both theoretical and practical work. Many other international journals publish work that involves the use of PGR but some of these have additional areas of research focus (e.g. animal genetics.

Two journals, *Genetic Resources and Crop Evolution* and *Plant Genetic Resources Newsletter* are also publishing plant genetic resources researches, however, they were not considered for case study.

For this study, they focused on one year's worth of the four journals (*Crop Science, Euphytica, Plant Breeding*, and *Theoretical and Applied Genetics)* and analyzed all issues published in 1996. They considered and examined 859 articles for evidence of the use PGR and made three following categories:

1. From an institute with a known *ex situ* collection.
2. Material collected directly in the field.
3. More than 20 lines, source not specified, which could be presumed to originate in a genebank.

Table 3.2: Articles involving use of PGR in four journals in 1996.

Journal	PGR articles	Total number of articles	Percentage of year's articles involving PGR
Crop Science	46	257	18
Euphytica	45	170	26
Plant Breeding	23		
Theoretical and Applied Genetics	77	337	23

They analyzed species, source and type of material, authors, and research topic of these articles. Sources of material were verified by means of the IPGRI online *Directory of Germplasm Collections.* They used the SNIGER database of CGIAR Center genebanks for comparative statistics on use of the material in these collections.

Table 3.3: Distribution of authors of PGR articles by geographic location of home institution. Numbers refer to number of articles involving at least one author from the country

Developing country	Number of articles+	Developed Country	Number of articles+	CGIAR Center	Number of articles+
Argentina	3	Australia	10	CIAT	5
Brazil	4	Austria	2	CIMMYT	1
Bulgaria	1	Canada	6	CIP	1
Chile	1	Finland	1	ICARDA	5
China	8	France	11	ICRISAT	1
Costa Rica	2	Germany	10	IITA	1
Cyprus	1	Greece	1	IRRI	2
Czech Republic	2	Israel	3		
Egypt	1	Italy	10		
Ethiopia	2	Japan	9		
India	4	Netherlands	8		
Malaysia	2	New Zealand	2		
Mexico	3	Norway	1		
Morocco	1	Spain	12		
Myanmar	1	Sweden	6		
Nepal	1	UK	11		
Pakistan	1	US	81		
Poland	1				
Russia	3				
Slovenia	2				
Thailand	1				
Trinidad	1				
Ukraine	1				
Yugoslavia	2				
Total	49 (19.67%)	Total	184 (73.89%)	Total	16 6.4%)

+ In each paper, multiple authors from one institution were counted together, but all institutions represented were counted separately.

The results of the survey indicated a high rate of use of conserved germplasm. Of the 859 articles, 191, or 22% involved use of PGR. The highest frequency of PGR use was found in *Euphytica*, but the value was very similar for all four journals. They further indicated that over half (57%) of the institution in developed countries were universities; 40% were national research centers and only 3% represented private industry organizations. In developing countries, 65% of the institutions were national or other research institutes and 30 % were universities; three private corporations (5% of the total) were all located in Brazil.

Table 3. 4: PGR use sorted according to source definitions

	1+	2	3	Both 1 and 2	Theoretical use of PGR
Number articles	152	30	3	4	3
Percent of PGR articles	79.17	15.63	1.56	2.08	1.56
Percent of total articles checked	17.69	3.49	0.35	0.47	0.35

+1" *Ex situ* collected material, 2: Material collected in the field, 3: More than 20 genotypes from an unlisted source, assumed to drive from genebank. Theoretical use: articles which do not involve manipulation of plant material, but are concerned with theoretical aspects of germplasm management.

What Sources of PGR Are Being Tapped?

The worlds *ex situ* germplasm collections are estimated to contain over 6 million accessions. The majority (79.2%) of the studies found in our survey used material explicitly obtained from *ex situ* collections and an additional 1.5% did not give a source but were assumed to be genebank accessions. There is a fair level of use of material collected directly in the field-31 cases or just over 15%.

According to the 12 countries hold more than 45% of the germplasm accessions kept in national collections. These countries are Brazil, Canada, China, France, Germany, India, Japan, Republic of Korea, Russian Federation, Ukraine, UK, and USA. In our study, germplasm from these countries was used in just under 60% of the studies, and 40% of the studies used material obtained from U.S. sources (either the USDA) genebank system, or universities and botanical gardens). This last figure is in keeping with the large number of U.S.-authored papers.

Dudnik et al. (2001) indicated that it is difficult to quantify the total number of accessions or genotypes used for research observations.

They further indicated that the 16 centers of the CGIAR hold in trust accession of over 3000 species, including major crops such as maize (*Zea* spp), rice (*Oryza* spp.), and wheat (*Triticum* spp.), which were freely available to researchers.

Table 3. 5: Sources of *ex situ* conserved material used.

Developing country institutions		Developed Country Institutions		CG Centers	
Brazil	1	Australia	9	CIAT	5
China	1	Canada	3	CIMMYT	4
Costa Rica	1	France	5	CIP	2
Czech Republic	1	European genebanks	1	ICARDA	8
Ethiopia	1	Germany	12	ICRISAT	3
India	1	Italy	5	IITA	2
Malaysia	1	Japan	4	IRRI	4
Niger	1	Netherlands	4		
Pakistan	1	Spain	7		
Poland	1	Sweden	1		
Senegal	1	Switzerland	1		
Slovenia	1	New Zealand	1		
Trinidad	1	UK	7		
Yugoslavia	2	USDA	53		
		US other	14		
Total	**15** **8.8%**	**Total**	**127** **74.7%**	**Total**	**28** **16.5%**

Which Species are Being Used?

Just as important as the degree of use of PGR is the range of species being tapped for research. Certain crops and species predominate in world collections, either because of their high levels of variation, or because of their importance in agricultural production. Their use in research, however, may reflect other factors as well as conservation priority and agricultural importance.

According to FAO estimates, 48% of the world's collections consists of cereals, 16% are food legumes, and 8% are vegetable species. Dudnik et al (2001) studied 112 different species, with a slightly different distribution into the various categories. The majority of studies did involve cereals but at a lower level (28% rather than 48%), while there was nearly equal use of vegetable species and food legumes. The major vegetable crops studied were

Brassica, Lycopersicon, and *Beta* species. While these are not the subjects of very large collections worldwide, the especial predominance of *Brassica* and *Lycopersicon* may reflect that they have become model crops for plant genetic research. Soybeans were the most commonly studied legumes, with the use of numerous and wild species. The level of use numerous cultivars and wild species.

The level of use of forage species in the research studies was equal to the FAO estimate of their predominance was in world collections. Research involving tuber species was very low.

The other category included industrial crops such as sugarcane (*Saccharum* spp. and *Erianthus* spp.) coffee (*Coffee* spp.) and tobacco (*Nicotiana* spp.), as well as few aromatic and parasitic species.

What are Plant Genetic Resources being Used to Study?

Dudnik et al. (2001) categorized the research in the articles according to what are interpreted as their primary objectives. Some of the categories used, such as breeding or cytology, are more thematic while others, agromorphology or biotic resistances, are more concerned with areas of agricultural performance. They used this classification because often the scope of an article has a thematic relevance broader than the involvement of a particular technique or character.

Not surprisingly, the largest single category was the measurement of genetic diversity, typically studied by means of a broad group of accessions. This usually involves comparison of diversity within and between different groups of accessions using morphological, electrophoretic, or molecular markers.

Work on biotic stress resistance constitutes the next largest category with many studies on the section of new useful resistance and its inheritance.

Work on breeding included studies of crossability, interspecific hybridization and production of somatic hybrids, and responses to selection for agronomic properties in breeding programs.

Other important areas of work were taxonomic studies where large amounts of diverse material are required and cytological studies. Work on abiotic stress resistance was surprisingly low with only six studies in the year investigated. Topics such as QTL identification, molecular mapping, and technology applications tended to be carried out with research lines and other types of elite cultivars rather than the accessions that fit our definition of PGR.

Dudnik et al (2001) concluded that the initial focus of the plant conservation community is to conserve biodiversity in respect to its threatened disappearance. More recently, the importance of using the material stored in genebank collections has been increasingly emphasized.

Use is often thought of solely as the incorporation of a few accessions into breeding programs, but genebank accessions are the ideal material for scientific research-agricultural but also cellular and molecular. PGR, however, provides a clear indication of the importance of PGR in developing the knowledge base of improved agricultural production through internationally recognized research.

4

Biotechnological Techniques for Biodiversity Conservation

PLANT TISSUE CULTURE

Plant tissue culture is the technique of growing plant cells, tissues and organs in an artificial prepared nutrient medium static or liquid, under aseptic conditions. It has advanced the knowledge of fundamental botany, especially in the field of agriculture, horticulture, plant breeding, forestry, somatic cell hybridization, phytopathology and industrial production of plant metabolites, etc.

The term *tissue culture* is actually a misnomer borrowed from the field of animal tissue culture. It is a misnomer because plant micropropagation is concerned with the whole plantlet and not just isolated tissues, though the explant may be a particular tissue. The term's *plantlet culture* or *micropropagation*, therefore, are more accurate. However, whether we call it cloning, tissue culture, micropropagation, or growing *in vitro*, the process remains the same; it is a vegetative method for multiplying plants. The nursery trade commonly accepts this method of propagation and it has had a significant impact on commercial horticulture.

Tissue culture is the ever-ready tool for specialists who hybridize plants by either sexual or asexual means. It is a clean and rapid way for genetic engineers to grow material for identifying and manipulating genes or to transfer individual characteristics from one plant to another. It plays a role in a wide array of fields, such as botany, chemistry, physics, genetic engineering, molecular biology, hybrid development, pesticide testing, and food science.

Callus Tissue and Organogenesis

Callus (pl. calli) *on a wounded plant parts or on a culture medium is made of an amorphous aggregate of loose parenchyma cells which proliferate from the mother cells.* Callus is either homogenous parenchymatous mass or treachery elements or sieve elements or submerized cells or secretory cells or the trichomes. Callus formation has been found in angiosperms, gymnosperms, pteridophytes, and bryophytes. Callus contains no organised meristems. Callus is somewhat an abnormal tissue which has the potentiality to produce normal roots and embryoids and in turn it develops into plantlets. Callus may be hard (due to lignification of cell walls) or brittle and sometimes soft.

Organogenesis is the development of adventitious organs or primordia (embryoid) from undifferentiated cell mass (callus) in tissue culture. It is controlled mostly by a balance between cytokinin and auxin. A relatively high ratio of auxin: cytokinin induces root formation in callus tissues whereas, a low ratio induces shoot formation. **Caulogenesis** is a type of organogenesis by which only adventitious shoot bud initiation takes place in the callus tissue. When it is applicable for root, it is known as **rhizogenesis**. Anomalous structures when develop during organogenesis is called **organoids**. The localized meristematic cells on a callus which give rise to shoots and/or roots is termed as **meristemoids**.

Callus Tissue and Organogenesis

Callus (pl. calli) *on a wounded plant parts or on a culture medium is made of an amorphous aggregate of loose parenchyma cells which proliferate from the mother cells.* Callus is either homogenous parenchymatous mass or treachery elements or sieve elements or submerized cells or secretory cells or the trichomes. Callus formation has been found in angiosperms, gymnosperms, pteridophytes, and bryophytes. Callus contains no organised meristems. Callus is somewhat an abnormal tissue which has the potentiality to produce normal roots and embryoids and in turn it develops into plantlets. Callus may be hard (due to lignification of cell walls) or brittle and sometimes soft.

Organogenesis is the development of adventitious organs or primordia (embryoid) from undifferentiated cell mass (callus) in tissue culture. It is controlled mostly by a balance between cytokinin and auxin. A relatively high ratio of auxin: cytokinin induces root formation in callus tissues whereas, a low ratio induces shoot formation. **Caulogenesis** is a type of organogenesis by which only adventitious shoot bud initiation takes place in the callus tissue. When it is applicable for root, it is known as **rhizogenesis**.

Anomalous structures when develop during organogenesis is called **organoids**. The localized meristematic cells on a callus which give rise to shoots and/or roots is termed as **meristemoids**.

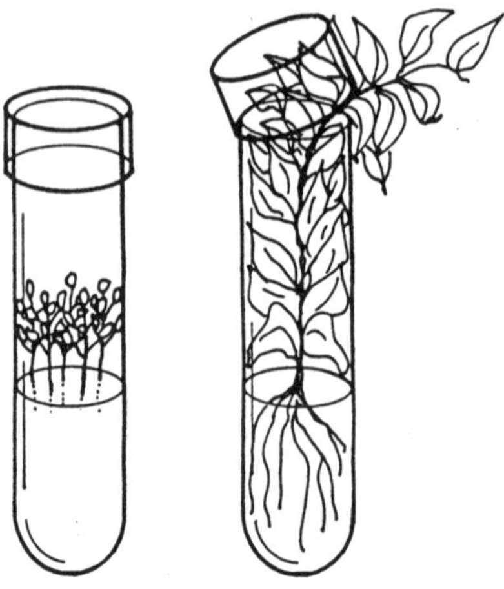

Fig. 4.1: If tissue culture plants did not revert to a juvenile state, most material would never fir into a test tube or jar and they would be too unwieldy to micropropagate.

When plants are multiplied vegetatively-as distinguished from those grown from seeds-whether by tissue culture or by cuttings, all the offspring from a single plant can be classified as a *clone*. This means that the genetic make-up of each offspring is identical to that of all the other offspring and to that of the single parent. On the other hand, plants propagated by seed, resulting from sexual reproduction, are not clones because each seed (and the resultant plant) has a unique genetic make-up-a mixture from two parents, different from either parent and different from one seed to another.

The term **cloning**, with respect to tissue culture, refers to the process of propagating in culture large numbers of selected plants with the same *genotype* (the same genes or hereditary factors) as their respective parent plant.

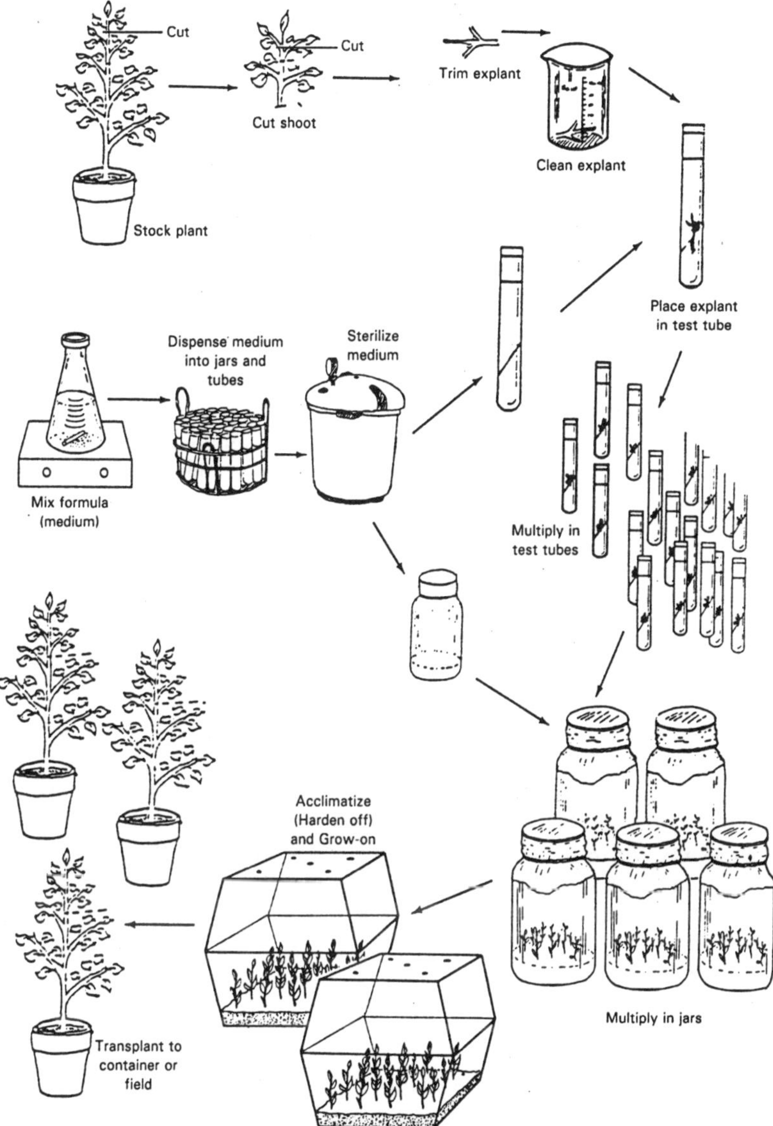

Fig. 4.2: Sequence of shoot tip micropropagation.

PRINCIPLES OF GROWTH

Growth is the self-multiplication of living material, the protoplasm it self. Growth is an increase in size (volume or length) due to cell divisions and subsequent enlargement. It is an increase in dry weight of bulk of an organism associated with development. **Development** is defined as *an ordered change* or *progress, often towards a higher, more ordered or more complex state*. Development may take place with growth and growth may take place with development. However, these are often quite integrated processes.

In most organisms, growth involves **cell division, expansion, differentiation** and **morphogenesis**. Increases in height and dry weight are the most obvious manifestation of growth, e.g., in forage and pasture crops. In grain crops, dry weight increases in nongrain components are important only in producing a plant capable of high grain yield. Growth is also expressed as the advance of a plant from one development stage to another. A seed germinates, the seedling develops roots, leaves, and stem, and the plant grows to maturity. The plant flowers and produces seed to complete its life cycle.

Morphological and chemical differentiation takes place during the development. These processes of cells are complicated and involve synthesis of many organic compounds like protein, cellulose, and nucleic acids and requiring physical forces that cause cell enlargement. The process becomes more complicated because it involves division and expansion of many different cell types. Growth includes assimilation and the formation of new protoplasm, permanent change in size, and increase in weight, either of the plant as a whole or of some organ or tissue. The growth pattern and division of the cells in root types, leaf buds, and developing seeds occur in the same way, but they serve vastly different functions for the plant. Several abiotic and biotic factors influence growth of an organ (leaf, fruit, root, etc.) or whole plant at the cellular level.

Why Growth Occurs?

Growth is expressed as the division of a cell to form two cells and the enlargement of the newly divided cells. When we say that cells double in all their constituents and then divide in half, we are obviously describing the average case. When we observe individual cells, however, we find deviations from the average. In some instances, the growth rate is constant rather than accelerating. In others, the growth rate is high at first and then decreases. A mass of cells doubles and halves only in an average way and not in an exact way. Cells divide when they are **ready**, and they are ready only when they have completed certain preparations for division.

Speculation and experimentation have now focused on other cellular factors as possible cell cycle controls. These include changes in the concentration of cyclic nucleotides, cAMP and cGMP.

Kinetics of Growth or Measurement and Pattern of Growth

Growth of a plant is usually measured in terms of increase in dry weight, length or height and such measurement indicates that growth-rate varies with the age of the organism. The growth of most plants follows a similar pattern, generally S-shaped or sigmoid shape curve. The size versus time plot is the most direct way of handling growth data. One can also plot the logarithm of the size as a function of time. In the size plot, the units are length, volume or weight, but in the rate curve, the units are length, volume or weight per unit time. A relatively slow growth rate characterizes early or seedling growth. The rate of growth increases as the plant becomes larger, is greatest just before or in early stages of flowering, and then decreases as the plant matures.

The plotted growth curves have following phases:

1. **First rapid phase:** During this phase the rate curve as well as size curve increases. It is also known as logarithmatic or exponential phase
2. **Maximum growth rate phase:** Growth curves of many plants also show another evident phase called maximum growth rate or the linear phase. Sachs called it the grand period of growth, and it lies between the logarithmatic phase and the senescence phase.
3. **Last phase:** During this phase, the size continues to increase but more slowly so that rate decreases. This phase is known as senescence or decreasing phase.

What Happens, When Growth Occurs?

Growth of a plant is initiated in the meristems and the cellular changes in these tissues. The three main changes in the complete development of a cell are **cell division**, **cell elongation** and **cell differentiation**.

After cell division, daughter cell enlarges to the size of the mother cell before they again divide. During cell growth, uptake of water takes place, which produce turgor pressure sufficient to overcome the force of attraction between the cell particles in cell wall. Consequently, wall stretches and become thin and the osmotic potential increases because of water uptake, which dilutes cell contents. The consequent increase in the cell's water potential brings water uptake (and therefore growth) to a stop. Besides water and solute absorption, new cell wall material is also added and cell wall became thick due to opposition (accumulation of new material on the inner surface of the wall). This happens during cell maturation.

Biochemical Changes: The important biochemical changes that occur during growth are: synthesis of protein, nucleic acid, phospholipids, multiplication of organelles and utilization of energy in the form of ATP, etc.

Dry Matter Accumulation: Dry weight accumulation pattern of plants in the early stages is like the accumulation of compound interest in bank account. The weight of the plant at any given time is likened to the size of the bank account and the increase in weight equated with the interest. In the early stages when plants are very small, the actual dry weight increase per day (interest) in small (first 4 weeks), but at the plant becomes larger, its weight gain per day increases (first 10 weeks).

THE BOTANICAL BASIS FOR TISSUE CULTURE

The diversity of naturally occurring vegetative reproduction is the landmark and reflects the amazing capability and potential of plants for multiplication. The same factors involved in multiplication and growth initiation in nature are involved in the greenhouse and in tissue culture. The natural capability of plants to multiply by asexual means is the basis for multiplication in vitro. Vegetative reproduction, whether occurring naturally or through human intervention, is initiated in stems, buds, roots, or leaves. We take cuttings, we make divisions, we layer, we make grafts, and we tissue culture-in short, we promote vegetative reproduction, a natural phenomenon.

Power of Regeneration in Plants

The multiplication of plants in vitro does not establish any new processes within the plants. Tissue culture simply directs and assists the natural potential within the plant to put forth new growth and to multiply in a highly efficient and predictable way. In contrast to plants produced from seeds as a result of sexual reproduction, new plants produced through vegetative reproduction are basically severed extensions of the original plants.

Fig. 4.3: A natural clone. Nature has been **cloning.** Any time a plant reproduces itself vegetatively it produces a clone.

Plant stems have tremendous potential for regeneration because they can grow in various forms and habits: long or short, slender or stout, round, flat, or square, above ground or underground, trailing or upright. One of the methods of vegetative propagation used most often by growers is that of growing new plants from stem cuttings. Stem cuttings are shoots or sections of stems that root when they are inserted into a growing medium (potting mix), which may be a mixture of peat or bark with sand or perlite, or simply sand and perlite.

Roots grow from the nodes (the part of the stem from which buds or leaves originate) when placed below the surface of the medium. Frequently cuttings are encouraged to root by wounding, and often an auxin (a growth regulator) is applied to the base of the shoot. To wound a cutting, a thin slice of epidermis (outer layer of tissue) is peeled away with a knife on 1 or 2 sides of the lower part of the stem. This will often cause the stem to grow callus, which can help root initiation.

Small **stem cuttings** are frequently used as starting material for tissue cultures, and microcuttings, tiny cuttings taken from tissue cultured material, are a product of tissue culture that can be planted and grow on to mature plants. The size of microcuttings can be critical to the success of growth; 1 to 2 in (2.5 to 5 cm) is usually best.

Layering is another form of vegetative reproduction that occurs frequently in nature and is widely used by growers as well. A branch is said to layer when it comes in contact with the soil, forms roots, and grows on to become a new plant. Wild blackberries (*Rubus*) rapidly expand their territory by layering. Layering is used commercially to propagate filberts (*Corylus*), grapes (*Vitis*), black raspberries and trailing blackberries (*Rubus*), currants (Ribes), apple (*Malus)* rootstocks, and some ornamentals. Air layering is a method whereby growers induce a stem to root without it being in contact with the soil.

Fig. 4.4: A grower assisted clone by use of stem cuttings. Growers have been cloning by cuttings for centuries.

In **grafting**, growers attach a scion (a shoot or bud) of one plant onto the understock or rootstock (rooted stem) of another plant to obtain the desired traits of both. Various methods can be employed in grafting, but whatever method is used, it is important that the cambial layers match-the cambium is the thin layer of tissue between the wood and the bark of a stem. Micrografting involves the grafting of tissue cultured material in aseptic conditions, a craft that requires great skill.

True **bulbs**, such as those of lilies (*Lilium*), tulips (*Tulipa*), and daffodils (Narcissus), are underground storage organs that function as modified stems. A bulb consists of scales (modified leaves) attached to a basal plate (a flat modified stem at the base of the bulb and the source of roots).

Corms are swollen underground stems, complete with nodes, internodes (the section of stem between 2 nodes) and lateral buds; they do not have scales like true bulbs. Plants that produce corms, such as Gladiolus and Crocus, multiply naturally by producing cormets (miniature corms).

Rhizomes, horizontal underground stems, and tubers, the swollen fleshy part of a rhizome. Potatoes (*Solanum tuberosum*) Caladium, and gloxinias (*Sinningia*) produce stem tubers. A piece of tuber or fleshy rhizome generates a new plant if it contains an eye, or bud.

Stolons or **runners** are long, prostrate, aboveground modified stems. Examples of stoloniferous plants are creeping dogwood (*Cornus canadensis*), strawberry (*Fragaria*), Bermuda grass (*Cynodon dactylon*) and white clover (*Trifolium repens*). The sole purpose of stolons or runners is the production of new plants.

Root cuttings are also used for propagation of gooseberries (*Ribes*), raspberries (*Rubus*), horseradish (*Armoracia rusticana*, syn. *Cochlearia armoracia*), apples (*Malus*), flowering quince (*Chaenomeles*), bayberries (*Myrica*), aspens (*Populus*), and roses (*Rosa*), to name a few.

Leaves can occasionally produce new plants. Leaf cuttings are made from such plants as *Bryophyllum*, *Begonia rex*, *Sedum*, African violets (*Saintpaulia*), and *Sansevieria*, among others.

WHY PLANTS REGENERATE?

Meristematic cells are located at the tips of stems and roots, in leaf axils, in stems as cambium, on leaf margins, and in callus tissue. Under the influence of genetic make-up, location, light, temperature, nutrients, hormones, and probably many other factors, meristematic cells differentiate into leaves, stems, roots, and other organs and tissues in a organized fashion. Meristematic tissue is the basis of plant growth and development.

Parenchyma cells, the most common type of plant cell, are thin-walled cells that have the capacity to regenerate and differentiate, to initiate the growth of new and varied tissues or organs for specialized functions

Meiotic division, or reduction division, is the process of forming sexually reproductive cells. In meiosis, each chromosome of a 2n cell splits in 2. The chromosomes segregate such that one chromosome from each set of 2 goes to each of the 2 new sex cells, or gametes, each of which has, thus, only one set of chromosomes (n).

Whenever cells divide there is the possibility of genetic variability. If a mutation (a change in genetic make-up) occurs during cell division, and assuming the cell survives the mutation, the mutation is carried in all future divisions. A mutation is a sudden abnormal change in genetic order that will alter some characteristic, or it can be a change in chromosome number. The effects of mutations are not always noticeable. Most mutations will cause a plant to die or to produce undesirable qualities, such as misshapen fruit or abnormal shoots. Growers expect a small percentage of plants to undergo mutation, and they will discard them if they see them.

PLANT TISSUE CULTURE: PRINCIPLES

The principles of tissue culture are all around us-in nature, in the field, and in the greenhouse.

The technique has developed around the concept that a cell is **totipotent** that is has the capacity and ability to develop into whole organism. The principles involved in plant tissue culture are very simple and primarily an attempt, whereby an explant can be to some extent freed from inter-organ, inter-tissue and inter-cellular interactions and subjected to direct experimental control.

Cell culture is the cultivation of cells on a solid gel medium or in a liquid medium, the latter commonly known as cell suspension culture. *Callus culture* is the multiplication of *callus* (a mass of disorganized, mostly undifferentiated or undeveloped cells), usually on a solid medium.

The **apical meristem** is the new, undifferentiated tissue at the microscopic tip of a shoot. It is often virus free even in diseased plants because these meristematic cells are not yet joined to the plant's vascular system, and perhaps they grow faster than the viruses. Thus, if the few virus-free cells that make up the microscopic dome of apical meristem are removed from the plant and placed in a culture, they can grow and produce healthy, disease free plants. This technique is known as *meristem culture*, a term sometimes wrongly used to broadly denote micropropagation or tissue culture, but which should be limited to indicate cultures started from an apical meristem.

Callus Culture

For raising the callus tissues, a tissue culturist must have clear understanding of some basic principles. A cell from any part of the plant like shoot apex, bud, leaf, mesophyll cells, epidermis, cambium, anthers, pollen, fruit etc., when inoculated in a suitable medium under aseptic laboratory conditions can able to differentiate and multiply. This results into the formation of an amorphous mass of cells known as **callus**, which can be induced to re-differentiate on appropriate medium to develop **embryoids** which directly develop into the plantlets, eventually giving rise to a whole viable plant.

The term **clone** (from the Greek *klon,* meaning: a slip or twig suitable for plant propagation) was suggested by Webber (USA) in 1903 to explain those plants which were obtained by a sexual reproduction; it is even applied to DNA multiplication (cloning of genes in bacteria). In strict scientific sense, cloning means *an organism obtained from a single cell through mitotic divisions*.

Meristem Culture

When a meristem is cultured *in vitro*, then it produces a small plant bearing 5 or 6 leaves. This could be obtained within a few weeks. Then the stem is cut into 5-6 small micro cuttings, which under favourable conditions, become fully-grown plants.

Organ Culture

A body of higher plants has complex inter-relationships between different organs like root, shoot, apical meristem, leaf primordia, floral buds, ovary, ovule, anther lobs, pollen grains, fruit, seed, etc. In this method a particular organ is isolated and cultured under laboratory conditions in a chemically defined medium where they retain their characteristic structures and other features and continue to grow as usual. In organ culture, organs are not induced to form callus, therefore, it differs from the callus culture where the organization of the intact tissues is lost.

This technique provides an experimental system to define the nutrients and growth factors that are usually received by the organ from other organs of the plant body and from surrounding environment. It also helps us in understanding the inter-dependence of organs with respect to various physical and chemical growth factors including growth hormones. Organ culture technique also provides the knowledge about the various problems of morphogenesis and the sites of biosynthesis of specific metabolites and growth compounds. It may be used as a tool for improvement of various economically important crops.

Embryo culture, cell culture, and callus culture are usually designated as such, but they may also fall under the broad term of tissue culture. *Embryo culture* can mean the *rescue* of an embryo from a seed and fostering its plantlet development and multiplication in a culture, or it can refer to *somatic embryogenesis*, whereby embryos are induced to form from *somatic* (vegetative, or asexual) cells.

Organ culture may be grouped into two major categories: **vegetative organs** (root culture, leaf culture, and shoot tip culture) and **reproductive organs** (complete flower culture, isolated ovary culture, isolated ovule and embryo culture, pollen mother cell culture, seed and fruit culture).

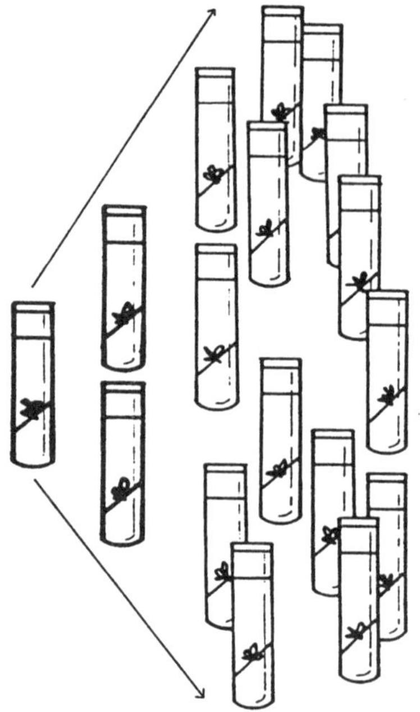

Fig. 4.5: Tissue cultured clone. Potentially, one explant can produce an infinite number of plants.

Anther and Microspore Culture (Production and Uses of Haploids)

Haploids are the sporophyts with gametophytic chromosome number. They are produced in a variety of plant species using varies methods. Haploids have their significance in genetics and plant breeding. Currently, haploids are the important tools for biotechnological programmes. They are being produced either by ovule/anther/microspore culture on artificial culture medium or by chromosome elimination.

Anther and Microspore Culture

In *Datura*, Guha and Maheshwari (1966, 1967) successfully produced haploids from anther culture. Use of microspores as the basic source material is very common, specially in plants like *Datura inoxia, Nicotiana sylvestris, N. tabacum, Oryza sativa*, etc. Various conditions are important fo such sulture, however, the important are: genotype of donor plant, the placement , growth and development stage of donor plant and type of culture medium used. Though, the culture includes various steps, the important are:

1. The flower buds from selected plant species are collected and brought to a liminar flow chamber. They are sterilized by appropriate chemical treatment, before dissection.
2. Avoid injury while removing anthers from flower buds. The injury will effect growth and development of callus and may produce variations in chromosome number. Injury may produce diploids, haploids or aneuploids.
3. Culture the anthers on solid agar medium. This will directly gives embryoids or may directly develops in callus, before differentiation.
4. Embryoids develop into plantlets.
5. Colchicine treatment gives rise to diploid homozygous plantlets, which may be directly used for field-tested for selection.
6. Microspore culture is more preferred than anther culture.

Collection of Microspore

The anthers are first dissected and then transferred to beaker containing liquid medium. Press them with a glass rod or syringe piston for squeezing out the microspore. Use nylon sieve to filter the suspensions with anther which allows only the microspores to pass through. Centrifuge the filtrate for three times at 600-800 rpm. The pellets get resuspended in a fresh medium every time. Then the microspores are inoculated on a solid or in a liquid medium. The cultures are maintained at 16/8 hr photoperiod and 25° C. In about 15 days, the microspores may directly develop into embryoids. In anther culture spontaneous doubled haploids (SDH) are also obtained. This does not requires colchicine treatment. SDH are used directly for field test selection.

Ovule Culture

Female gametophyte cells ate also a source for haploid production. First success was obtained from gymnosperms plants e.g., some cycads, *Ephedra*, *Zamia*, etc. Similar haploids have also been obtained from crop species like barley, wheat, tobacco etc. For haploid production from female gametophyte it is necessary to know: (1) the events related to the induction of haploidy in these tissues, (2) factors that control in vitro development of proembryo into the fully organized plants, and (3) major differences in the growth patterns of in vitro development of unfertilizes ovule cells (female gametophyte) and in pollen cells (male gametophyte.

THE CONCEPT OF TOTIPOTENCY OF CELLS

Some concepts in science become inherently acceptable long before their practicality is demonstrable. This was so in the concept of the totipotency of cells of higher plants. Even in the mid-twenties, one encountered the tacit view that, apart from inherent practical difficulties, there was no theoretical reason why one should not rear a begonia plant from a single leaf hair cell. This view was traceable **first** to the then well recognized principle that as cells divide mitotically, they do so equationally to produce daughter cells in facsimile.

Secondly, the concept was due to G. Haberlandt's historic but then unsupported claim, expressed in 1902, that one day it should be possible to rear plants from isolated surviving cells of flowering plants. Haberlandt's insight was, in fact, the more remarkable because he even stated that out of surviving somatic cells artificial embryos would be reared asexually. In higher plants, embryos do develop in situ from appropriately stimulated somatic cells but without any prior sexual act. At the outset, the special significance of the fertilized egg in plants should be seen as restricted to its role as the genetically unique product of a fusion of male and female gametes. Having established the genetic constitution of a given individual, the zygote (or fertilized egg) really behaves developmentally as a very general kind of living cell- a cell with a built-in capacity to grow in an organ, the ovule, that fosters that growth. Moreover, as will be shown, and in a satisfying number of cases, isolated somatic cells may return to a simulated zygotic state and grow into embryos and to plants under the appropriately applied conditions which furnish the requisite nutrients and stimuli.

The Autonomous Organelles: Their Behaviour During Growth Induction and Morphogenesis

The stimuli that cause quiescent cells of carrot to proliferate and grow also affect all their organelles (mitochondria, plastids, dictyosomes, reticulum and polysomal aggregates of ribosomes in the ground cytoplasm) in various ways. The same stimuli affect the metabolism of the tissue and its ability to form residual or secondary biochemical products.

Genetic Approach of Plant Tissue Culture

1. Somatic embryogenesis is studied because it is an interesting process, liable to produce results of economic interest.
2. The field of plant embryogenesis is not well developed and somatic embryogenesis looks a promising model system.
3. Somatic embryos provide an interesting possibility as it is possible to mutagenize a cell population, make it to embryogenize and look for mutants of conditional type, e.g. temperature-sensitive mutants.
4. The seed embryos, being wrapped in various coats, are not amenable to microsurgery as in case of animal embryos. Moreover, for biochemistry, a large number of synchronized embryos are needed and this is not so easy to obtain, particularly if one is interested in early embryonic stages.

Importance of Tissue Culture Technique

All the cells in an organism carry the same genetic information, yet show variations in expression. Our knowledge of cell and tissue cultures has been developing with full swing, specially in biotransformation, forestry, genetic engineering, morphogenesis, somatic hybridization, secondary metabolite production, hybridization, variety development and their conservation, maintaining pathogen free plants and rapid clonal propagation, totipotency, differentiation, cell division, cell nutrition, metabolism, radio biology, cell preservation, etc.

It is now possible to cultivate cells in quantity, or as clones from single cells; to grow whole plant from isolated meristems and to induce callus or even single cell to develop into complete plant either by organogenesis or directly by embryogenesis *in vitro*.

The production of haploid through tissue culture from anthers or isolated microspores and of protoplasts from higher plant cells has served as the basic tools for genetic engineering and somatic hybridization. Tissue culture technique helps to propagate plants of economic importance such as orchids and other ornamental plants in large numbers by their meristem culture or by other in vitro methods. This provides them virus-free plantlets. Propagation of valuable economic plants through tissue culture based on the principle of **totipotency** (*every cell within the plant has potential to regenerate into a whole plant*).

In plant breeding, embryo, ovary and ovule culture as well as in vitro pollination have been employed to overcome morphological and physiological sterility and incompatibility. In recent years, plant tissue culture technique is in increasing use for producing haploids from anthers or isolated microspores, and of protoplasts from higher plant cells and the recognition of the potential of these materials in genetics and plant breeding. One of the most significant developments in the field of plant tissues culture during recent years are the isolation, culture and fusion techniques which have their special importance in studies of plant improvement by cell modification and somatic hybridization.

Plant tissue culture technique is a boon in the studies of the biosynthesis of secondary metabolites and provides an efficient means of producing economically important plant products (fine chemicals).

Tissue culture can serve a number of purposes, and growers have started their own commercial production laboratories for a variety of reasons. Growing plants from seeds or cuttings can be unacceptable or impractical due to some of the following (and other) factors:

1. Seed-grown products lack uniformity
2. Seed-grown products are not true to type
3. Seeds take too long to grow to mature plants
4. Seeds are difficult to handle
5. Seeds are not available
6. Cuttings grow too slowly
7. Cuttings have poor survival rate
8. Cuttings require too much care
9. Cuttings are too vulnerable to disease
10. There is a shortage of stock plants from which to take cuttings because there is:
11. Only one hybrid
12. Only one virus-free plant
13. Only one desirable mutant

14. There is insufficient room for stock plants
15. It is not cost effective to maintain the stock plants

The simplest tissue culture hobby is the multiplication of easy, fast-growing plant material, such as *Kalanchoe*, Boston ferns (*Nephrolepis*), African violets (*Saintpaulia*), or *Begonia*; next in order of complexity are carnations (*Dianthus*), strawberries (*Fragaria*), or *Syngonium*. The chemist looking for a tissue culture hobby may be challenged to explore the field of plant by products; dyes, flavorings, medicinals, and oils are just some of the by-products of certain plants.

5

Biodiversity and Germplasm Conservation

The conservation of biodiversity is widely recognized as a high priority area for attention in the ongoing debate linking environment and development. In particular, the conservation of plant genetic resources for food and agriculture, one sector of biodiversity, is considered to be a major element of any strategy to achieve sustainable agricultural development, along with the conservation of other natural resources.

Genetic resources are renewable, provided they are well managed. Conservation and management of biodiversity, closely linked to the conservation and wise use of natural ecosystem, is of major importance for sustainable development. The tropical and sub-tropical areas are rich in term of plant genetic resources. India is particularly rich in biodiversity, including the ecosystem, species and genetic diversity, species and genetic diversity, by virtue of its varied climates and physical features.

In view of the unique importance of the grass covers in terms of total agro biodiversity scenario, it is appropriate to the realistic and assume a the following as priority concerns in the future context. The updating of grass cover/grazing resources of India is required to be done on priority. Almost three decades have passed when grass cover of India was documented. The environmental degradation has caused enormous impact not only on the flora and fauna but the area represented activity. Remote sensing and GIS approach have been refined and made more widely available in the country during this period. These can be suitably used for the required updating.

1. Collection, evaluation and exploitation of promising grasses/legumes, which have not yet been tested for their prudential as forage species.

2. Inventory of grazing routes, grazing system and designing sustainable production system for migratory grazers of Himalayas as well as of the Thar Desert.
3. Soil and water conservation may be taken up an integral part of grassland development.
4. Greater thrust is required to be made for selection of native types from stress prone habitats i.e. ravines, riverine, alkaline, acidic, hot desert, cold desert and steep slopes etc, their characterization, conservation and value addition.
5. In situ conservation of genetic wealth of above mentioned stress prone areas through botanical gardens can help in preserving several species which are facing extinction are less important today but may be of much value in future.

Ever since primitive man learned the art of farming and realized the economic utility of plants, he started saving selected seeds or vegetative propagules from one season to the next. Infact, he was practicing a type of germplasm conservation and management. The primitive cultivars and their wild relatives constitute a poll of genetic diversity (germplasm or gene bank) which is invaluable for breeding programmes. However, the population and needs of man have steadily increased over time leading to persistent exploitation of natural resources. The pressure thus created led to the way for gradual exclusion of traditionally used genotypes. In this way, some of the valuable gene pools might be lost unless coordinated efforts are made towards the conservation of genetic stocks all over the world.

Realizing the danger of erosion of genetic resources, the objective of providing necessary support to collection, conservation and utilisation of the plant genetic resources is concern of the world now.

Diverse conservation methods are pursued according to the situation at hand. These methods can be divided broadly into ex situ and in situ. The latter cover conservation in wilderness areas, reserves, protected areas, and within traditional farming systems (so-called, on-farm conservation). Ex situ conservation involves removing the plant genetic resources from their natural habitat and placing them under artificial storage conditions. The following sections examine the different ex situ options available.

GERMPLASM CONSERVATION

Germplasm (i.e., the self-contained units of propagation in plants (seed and vegetative propagules) and reproduction in animals (embryos) is, in fact, a *concentrated package of genes.* It is thus obvious why biotechnological techniques are especially relevant to commercialization of unique germplasms. This includes biotechnology of hybrid and artificial seeds, germplasm banks and cryopreservation, in vitro fertilization, and embryo implantation.

Both *in situ* and *ex situ*, the ICFRE and ICAR Institutions have put conservation methods into practice. In situ conservation is preferred option in case of grassland species because it allows for continuing adaptation of wild populations and the natural evolutionary processes. Following the *man and biosphere* approach advocated by the UNESCO, 14 sites have been identified as Biosphere Reserves. They have been set up in different parts of the country starting with Nilgiri biosphere reserve in 1986 covering an area of 5520 sq. Km. Each biosphere reserve is intended to fulfill three complimentary functions, namely:

1. **Conservation function**, to preserve genetic diversity, ecosystem and landscapes.
2. **Development function**, to foster sustainable economic and human development.
3. **Logistic support function**, to support demonstration projects, environmental education and training, research and monitoring related to local, national and global issues of conservation and sustainable development.

Germplasm Conservation Using Cell and Tissue Culture Objectives

1. To ensure the maximum level of genetic stability of the conserved material that is consistent with practical storage periods and with the available technology and resources.
2. To maintain accurate and detailed records concerning the biology of the conserved material, and its know history prior to collection. It is important to known the disease status of the material, and its full culture history including media details, the use of growth regulators, frequency of transfer etc. Cultures and derived plants should be characterised at some level of morphology performance and any known genetic characteristics, including analysis of .gene products and DNA fingerprinting where this is possible and appropriate. There is a responsibility to disseminate this information as widely as possible.

3. To develop and improve technology at all levels, and to increase the efficiency and utility of in-vitro techniques in the conservation of plant germplasm.
4. To construct and maintain a policy for the in-vitro conservation of those plant genetic resources appropriate to the collection with due regards to Internationals. National and Regional/Local priorities.
5. Conservation of somaclonal and gemetoclonal variations in cultures.
6. Maintenance of recalcitrant seeds.
7. Conservation of cell lines producing medicines.
8. Storage of pollen for enhancing longevity.
9. Conservation of rare germplasm arising through somatic hybridization or other methods of genetic manipulations.
10. Delaying the process of ageing.
11. Storage of meristem culture for micro propagation, micro grafting and production of disease free plants.

Germplasm Facilities

Recognizing the need for sophisticated facilities for research and development and providing services, the following additional germplasm facilities have been set up:

1. The National Facility for Microbial type Culture Collection at the Institute of Microbial Technology, Chandigarh, with over 1,600 culture in its stock.
2. The National Facility on Blue green Algal Collection at the Indian Agriculture Research Institute, with over 500 strains and several pure cultures as well as soil-based cultures, which have been supplied to farmers for production of bio-fertilizers.
3. The National Facility for Marine Cyanobacteria at the Bharatidasan University, Tiruchirapalli, which is coordinating extensive surveys on the southern coast.
4. The National Facility for Plant Tissues Cultures Repository at NBPGR, New Delhi, which has undertaken in vitro conservation of germplasm (cell, tissue, organ for medium and long term, particularly for those species for which conventional methods are inadequate. It has 850 accessions of crop species and employs molecular methods of characterization and classification.

5. The National Facility for Laboratory Animals at the Central Drug Research Institute, Lucknow and the National Institute of Nutrition, Hyderabad have made available quality animals for biomedical research and industry in the country.
6. The National Facility for Animal Tissue and Cell Culture, Pune, an autonomous institution under Department of Biotechnology (DBT) has 1127 stock cultures comprising 594 different cell strains. The facility has supplied 401 culture consignments to 84 institutions throughout the country. It also has 50 rectors, plasmids and genomic libraries.
7. Three National Gene Banks for Medicinal and Aromatic Plants at the Central Institute of Medicinal and Aromatic Plants, Lucknow and NBPGR, New Delhi, for the northern region; and the Tropical Botanical Garden and Research Institute, Trivandrum, for peninsular India have been established. These banks will conserve important species of proven medicinal value, which are categorised as endangered, threatened or rare and are used extensively in traditional systems of medicine, difficult to propagate, have significance for R&D for the future, and are of commercial values. India is the regional coordinator for Asia and also the overall coordinator for the establishment of Gene Banks of Medicinal and Aromatic Plants among G-15 countries.

Limitations of In vitro Conservation

There are practical limitation to any conservation system that does not provide a maximum genetic resource whilst, at the same time, conserving the potential of the system to produce whole plants. The level of limitation that is acceptable depends upon individual circumstances. However, where resources are not themselves limiting and the pressure of the time allow for significant research and development, the technique of cyropreservation must be the current in vitro conservation method of choice.

Yet, application of drying to somatic embryos, and the construction of artificial seeds, presents and exciting prospect for the future. Once successfully processed, the dried somatic embryos might be effectively stored at -20° C using a simple domestic refrigerator, which is globally available technology. However, the pre-treatments necessary to achieve desiccation without loss of viability are often complex and currently it is only the tissue of somatic embryos that can be treated in this way.

It is likely that the somatic embryo shares much of its genetic expression and developmental pathways with the zygotic embryo. Both have a potential to withstand extremes of desiccation when the zygotic embryo, and that both have a potential to withstand extremes of desiccation when the appropriate sequence of genes are in operation. In a developing orthodox seed (Roberts, 1973) the hydrated, zygotic embryo will respond to chemical stimuli from the parent plant to develop desiccation resistance and thus dry to the low water content characteristic of the seed. It appears that a somatic embryo can be triggered artificially in much the same way (Ammirato, 1989). This technique has obvious potential in circumstances where embryogenesis is both possible and desirable, and where the zygotic embryo does not show recalcitrant characteristics (Grout, 1986). However, it is not clear whether the techniques for drying could be feasible for other somatic cell and tissue types, that have patterns of development and differentiation far removed from that of the embryo whilst it is true that leaf callus cell for instance, contain the genes used in the developing embryo to tolerate desiccation during seed formation. It is another matter to be able to exploit this aspect of the genome independently of the rest of the embryo development. (Morris and Grout, 1990).

As cryopreservation has developed as a technology, and surviving material examined more critically, the concepts of optimized procedures and modified culture conditions by support recovery growth following thawing have been considered in more depth.

It is probably an exceptional occurrence when a cryopreserved cell is thawed to physiological temperatures with structure and functions completely unaltered, ready to resume typical activity. More commonly, surviving cells will have varying cells will have varying degree of damage that may will be able to repair, depending upon available reserves and metabolic integrity, and may require specific *recovery media* or past thaw conditions to ensure a reliable return to a known and predictable performance. Those that cannot effect satisfactory repair will die, sometime after thawing.

Different circumstances of cooling and warming may well generate different lesions and consequently differing requirements for repair, and it is not improbable that, even after repair, recover ells might show a range of alteration in both structure and function.

CLASSIC APPROACHES TO EX SITU CONSERVATION

The most familiar approach to ex situ conservation is seed storage. A large proportion of agricultural crops produce seeds that can be dried to a sufficiently low moisture content that they can be stored at low temperatures. There is an interaction between moisture content, temperature, and survival in storage and longevity, so that drying to lower moisture content permits storage at relatively higher temperatures. This principle underlies seed storage that should greatly reduce the constraints imposed by difficulties in maintaining sufficiently cold seed stores. This can be an especially serious problem in developing counties and, worldwide, is a factor in the cost of operating seed stores. Nevertheless, for crops that produce seeds amenable to drying and cold storage (i.e., *orthodox* seeds), this approach to conservation is convenient, is easily adopted, and is secure. Its drawbacks relate to biological, rather than practical, features that prevent its wider application beyond orthodox seeds.

1. There are those that do not produce seeds at all, and are propagated vegetatively, for example, banana and plantain (*Musa* spp.).
2. There are crops, including potato (*Solanum tuberosum*); other root and tuber crops, such as yams (*Dioscorea* spp.), cassava (*Manihot esculenta*), and sweet potato (*Ipomoea batatas*); and sugarcane (*Saccharum* spp.), that have some sterile genotypes and some that produce orthodox seed. However, similar to temperate fruits, including apple (*Malus* spp.), these seeds are highly heterozygous and, therefore, of limited usefulness for the conservation of gene combinations. These crops are usually propagated vegetatively to maintain clonal genotypes.

Those crops that produce *recalcitrant* seeds like coconut (*Cocos nucifera*), avodado (*Persea Americana*), mango (*Mangifera indica*), cacao (*Theobroma cacao*), and members of the Dipterocarpaceae family, cannot tolerate desiccation to moisture contents that would permit exposure to low temperatures. They are often large, with considerable quantities of fleshy endosperm. Although there are clear groups of species that can be classified categorically as orthodox or recalcitrant, there are also intermediate types for which seed storage is problematic. Traditionally, the field gene bank has been ex situ storage method of choice for these problem materials. In some ways, it offers a satisfactory approach to conservation.

The genetic resources under conservation are readily accessed and observed, permitting detailed evaluation. However, there are following certain drawbacks that limit its efficiency and threaten its security:

1. The genetic resources are exposed to pests, diseases, and other natural hazards, such as drought, weather damage, human error, and vandalism. Nor are they in a condition that is readily conducive to germplasm exchange.
2. Field gene banks are costly to maintain and, as a consequence, are prone to economic decisions that may limit the level of replication of accessions, the quality of maintenance, and even their very survival in times of economic stringency.

3. Even under the best circumstances, field genebanks require considerable input in the form of land (often needing multiple sites to permit rotation), labor, management, and materials.

NEW APPROACHES TO CONSERVATION

In vitro techniques features in the conservation strategies of microbes, and plants, but it is probably fair to say that the potential for exploiting these techniques and integrating them into wider practices, including genetic improvement, is greatest for higher plants. This relates largely to the ease with which plant material can be manipulated in vitro, in particular the phenomenon of totipotency (capacity to regenerate whole plants from single cells through tissue culture techniques).

Early efforts in the development of new approaches to conservation focused on storage *per se*, but applications of biotechnology have been demonstrated in all aspects of conservation and use, from germplasm collecting and exchange, to multiplication, disease indexing and eradication, characterization and evaluation, storage, stability monitoring, distribution, and utilization. It is instructive to explore the broader context of conservation and use of plant genetic resources.

One of the most important aspects in vitro culture, particularly for the problem crops identified earlier, is mass propagation.

In vitro Propagation

In vitro propagation not only facilitates the agricultural production of the crop, but also underpins the use of all other biotechnologies in conservation and use. The advantages of being able to multiply a given genotype with relative ease, with a low risk of introducing or reintroducing pathogens, and with a low risk of genetic instability, need not be emphasized. However, the latter point of genetic instability bears further examination. Genetic stability in culture is not a given. There are clear links between the culture system in use and risk of instability through Somaclonal variation. As a broad generalization in the context of genetic conservation, the more instability-prone culture systems, such as protoplast and cell cultures. Somatic embryos present some attractive options here, being both relatively amenable to storage cryopreservation and manipulability as synthetic seeds. Genetic stability under conditions in vitro conservation is dealt with later, as is the amenability of different culture systems to in vitro storage.

Molecular Technologies

Molecular technologies based on DNA extraction and storage offer new ways of conserving may be some way off in the future, it is possible to envisage application for the storage of specific gene sequences within a broad complementary conservation strategy. Similarly, in vitro conservation techniques can facilitate the application of genetic manipulation procedures by, for example, providing a simple way of storing experimental material in the form of in vitro culture. More importantly perhaps, new storage techniques can relieve the burden placed on all in vitro-based procedures imposed by the need to maintain stock cultures.

COLLECTION AND EXCHANGE OF GERMPLASM

Problems with Conventional Methods

In clonally propagated crops, the material of choice for collecting is often a vegetative propagule (a stake, piece of bud-wood, a tuber, corm, or sucker). In some cases the materials adapted for survival are once excised from the parent plant are used. They have plant health risk owing to their vegetative nature and contamination with soilborne organisms. The collector can compensate for these problems, to some extent, by good planning, careful selection of material, and observation of basic plant health precautions. Nevertheless, fundamental and unavoidable risks remain.

Problems for recalcitrant seeds

1. The recalcitrant seeds prone to microbial attack or deterioration if exposed to unsuitable environmental conditions.
2. If they are stocked for too long period during transition for the gene bank, then there are very limited options to handle them, once they do arrive at their destination.

Problems for orthodox seeds

1. Collecting the germplasm of orthodox, seed-producing species can also be problematic.
2. The material available for collecting may be sparse, immature, past its optimal state of maturity, shed from the plant, or even eaten by grazing animals.

The adaptation of in vitro techniques to the collecting environment, illustrates one of the simplest, but most effective, applications of biotechnology to plant genetic resources work.

In Vitro Field Collection

The first coherent and comprehensive examination of its potential was made by the International Board for Plant Genetic Resources (IBPGR). This took the form of a meeting of scientists with expertise and proposed four crop models serve to illustrate the general potential and flexibility of in vitro collecting. These are coconut, cacao, forage grasses, and *Musa* spp. models (model for grasses has not been described here).

Coconut

The coconut produces large, recalcitrant seeds. However, the key to finding a solution lies in recognizing that only the small embryo is needed to propagate a coconut palm, given adequate handling techniques. The principle used in the field collecting was the idea that, with minimal, but dexterous aseptic precautions, embryos could be isolated from nuts in the field, surface sterilized, and then either dissected at the field location or transported in endosperm plugs held in sterile salt solution for subsequent dissection in the laboratory. Table 5.1 provides details of the different approaches that have been taken with coconut and the other model crops examined.

The success that has been demonstrated with coconut can be repeated with other recalcitrant seeds that are physiologically similar but structurally very different, such as cacao, avocado, and *Citrus* spp. Embryos of these have fleshy cotyledons that can be dissected away to reveal the embryo axis. IN these cases, it is easier to keep the embryo axis sterile, and it can suffice to flame sterilize the fruit and, by using frequently sterilized instruments, maintain the inherent sterility of the interior of the seed.

Cacao

The seeds of cacao are recalcitrant. In this method, bud wood is often the target for collecting. This is vulnerable owing to the potentially long transit period involved in collecting from, for example, the Amazon basin, which is the center of origin of cacao and a valuable source genetic resources. Experimentation to develop an in vitro collecting method for cacao bud wood sought to minimize the materials and equipment to be carried to the collecting site. It was based on the premise that absolute sterility would be difficult to achieve in the field and would not necessarily be essential for robust, woody material. The technique involves the use of the drinking water-sterilizing tablets and culture medium supplemented with fungicides (Table 5.1). Antibiotics can also be used, but this must be weighted against the inherent disadvantages in their sue from the point of view of hazards to the user and the maintenance of low levels of persistent infection. The general approach used for cacao has been used successfully for other similar materials, such as woody shoots of coffee (*Coffea* spp.), *Prunus* spp., and grape *Vitis* spp.

Musa spp.

Musa spp. is collected in the form of stem suckers, which are large, fleshy, and likely to be covered in soil. The method includes the surface sterilization, combined with extensive dissection to reach naturally aseptic inner tissues. A sucker of some 30 × 8 cm is reduced to a shoot tip less than 1 × 1 cm.

The above methods illustrate the flexibility of in vitro collecting. There is no one formula to be followed, nor need there be. The approach should be based on :

1. Prior knowledge of the requirements of the species and explant in question.
2. Collective experience gained with diverse species in different collecting environments.
3. During any germplasm transfer operation, particular attention should be given to phytosanitary considerations.
4. In vitro collected explants should be treated with the same care and observance of regulations as any other type of collected material.

Table 5. 1: Summary of conditions used In vitro collecting diverse specimens

Species	Explant	Surface sterilization	Initial handling	Laboratory treatment
Coconut (*Cocos nucifera*)	Embryo in endosperm plug, extracted with cork borer	Calcium hypochlorite at 45 gL^{-1}	Endosperm plug inoculated into sterile solution of KCl at 16.2 gL^{-1}	Repeat sterilization if necessary. Embryo dissected, inoculated onto semisolid medium, cultured under standard conditions, transferred to the nursery.
Coconut (*Cocos	Embryo in endosperm	Commercial bleach (8% Cl)	Embryo dissected at	Embryo cultured under standard

Species	Explant	Surface sterilization	Initial handling	Laboratory treatment
nucifera)	plug, extracted with cork borer		field work bench and inoculated onto semisolid medium	conditions, transferred to the nursery.
Coconut (*Cocos nucifera*)	Embryo in endosperm plug, extracted with the cork borer	None	Endosperm plugs placed in a bag of freshly gathered coconut milk, held in a cool box	Endosperm plug surface sterilized, embryos dissected and inoculated onto standard culture medium
Coconut (*Cocos nucifera*)	Embryo in endosperm plug, extracted with cork borer	Inoculation is carried out in an inflatable glove box sterilized with alcohol. Sterilization with 5% calcium hypochlorite for 15-20 min. embryo is then excised and sterilized with 2% calcium hypochlorite for 2-5 min, followed by washing in sterile water.	Embryos are placed on standard culture medium in screw-top flasks	Standard culture procedure
Cacao (*Theobroma cacao*)	Stem nodal cutting	Drinking water sterilizing tabl[illegible] containing "Halozone" (*p*-carboxybenzene-sulfondichloroamide), 4 mg/tablet: 10 tablets dissolved in 100 mL of boiled water, plus 0.05% FBC protectant fungicide	Inoculation onto semisolid medium containing funicide Tilt MBC at 0.1% with or without antibiotics rifamycin at 30 mgL^{-1} and trimethoprim at 30 mgL^{-1}	Continued culture or resterilization using standard treatments, or grafting
Digitaria decumbens (forage grass)	Stem cutting	Drinking water-sterilizing tablets containing "Halozonc" (*p*-carboxybenzene-sulfondichloroamide) at 1 g tablets/L of boiled water	Inoculation onto culture medium containing 1.5 gL^{-1} benlate and 0.1 mgL^{-1} rifamycin	Standard culture conditions; transfer to soil after 14 weeks

Species	Explant	Surface sterilization	Initial handling	Laboratory treatment
Cotton (*Gossypium* sp.)	Stem nodal cutting	20% commercial bleach in 30% ethanol for 45 s; no washing	Inoculation onto solid culture medium containing half-strength salts, 1% glucose, antibiotics rifamycin at 15 mgL^{-1}, fungicide Tilt MBC at gL^{-1}, NAA at 1 mgL^{-1} and casein hydrolysate at 0.5 gL^{-1}	Resterilize with 4% bleach, treat with rooting hormone and plant in soil/sand/vermiculite mix with lime and slow-release nutrients

In vitro Germplasm Exchange

In vitro techniques are in use for several years for the international distribution of plant genetic resources. The techniques used are based on standard mass propagation procedures, with minor but following important modifications of detail to increase structural stability in transit.

1. The concentration of agar or other gelling medium used for preparing the culture medium may be raised to increase its firmness.
2. The plantlets may be transferred in sterile heat-salable polyethylene begs, rather than the more fragile glass or plastic containers.
3. Those species that produce storage propagules (stem tubers, capable of regenerating plants) a further option should be used. This approach has been used successfully in potato and yam, the tubers being more resilient with the result of producing a more robust system for germplasm exchange.

SLOW-GROWTH STORAGE

Classic Techniques

Standard culture conditions can be used only for medium-term storage of naturally slowly growing species. These techniques have been developed for reducing the growth rate of cultures. Slow-growth storage techniques involve modification of the physical environmental conditions or culture medium, or both. The most successful and widely applied technique is temperature reduction. A decrease in light intensity or culture in the dark is often used in combination with temperature reduction. Strawberry (*Fragaria* x *ananassa*) plantlets have been conserved in the dark at 4°C. Regular addition of a few drops of liquid medium to the cultures maintained the plantlets viable for up to 6 years.

Apple (*Malus domestica*) and *Prunus* shoots survived 52 weeks at 2°C. Temperatures in the range of 0°-5°C can be employed with cold-tolerant species that are often cold-sensitive. The cassava shoot cultures have to be stored at temperatures higher than 20°C. Oil palm (*Elaeis guineensis* Jacq.) somatic embryos and plantlets cannot withstand even short-term exposure to temperature lower than 18°C, while banana in vitro plantlets can be stored at 15°C without transfer for up to 15 months.

A reduction in the concentration of mineral elements and elimination of sugar allowed the conservation of *Coffee* plantlets for 2 years. Addition of osmotic growth inhibitors (e.g., mannitol) or hormonal growth inhibitors (e.g., abscisic acid) is also an efficient way to achieve growth reduction.

The type of culture vessel, its volume and the volume of medium, and the closure of the culture vessel influence the survival of stored cultures. Keeping shoot cultures in larger vessels improved their condition and maintenance of viability during storage. Replacing cotton plugs by polypropylene caps, thereby reducing the evaporation rate of the culture medium. It also increased the survival of *Rauvolfia serpentina* during storage. The use of glass and plastic vessels, the use of heat-salable polypropylene bags has been reported.

Alternative Techniques

Alternative slow-growth techniques include modification of the gaseous environment and desiccation or encapsulation of explants. Growth reduction can be achieved by lowering the quantity of oxygen available to the cultures. The method includes covering of the tissues with paraffin oil, mineral oil, or liquid medium.

This techniques was first developed by Caplin (1959). He stored carrot (*Daucus carota*) callus under paraffin oil for 5 months. It was employed more recently by Augereau et al. (1986) with *Catharanthus* calluses and by Moriguchi et al. (1988) with grape. Florin showed that 86 and 50% of a collection of 313 different callus lines could be stored with the same technique for 6 and 12 months, respectively. Similarly, 13 of 20 cell suspensions from eight different species survived after 6 months of storage under liquid medium without shaking.

Attempts to store microcuttings under mineral oil have been performed with pear (*Pyrus communis*), coffee, and several ginger (*Zingiber officinale*) genotypes. Growth reduction was achieved in all cases, but hyperhydration of explants was often observed during storage. After return to standard conditions following four months in storage, re-growth of surviving cultures was very slow for coffee, and partial or complete necrosis of explants was noted with pear. However, this storage technique was very efficient with some ginger genotypes, which could be conserved under mineral oil with viability for up to 2 years.

Reduction in the level of available oxygen

Oxygen plays important role in storage. Reduction in Oxygen be achieved by decreasing the atmospheric pressure of the culture chamber or by using controlled atmosphere. Tobacco (*Nicotiana tabacum*) and chrysanthemum (*Chrysanthemum morifolium*) plantlets were stored under low atmospheric pressure (with 1.3% O_2) for 6 weeks. Oil palm somatic embryos could be conserved for 4 months at room temperature in a controlled atmosphere with 1% O_2, and proliferated rapidly after subsequent transfer to standard conditions (Engelmann, 1990).

Desiccation

It is a means of storage of embryogenic cultures. It was first described by Jones (1974). Embryogenic cultures of carrot were left on semisolid medium for up to 2 years at 25°C. Supply, of a sucrose solution to the cultures resulted in "germination" of the embryos that, on planting out, produced healthy individual plants. Senaratna et al., (1991) have shown that alfalfa (*Medicago sativa*) somatic embryos, dehydrated progressively using saturated salt solutions, could be conserved with 10-15% moisture content for 1 year at room temperature. They displayed only a 5% decrease in their conversion rate after storage. Lecouteux et al. (1992) stored carrot somatic embryos for 8 months at 4°C without viability loss.

Large-scale propagation by means of somatic embryogenesis

This technique has been developed for elite genotype of numerous crop species, leading to the production of large numbers of synchronously developing embryos. These embryos can be encapsulated in a bead of gel (e.g., calcium alginate), containing nutrients and fungicides, thereby forming synthetic seeds which, theoretically, can be stored and sown directly in vivo similar to true seeds. The production of synthetic seeds has been developed for many plant species. The application of synthetic seed technology to somatic embryos or shoot tips also appears of interest in a germplasm conservation context. However, only a limited number of short-to medium-term storage experiments have been performed with encapsulated material.

Encapsulated axillary buds of mulberry (*Morus indica* L.) and somatic embryos of sandalwood (*Santalum album*) can be stored for 45 days at 4°C and somatic embryos of interior spruce (*Picea glauca*) for only 1 month. Storage for longer periods was achieved if beads were placed in liquid medium at low temperature. Under these conditions, Machii (1992) conserved mulberry apices for 80 days and Shigeta et al., (1993) carrot somatic embryos for 3 months. Redenbaugh et al. (1991) mention that the rapid survival loss of encapsulated material that is generally observed is mainly due to the encapsulating matrix, which dehydrates rapidly and limits the respiration of the embedded embryos.

CRYOPRESERVATION

History

Cryopreservation is the storage at ultralow temperatures in a cryogenic medium, such as liquid nitrogen. It has the potential to achieve the goal of suspending metabolism and, to all intents and purposes, suspending time. Cryopreservation has a relatively long history in microbiology for the storage of stock cultures, and in livestock husbandry for the storage of semen of elite male cattle. In higher plant systems to cooling to ultralow temperatures has been carried out over the past 40 years, following two main themes:

1. To gain an understanding of the physiological and biochemical processes involved in the transitions to and from the frozen state, including cold acclimation.
2. To preserve plant material in a viable state. The most successful cryopreservation work has involved attention to the underlying processes of cryoinjury and cryoprotection, rather than an empirical approach alone.

The first report of exposure of cultured plant material to the temperature of liquid nitrogen was made by Quatrano in 1968, using cultured cells of flax (*Linum usitatissimum*). This methods adopted closely followed the classic procedures found to be successful with other living systems; namely; chemical cryoprotection, slow, dehydrative cooling, storage in liquid nitrogen, rapid thawing, washing, and recovery. In recent years there have been various diversification in cryopreservation technique, specially by varying biological requirements and varying infrastructural situations, thereby extend cryopreservation to a wide range of users.

Assuming that differentiated plantlets are to be stored in-vitro, with all that this implies for genetic stability, and then lowering growth temperature to slow metabolism and development has previously reduced the labour and expense for repeated subculture. The security of the cultures is also improved as the frequency of the physical interventions of subculture is also reduced. Similarly, the use of osmotic stress and the addition of growth retardant to the culture medium have effectively prolonged the interval between culture transfers (Withers 1987, Pritchard et. al., 1986).

The greatest genetic stability is achieved by storage of propagules, capable of regeneration in-vitro liquid nitrogen i.e. **cryopreservation** (Kartha 1985, 1987, Grout and Morris 1987, Withers 1987) where possible these propagules will be excised meristems, capable of direct regeneration into plantlets, but they may, of necessary, by isolated cells or protoplasts.

An *in vitro* active genebank (IVAG) has been developed at the Centro Internacional de Agricultural Tropical (CIAT), Cali, Colombia, which maintains approximately 5500 clones of world collection of *Manihot* under reduced growth. The other important centres maintaining *in-vitro* genebanks of root and tuber crops are the CIP (Lima, Peru), IITA, (Nigeria), CATIE (Cost Rica), INRA (Guadalope).

Cryostorage or Long-term Storage

Cryostorage or long-term storage provides indefinite storage system for organ, tissue or cells at ultra low temperature (at – 196°C using liquid nitrogen) by reducing metabolism significantly with no biological deterioration. Long term·storage lessens the risks associated with slow growth storage. Despite the genetic stability inherent in organized plant structures such as meristem, there are risk to DNA structure, integrated metabolism and viability under of slow growth such as resulting from the oxidative activity of free radicals (Benson 1990). Cryopreservation is an economical, space saving method for a long-term storage (Kartha 1985; Withers 1987).

To avoid periodic sub-culturing and maintain genetic integrity of the collection, Escober *et al.* (1994) developed cryopreservation protocol to recover shoot tips from liquid nitrogen with cassava M Col 22 as a model having 60-70% plant recovery.

Preculture of shoots followed by cryoprotectionwith sorbitol and dimethyl sulphoxide (DMSO) are considered to be important for high rate of survival. Protocols have also been developed for cryostorage of seeds and zygotic embryos and somatic embryos (Marin *et al.,* 1990; Sudarmonowati and Henshaw, 1990). In sweet potato cryopreservation of embryogenic tissue has been attempted by Blakesley *et al.,* (1995) and Bhatti *et al.,* (1997). Their protocols were based on encapsulation sucrose pre-treatment and evaporative dehydration prior to freezing. Survival of embryogenic tissue was 4%-38% depending on the genotype.

Cryogenic gene banks are relatively easy and inexpensive to operate. Many people do not realize that cryogenic banks involve such simple techniques. For them, cryogenics conjures up images of technicians in white coats wheeling gleaming vats of liquid nitrogen. As the simplicity of these banks becomes widely known, more and more facilities will be developed. Already, some of the institutions and groups that currently use gene banks without sub-freezing temperatures have been persuaded to adopt the more efficient, cryogenic techniques.

Gene banking is the practical and effective method to combat plant extinctions. It is a kind of freezer that preserves seed and pollen. This technology does not require extensive knowledge or specialized training, and expensive equipment. Gene banks are generally easy to construct and maintain, although a few problems may arise during or after construction of the bank.

Classic Techniques

Most of the early work on the cryopreservation of in vitro plant cultures focused on a two method based on chemical cryoprotection and dehydrative cooling. This was particularly successful with cell suspension cultures, which is not surprising when the underlying biophysical events are explored. The vast majority of higher plant somatic cells, be they in vivo or in vitro, are not inherently freeze-tolerant. The transition of extra- and intracellular water into ice causes damage of a physical or biochemical nature.

Freezing Techniques

Extracellular freezing commonly occurs first, causing a flow of water from the cytoplasm and vacuole to the extra-cellular space where it freezes. Depending on the rate of cooling, different amounts of water will leave the cell before the intracellular contents solidify. Rapid cooling will result in more water remaining within the cell and causing potentially damaging ice than in slow cooling.

1. Ice causes damage when formed in the freezing process *per se.*
2. It can also cause damage during rewarding owing to the phenomenon of recrystallization, in which ice melts and reforms at a thermodynamically favorable, larger, and more damaging crystal size. This can be mitigated by rapid thawing.
3. Shrinkage of the protoplast and loss of surface area in the plasmalemma can render the protoplast incapable of resuming its original volume and surface area after thawing, resulting in rupture.

Slow cooling reduces this risk, but can incur different damaging events owing to the concentration of intracellular salts and changes in the cells membrane. Light and electron microscopic studies of cell suspension cultures and isolated protoplast systems have helped clarify the nature of damage under different cooling regimens. They have also revealed the mitigating effects of cryoprotectants. Cryoprotectants facilitate the flow of water across the cell membrane, and protect both molecular and gross structures through a range of modes of action, including colligative effects and free radical scavenging.

The effect of cooling rates on survival in cryopreserved cell suspension culture systems clearly illustrate the existence of an optimum cooling rate, commonly in the region of -1°C or -2°C min^{-1}. This method provides central strategy of the classic approach to cryopreservation. The effect of culture conditions before cryoprotection, the age of the cells at the time of harvest for cryopreservation, immediate post thaw treatment, and recovery growth conditions, as well as looking more closely at the temperature excursions are also important because of:

1. the age of cells at the time of harvesting for storage can affect their survival,
2. this is linked to cell size and water content,
3. rapidly growing cells are small and have a relatively low water content, and
4. modification of the pregrowth medium used for the passage before cryopreservation by, for example, the addition of osmotically active compounds, such as mannitol and sorbitol, can lead to reduction in cell size and an increase in freeze tolerance.

For cell suspension cultures

For cell suspension cultures, in particular, mixtures of cryoprotectants are much more effective than single cryoprotectants, and preparation in culture medium is usually beneficial. Removal of cryoprotectants after thawing has not been demonstrated to be essential, and there is clear evidence for a detrimental effect of washing.

Recovery of growth on solid medium

Similarly, recovery growth on solid medium is generally much more effective than dilution in liquid medium. If toxicity is suspected, precautions can be taken, such as moving cells on a supporting filter paper through a series of dishes of solid culture medium. Table 5.2 provide examples of the application of the classic approach to cryopreservation for a range of in vitro plant cultures systems.

Table 5.2: Examples of (a) the Classic, Slow, Cooling-Based Approach to Cryopreservation of In Vitro Plant Cultures, and (b) Rapid Cooling

Species	Culture system	Pre-growth	Cryoprotection	Cooling, storage, warming	Recovery
(a)					
Sycamore (*Acer psedopatanus*)	Cell suspension	Culture for 3-4 days in medium containing 6% mannitol	0.5 M DMSO + 0.5 M glycerol + 1 M sucrose	-1°C min-1 to -35°C, hold for 30 mins; transfer to liquid nitrogen; thaw in water bath at +40°C	Layer cells in suspending liquid over semisolid medium
Soyabean (*Glycine max*)	Protoplasts	Protoplasts isolated from exponentially growing cells	5% DMSO + 10% glucose	-10°C min^{-1} to -35°C; transfer to liquid nitrogen; thaw in water bath at +40°C	Wash in liquid medium; transfer to standard medium
Potato (*Solanum tuberosum*)	Shoot-tip	Shoot-tip dissected from glasshouse or in vitro plants; incubate overnight in standard liquid medium	10% DMSO	-0.20°C min-1 to -35°C; transfer to liquid nitrogen; thaw in water bath at +37°C	Wash twice with liquid medium; transfer to semisolid medium
Pear (*Pyrus* spp.)	Shoot-tip	Culture in vitro plantlets at 22°C/16 h day, -1°C/8 h night for 7 days; dissect shoot-tips and	10% polyethylene glycol + 10% glucose + 10% DMSO	-1°C min^{-1} to -40°C; transfer to liquid nitrogen; thaw in water bath at	Wash in liquid medium; drain; transfer semisolid medium

Species	Culture system	Pre-growth	Cryoprotection	Cooling, storage, warming	Recovery
		pregrow for 48 h in medium containing 5% DMSO		+40°C for 1 min; transfer to +23°C	
(b)					
Carnation (*Dianthus caryophyllus*)	Shoot-tip	Dissect shoot-tip from cold-hardened plant	5% DMSO	Place in ampule and plunge into liquid nitrogen	Culture under standard conditions
Potato (*Solanum tuberosum*)	Shoot-tip	Dissect shoot-tip from plantlet and culture on filter paper floating on liquid medium for 2 days.	10% DMSO	Collect shoot-tip on hypodermic needle and plunge into liquid nitrogen; thaw by plunging into liquid medium at 34°C-40°C	Culture on filter paper bridge over liquid medium
Oilseed rape (*Brassica napus*)	Shoot-tip	Dissect shoot-tip from in vitro plantlet; incubate for 24 h in medium containing 5% DMSO	15% DMSO	Collect shoot-tip on hypodermic needle and plunge into liquid nitrogen; thaw by plunging into liquid medium at room temperature (+40°C)	Transfer without washing to semisolid shoot-induction medium

NEW CRYOPRESERVATION TECHNIQUES

Principle

The new cryopreservation techniques are based on the phenomenon of vitrification. *Vitrification* is the process of transition of water directly from the liquid phase into an amorphous phase or glass, while avoiding the formation of crystalline ice (vitrification in the present context should not be confused with the phenomenon of hyperhydration).

Vitrification is a process in which cell dehydration is performed before freezing by exposure of samples to concentrated cryoprotective media or air desiccation. This is followed by rapid cooling. As a result, all factors that affect intracellular ice formation are avoided. Glass transitions (changes in the structural conformation of the glass) during cooling and re-warming have been recorded with various materials using thermal analysis.

Vitrification-based procedures offer practical advantages in comparison with classic freezing techniques. Vitrification is more appropriate for complex organs (shoot tips or embryos) which contain a variety of cell types, each with unique requirements under conditions of freeze-induced dehydration. By precluding ice formation in the system, vitrification-based procedures are operationally less complex than classic ones (e.g., they do not require the use of controlled freezers) and have greater potential for broad applicability, requiring only minor modifications for different cell types.

Luyet (1937) was the first to envisage the use of vitrification for cryopreserving biological specimens, but it is only somewhat recently that numerous reports on cryopreservation of plant material using vitrification-based procedures have appeared in the literature (Sakai, 1993). Four different procedures based on the phenomenon of vitrification can be identified: encapsulation-dehydration, desiccation, pregrowth-desiccation, and a procedure that actually goes by the name of vitrification

Encapsulation-Dehydration

The encapsulation-dehydration technique is based on the technology developed for the production of synthetic seeds, by which embryos are encapsulated in a bead of calcium alginate gel (see Chapter on Synthetic seeds). The encapsulation-dehydration technique permits freezing explants of large dimensions; pear shoot tips up to 5 mm in length, and heart to torpedo stage embryos (2-3 mm in length) have been successfully cryopreserved.

Before the cryopreservation procedure itself, plant material is often subjected to various treatments that increase survival potential. For cold-tolerant species, such as pear, apple, or mulberry, mother plants or apices can be placed at a low temperature (0-5°C) for several weeks. Scottez showed that this cold treatment resulted in an increased quantity of unsaturated fatty acids in the pear apices. Before encapsulation, apices of mulberry are transferred daily onto solid media progressively increased sucrose concentrations to initiate dehydration.

The stages of the process after encapsulation are pregrowth, desiccation, cooling, warming, and recovery growth. Pregrowth is performed in liquid medium enriched with sucrose (0.3-1 M) for periods of between 16 h and 7 days. Partial replacement of sucrose with other sugars (raffinose, maltose, glucose, or trehalose) did not improve the survival of cryopreserved grape (*Vitis vinifera*) shoot tips. For plant species that are sensitive to direct exposure to high sucrose levels, a progressive increase in sucrose concentration is used.

Table 5.3: List of plant species and specimens (apices, somatic and microspore embryos) that have been successfully cryopreserved using the encapsulation-desiccation technique

Specimen	Species
Shoot apex	Potato, Apple, Pear, Mulberry, Carnation, Grape, Chicory, Eucalyptus, Coffee, Cassava, Sugarcane
Somatic embryo	Carrot, Coffee Walnut
Microspore embryos	Oilseed rape

Encapsulated samples are desiccated either in the air current of a laminar airflow cabinet or by using silica gel. The latter method is preferred because it provides more precise and reproducible desiccation rates. The optimal water content of desiccated beads is about 20% (fresh weight basis), ranging from 13% with coffee somatic embryos to 30-35% with apple, grape, mulberry, and cassava apices.

Cooling

Cooling is usually carried out rapidly, by direct immersion of samples in liquid nitrogen. However, controlled slow cooling, down to -100°C, led to improved survival of grape apices. In contrast, the survival rate of sugarcane (*Saccharum* spp.) apices was higher after rapid than after slow cooling. These results suggest continued dependence on control of the residual water content in the specimen. Storage is usually performed at -196°C.

Recovery

For recovery, samples are usually placed directly under standard conditions. Recovery growth of cryopreserved material is usually direct and rapid, without callus formation. Histological studies performed with apices of several plant species revealed that the structural integrity of most meristematic cells is preserved after cryopreservation by encapsulation-dehydration. Therefore, recovery growth originates from the entire meristematic zone. This is contrary to what is generally observed after classic cryopreservation, during which many cells are destroyed, frequently leading to callus formation during recovery.

Desiccation

Cryopreservation using a desiccation procedure consists of dehydrating the plant material, then freezing it rapidly by direct immersion in liquid nitrogen. Desiccation has been applied mainly to zygotic embryos of a large number of species. Desiccation is usually performed by placing the embryos or embryonic axes in the air current of a laminar airflow cabined. However, more precise and reproducible desiccation can be achieved by placing plant material in a stream of compressed air or in an air-tight container with silica gel.

Duration of desiccation

The duration of desiccation varies with the size of the embryos and their initial water content. Optimal survival rates are generally noted when embryos are dehydrated down to 10-20% moisture content (fresh weight). Dehydration must be sufficient to ensure survival after freezing, but not so intense to induce extended desiccation injury. In optimal cases, no significant difference is observed in the survival rates of desiccated control and cryopreserved embryos, as noted with tea (*Camellia sinensis*), banana, and hazelnut (*Corylus* spp.).

Regrowth of plant material

Regrowth of plant material after warming is usually direct, but modified regrowth patterns are occasionally observed. Chin et al. noted the nondevelopment of the hustorium and more rapid leaf expansion of cryopreserved embryos of *Veitchia* and *Howea,* in comparison with unfrozen controls

Modified recovery conditions, notably of the hormonal balance of the culture medium, can significantly improve the survival rate of the cryopreserved material (e.g., coffee embryos).

Pregrowth-Desiccation

Cryopreservation using a pregrowth-desiccation procedure comprises the following steps:

This techniques has been applied to only a limited number of specimens: stem segments of in vitro plantlets of asparagus (*Asparagus officinalis* L.), somatic embryos of melon (*Cucumis melo*) and oil palm, microspore embryos of rapeseed, and zygotic embryos of coconut.

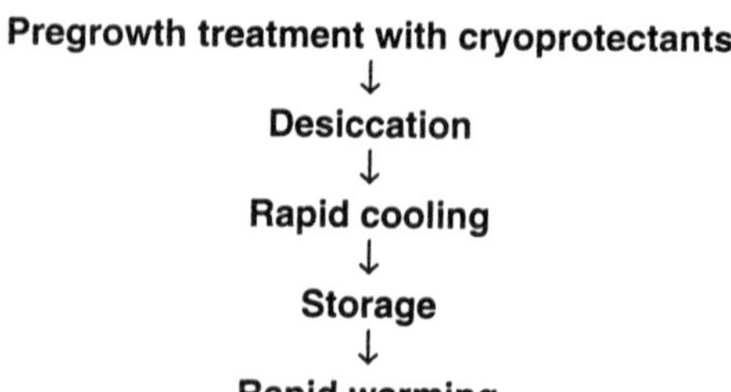

The application of cryoprotectants is usually performed before desiccation.

Table 5.4: Procedures adopted for dehydration/ desiccation

Plant	Dehydration/ desiccation
Coconut embryos	Ddehydration was carried out before preculture with cryoprotectants. The duration of treatment with cryoprotectants varies between 20 h for coconut. Various methods have been employed for desiccation embryos were placed either in the air current of a laminar airflow cabinet.
	Pregrowth-desiccation has been tested with four varieties of coconut, giving recovery rates of between 33 and 93%.
Oil Palm	Ddehydration was carried out before preculture with cryoprotectants. The duration of treatment with cryoprotectants for 7 days with somatic embryos. Somatic embryos were placed over a salt solution, ensuring a constant relative humidity.
	Storage is usually performed in liquid nitrogen. Experiments with oil palm somatic embryos have shown no modification in the recovery rate after 1 or 52 months of storage at -196°C.
	Dumet et al. have been able to conserve oil palm somatic embryos for 6 months at -80°C (i.e., below the glass transition temperature) without any modification in recovery rate compared with embryos stored at -196°C.
	Somatic embryos were cultured on media with a progressively reduced sucrose concentration and transitory supplement of 2,4-dichlorophenoxyacetic acid (2,4-D), to stimulate proliferation. Specimens are usually transferred directly onto standard medium for recovery.
	Pregrowth-desiccation (large-scale application) technique has been performed in the case of oil palm somatic embryos. Eighty clones are now routinely stored at -196°.
Melon	Sugars (sucrose or glucose) are generally employed for preculture. However, abscisic acid only was used for the pretreatment of somatic embryos.

Plant	Dehydration/ desiccation
Asparagus stem segments, rapeseed and oil palm	Embryos are placed in an air-tight chamber silica gel.

All materials cryopreserved using pregrowth-desiccation are cooled rapidly by direct immersion in liquid nitrogen. Warming is generally carried out rapidly except for stem segments of asparagus and rapeseed microspore embryos, which were rewarmed slowly at room temperature. Cryopreservation using pregrowth-desiccation has ensured satisfactory survival rates with all materials tested, and recovery is usually rapid and direct.

Vitrification

Vitrification procedures consist of the following steps:

1. Treatment (*loading*) of samples with cryoprotective substances.
2. Dehydration with a highly concentrated vitrification solution.
3. Rapid cooling and warming.
4. Removal (*unloading*) of the vitrification solution.

Vitrification solutions are complex mixtures of cryoprotective substances. The most commonly employed are derived from the solutions proposed by various groups are as follows:

Table 5.5: The most commonly employed solutions proposed by various groups

Group	Percentage of the chemical
Sakai's group	
Glycerol	22%
Ethylene glycol	15%
Propylene glycol	15%
Dimethyl sulfoxide (DMSO)	7%
Sorbitol	0.5 M
Steponkus' group	
Ethylene glycol	40%
Sorbitol	15%
Bovine serum albumin	6%

Vitrification procedures have been developed for about 20 species, using protoplasts, cell suspensions, shoot apices, and somatic embryos. The plant material is often submitted to various treatments before the cryopreservation procedure itself, to increase its survival potential. In the cold-tolerant species, the in vitro mother plants can be cultural at low temperature for several weeks. Examples have been placed for 1 or 2 days on a medium supplemented with a high sugar concentration or cryoprotective agents. Mint (*Mentha* spp.) shoot tips have been cultured thus for 2 days on a medium containing 0.75 M sucrose and 4% DMSO.

Explants are then loaded (i.e., suspended) in a medium containing cryoprotective substances (ethylene glycol, glycerol, sucrose) for a short period (5-90 min, depending on the material). This reduces their sensitivity to the vitrification solution. Survival of rye (*Secale cereale*) protoplasts after exposure to a vitrification solution increased from 4% without loading with 1.5 M ethylene glycol to 65% with loading.

The duration of contact between the plant material and the vitrification solutions is a critical parameter owing to their high toxicity. The period genetically increases with the size of explants treated. Rye protoplasts have to be dehydrated for only 60s, whereas the optimal dehydration period is 80 min for apple and pear shoot tips. Dehydration of samples at 0°C instead of room temperature reduces the toxicity of the vitrification solutions and increases the potential period of exposure to vitrification solution, thereby giving more flexibility for handling the plant material during this critical phase of a vitrification protocol.

Once dehydrated, samples are cooled rapidly by direct immersion in liquid nitrogen to achieve vitrification of intracellular solutes. Reduction in the quantity of suspending cryoprotectant solution and the use of containers of a small volume (e.g., 500 µL plastic straws) led to increased cooling rates.

Rewarming of samples has to be performed as rapidly as possible to avoid devitrification processes, which would lead to the formation of ice crystals that would be detrimental to cellular integrity. Thus, samples are immersed in a water bath or liquid medium held at 20°-40°C. However, Steponkus et al. have advised holding vitrified samples in air for a few seconds before pluging them in a thermostated water bath. This is to achieve slow rewarming through the glass transition region (ca. -130° C) to minimize mechanical fracturing of the glass caused by excessive thermal gradients.

After warming, the highly concentrated vitrification solution must be removed progressively to minimize osmotic shock. This is usually performed by diluting the vitrification solution in liquid medium supplemented with 1.2 M sucrose or sorbiol, before transferring the explants to standard medium.

Vitrification procedures generally lead to high survival rates, and direct and rapid recovery is usually observed. However, Towill mentioned callus formation and abnormal development of some mint apices after vitrification. Vitrification experiments involving a large range of genotypes are still infrequent.

GENETIC STABILITY OF IN VITRO CONSERVED GERMPLASM

During genetic conservation high priority should be given to genetic The facility of cloning in vitro does than offer, superficially, a very attractive means of perpetuating given genotypes, particularly for traditionally vegetatively propagated material. However, assumptions of clonal integrity in the in vitro situation may be unsafe. It is more pragmatic to consider supposed clones to be very tight populations with potential for deviation from the original distribution of genotypes. This then leads to a consideration of the factors that might contribute to such a deviation, namely creation and selection.

Genetic variation may arise by somaclonal variation, with obvious implications for the choice of culture system used for genetic conservation. It may be intrinsic in the cultured material, possibly linked to its genetic structure, such as in sugarcane or banana, for which polyploids are more prone to instability than diploids.

Selection may occur under conditions that either cause differential damage of a lethal nature or that favor the growth of one genotype over others in a mixture. The issues to be taken into consideration and the information now available on genetic instability in material conserved in vitro are rather different for slow growth and cryopreservation.

Slow growth

In cultures comprising a mixture of genotypes, the different components of the mixture may not grow at the same rate. Under the stressed conditions implicit in slow growth. The risk of selection must be considered to be greater than under standard growth conditions. Accordingly, it is important to minimize the initial risk of instability and take measures to minimize additional risks and monitor cultures at intervals to detect variation.

The risk of instability can be minimize at the outset and during slow growth storage through control of the culture system. If the cultures are maintained in a highly organized state (as shoots, plantlets, or embryos) the risk of somaclonal variation is much lower than if they were in the form of cells or calluses. The organized cultures need not incur unacceptable risks. The genetic integrity of cassava cultures maintained in slow growth at CIAT over a period of 10 years, was confirmed when tested by isozyme analysis, DNA analysis, and by monitoring their morphology when returned to the field.

Differences occur in species to species during their susceptibility to somaclonal variation. When this is combined with the clear differences in response to storage under slow growth conditions experienced not only between species but also between cultivars, it is clear that one cannot simply extrapolate from one fortunate example to all others. Thus, there is a pressing need for controlled experiments to test the genetic integrity of cultures stored in slow growth in comparison with controls maintained under normal growth conditions, and ideally, with cryopreserved specimens, to gain an insight into the relative risks.

Musa is an interesting model for experiment. Mass propagation in vitro is widely used for bananas and plantains, and a risk of somaclonal variation, even under optimal propagation conditions, is recognized. This is strongly linked to genotype. Studies of this phenomenon and parallel development of morphological, biochemical, and molecular methods for characterizing variants, have yielded a wealth of information on patterns of variation and potential markers to use in monitoring instability.

The risks of selection under slow-growth conditions should not be evaluated alone. Culture under any conditions carries risks. The risks of loss through human error are reduced by most storage measures. Risks caused by equipment failure are highly variable and, under some slow growth conditions, might be considered to be greater than in normal growth.

Cryopreservation

Cryopreservation involves a series of stresses that may destabilize that plant material and lead to modifications in recovered cultures and regenerated plants. Therefore, it is essential to verify the genetic stability of material recovered from cryopreserved samples before this technique is routinely used for the long-term conservation of plant germplasm. Even though freezing protocols have been developed for many species, only a limited number of studies have considered this aspect. No modification at the phenotypic, biochemical, chromosomal, or molecular level that could be attributed with certainty to cryopreservation has yet been reported. This correlates with observations from other biological systems.

In cell suspensions, numerous example are now available to illustrate that cryopreserved cells maintain their biosynthetic and morphogenic potential. The only published exception concerns lavender *Lavandula vera* cell suspensions submitted to repeated freeze-thaw cycles: the number of colonies recovered from cryopreserved cells increased with the number of freeze-thaw cycles, suggesting that the selection of more freeze-tolerant cells was taking place. However, no modifications were noted in the biosynthetic and regenerative capacities of cryopreserved cells, implying a change in population structure, rather than genetic change.

In the plants regenerated from cryopreserved apices no differences could be noted in the vegetative and floral development of several hundred oil palm plants regenerated from control and cryopreserved somatic embryos. The recovery of apices on certain media, plants failed to produce flowers in the first regeneration cycle. It is also suggested that use of tissue culture is better than that of cryopreservation.

METHODS OF CRYOPRESERVATION

Collecting, Cleaning and Drying

The first step in collecting pollen for the gene bank is to make certain that the pollen are alive. A variety of biochemical tests and stains can be used, but the most dependable method is also the easiest. Simply plant a known number of seed and see how many germinate. Only on rare occasions will all of the pollen begin to grow. Plant scientists usually strive for about 95 per centviability. Some plants have naturally low viability rates. For example, *Aloe albida* germinates at a rate between 25 and 30 per cent. This species is endangered, perhaps because of this highly inefficient rate of germination.

In this system stability is imposed by ultra low temperature and storage is at, or close to -196° C using liquid Nitrogen (or the vapour immediately above it), as practical and convenient oxygen. At such temperature normal cellular chemical reactions do not occur as energy level are too low to allow sufficient molecular motion to complete the reaction. Water exists either in a crystalline or glassy state under these conditions and such high viscosity (> 1013 poises) that rates of diffusion are insignificant over time spans measured at least as decades. The majority of the chemical changes that might occur in a cell are therefore; effectively prevented and so the cell is stabilized o the maximum extent that is practically possible.

Unfortunately, that is not to say that biological material successfully. cooled to -196° C is in state of complete suspended animation. Certain type of chemical reaction can still occur at these temperatures. Such as the formation of the free radicals and macromolecular damage due to ionizing radiation. The only real threat to genetic stability comes from such reactions, especially those that damage nucleic acids. Any damage that does occur will necessarily be cumulative as enzyme repair mechanism are also totally inhibited at these low temperatures.

While there are as yet few quantitative data on genetic stability at ultra low temperature for higher organisms, studied by Ashwood, Smith and Grant (1977) indicate that to reach a D10 level (Where D10 is the radiation dose resulting in 90% mortality of the population) a frozen cell population would have to be exposed to back ground radiation for some 32,000 years. It is also noteworthy that dimethyl sulphoxide, probably the most commonly used cryoprotectant in-vitro preservation and may aid in radiation damage. (Finkle, et. al., 1995).

The potential of conservation system for in-vitro material based upon cyrogenic storage is therefore, clear and the technique has become relatively widely used.

POLLEN AND SPORE GENE BANKING

Seeds are the most likely candidates for gene banks, but this cryogenic technique may be useful in preserving pollen from flowering or cone-bearing plants, and spores from non-flowering plants such as ferns and mosses as well. People sometimes ask why we bother with pollen, now that the cryogenic technology has been worked out for seeds. One reason is that pollen grains are so tiny that thousands and even millions of grains can be stored in a very small vial. Obtaining great diversity is quite simple. Botanists are able to use pollen grains from certain species to grow whole plants. This is still a difficult, tedious procedure with routinely perform these transformations. Pollen grains can also be used when needed to help create hybrids. When seed is used to develop hybrids, the seeds must be retrieved from the bank and then planted. It takes several years of maturation for the plant to be useful in hybridization. Because pollen can be used as is, it is much more economical and efficient. A pollen bank can be an extremely powerful tool in plant breeding since it frees breeders from the tyranny of time. The process can be streamlined and quickened by crossing pollen from plants that normally flower at the end of a season onto flowers that appear at the beginning of the season.

In Irvine we have found another use for pollen banks. Several specimens we grow at the arboretum are self-sterile- they set seed only if pollinated with pollen from another individual. We had managed to raise *Cytanthus obliquus*, a relatively rare African amaryllid, to maturity. This process took ten years and produced five plants, all of which became virus-infected. We decided to replace the diseased plants and also increase our stock from seed. Seedlings are usually virus-free. Here we ran into trouble. Each of the five plants flowered but never two at the same time. Even though each plant produced abundant pollen, none of the plants would set seed from its own pollen. We finally hit upon the perfect solution: storing pollen from this species in the bank. We now have no trouble pollinating the plants and we can generate as many seedlings as we require. The original five plants were replaced by several hundred offspring.

Fern spores, which are very similar in size to pollen, can be treated just like pollen. Even the spores of tropical ferns like staghorns, *Platycerium* species, can be germinated after drying and freezing.

The biggest problem with pollen is its susceptibility to water damage. Rain, heavy dew or water from any other source that comes into contact with pollen is liable to kill it. Pollen should be collected only from fresh flowers that have not been exposed too much to the elements.

We store pollen in gelatin capsules, the type used to hold medicine. These capsules can be purchased from the local pharmacist. The stamen, which contains pollen, is removed from the flower, inserted into the capsule and shaken several times to deposit the dust-like pollen on the capsule wall. The stamen is then withdrawn. The species name and date may be written directly on the capsule with a waterproof laundry marker. We dry the pollen by exposing the capsules to the air inside a frost-free refrigerator for 24 hours. The moderately dry air circulating in these refrigerators will cause sufficient water loss to permit the capsules to be frozen safely. Water is able to move out through the walls of a gelatin capsule. We store the capsule inside individual plastic containers that have a little DrieriteR in them to prevent moisture from moving through the gelatin and then place these containers in the bank. Relevant data are written on the wall of the plastic container. Pollen samples withdrawn from the freezer have limited viability and should be used within two days of withdrawal. Little is presently known about the longevity of pollen in cryogenic storage. We have tested pollen that had been frozen for eight years and found it to be viable, but no one knows yet whether pollen will be as hardy as seed seems to be.

Technique of Pollen Conservation

Erdtman (1952) and Nair (1970) described the method for preparation of pollen slides. Preparation of both Acetolysed (Ac) and Unacetolysed (Uc) grains could be made on the same slide. The detailed method is as under:

The polliniferous material is collected and placed in 70% alcohol n vials. After about 24 hours the material is crushed within plastic centrifuge tubes and the dispersion is sieved through a fine brass mesh into a glass centrifuge tube.

The whole quantity of the dispersion is divided into two halves, A and B. The part A is stained by safranin, warmed slightly over a flame, centrifuged and kept aside.

The part B contained in another centrifuge tube is centrifuged, the supernatant alcohol decanted, the sediment is covered by glacial acetic acid and centrifuged. Again after pouring out the glacial acetic acid the pollen sediment is covered by the acetolysis mixture composed of acetic anhydride and Concentrated sulphuric acid mixed in the ratio 9:1. The interaction between the anhydride and the acid produce heat, calculated at about 70°C and therefore the mixture contained in the centrifuge tube is placed in water bath and heated from 70°C to boiling point. The acetolysis mixture is stirred with a glass rod and the tube is left in the hot water for 2-3 minutes to enable the complete dissolution of the protoplasm.

The mixture is centrifuged, the liquid decanted and the sediment is again covered with glacial acetic acid followed by centrifuging and finally the sediment is subjected to many times centrifugation and brought to dilute 50% glycerin and left aside.

In order to prepare pollen slides, the dispersion of A and B is mixed and centrifuged, the glycerin decanted and the centrifuge tube is placed upside down on a filter paper so that the excess glycerin run down. A pellet of glycerin jelly, cut by means of blade is carried at the tip of a needle is taken inside the centrifuge tube kept inverted to touch the pollen sediment. The pollen were caught on the pallet of glycerin jelly, transferred to a microscope slide, warmed and left aside for one or two minutes so that the outer surface of the pallet of glycerin jelly slightly condenses. A cover glass is placed over the jelly and slightly pressed so that a round area of jelly is formed inside the coverslip. A piece of paraffin wax of melting point 60°C is placed on a side of the cover slip, warmed carefully so as to allow the molten wax to flow into the vacant area left by the glycerine jelly.

The slide is kept aside for a little while so that the wax solidifies. The extra wax is scraped by means of a blade, the slide is cleaned by xylol and made ready for microscopic examination.

Measurements

In taking the measurements of size, 10-20 grains have been studied and the average size is mentioned. In those which bear excrescences on the extine surface, the measurements given are exclusive of the excrescences. For radio-symmetric grains the polar diameter (P) is followed by two equatorial measurements (E and E_1). The size of pollen grain is measured from both acetolysed and unacetolysed grains, whereas, the other measurements have been made from acetolysed grains only. For delicate pollen grains like *Canna* and *Mussa* all measurements have been made from unacetolysed grains.

Data Presentation

In presenting the data, the families of plants are arranged according to the Bentham and Hooker (1862-83) system of classification, some times modified according to Hutchinson (1959). Within each family plants are grouped on the basis of the apertural characters of pollen grain. Within each such group, the alphabetical order is followed. In every palynological group, detailed and uniform pattern of description is adopted for all the species.

Glycerin Jelly Preparation for Pollen Slides

50 g, Genelin Powder

150 ml Glycerin

150 ml Distilled Water

Boil in water bath for 2 hours add phenol crystals (4 g)

For Aeroscope

Add 5 per cent safranin solution, 5-10 ml according to requirement.

CRYOPRESERVATION OF MICROORGANISMS

Reports are available from the studies with microorganisms that suggest sub-optimal cryopreservation procedures may cause genetic alternations. Chromosomal damage has been observed in *E. coli*, apparently caused by freeze thaw stresses during cryopreservation and subsequent recovery with the suggestion that DNA repair mechanisms are important in the recovery from atleast some aspect of freezing stress (Williams and Calcott, 1982). Studies with genetically manipulated yeasts also have shown that viability may persist at high level following cryopreservation. It is important, therefore, that in vitro conservation be based on cryopreservation procedure that is believed to be optimized.

PRESENT STATUS OF EX SITU CONSERVATION

To complement in-situ conservation, attention has been paid to ex-situ conservation measures. According to currently available survey, Central

Government and States Governments together run and manage 33 Botanical gardens; Universities have their own Botanic gardens. There are 275 zoos, deer parks, safari parks and aquaria etc. A Central Zoo Authority was set up by Government of India to secure better management of zoos. A scheme entitled assistance to Botanic garden provides onetime assistance to botanic garden to strength and institute measures for ex-situ conservation of threatened and endangered species in their respective regions.

6

Synthetic Seeds and Biodiversity Conservation

THE NATURAL SEED

The seed stage of seed plants represents a unique developmental phase of the spermatophyte life-cycle, and as such involves structures, not characteristic of other stages of development. The essential structure of seed is defined as a *ripened ovule consisting of an embryo and its coats*. Anatomically a seed consists of some old or parental sporophyte tissue viz. the seed coats, which are derived from the integument's and nucleus: some endosperm, which may be either gametophytic tissue or fertilized tissue: and the embryo, the new young sporophyte. The normal seed contains materials which it utilizes during the process of its germination. These substances are frequently found in the endosperm. Thus endosperm may contain variety of stored materials such as starch, oils, proteins etc. In some plants, however, the reserve food material is present in cotyledons.

DEVELOPMENT OF THE CONCEPTS OF TISSUE CULTURE AND ARTIFICIAL SEEDS

P. R. White is acknowledged as the father of tissue culture in the United States. He was the first to grow excised root tips of tomatoes (*Lycopersicon*) in continuous culture. When new material is started in culture, grown in vitro (literally, in glass), it develops very small juvenile shoots, which are reminiscent of seedlings. A plantlet continues to produce and maintain small stems and leaves throughout its duration in culture. This is fortunate because most mature material would be too unwieldy for micropropagation to succeed in a test tube. After multiplication in culture and when transferred to soil outside the laboratory, the plantlets will produce leaves of normal size and assume the mature features of the plants from which they originated.

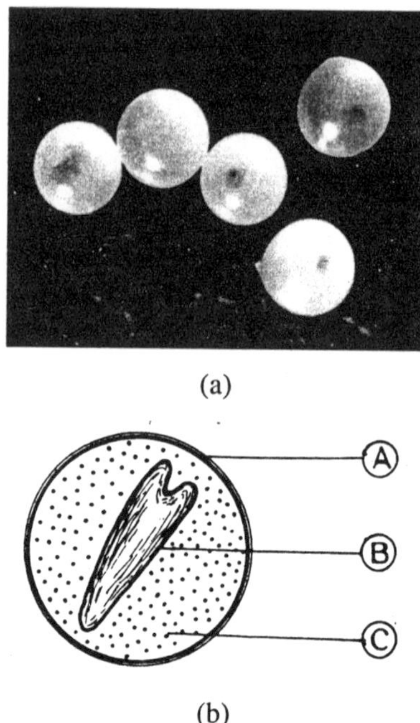

Fig. 6.1: (a) photograph of synthetic seed. (b) Capsule gel with hydrophobic membrane, the (A) artificial seed coat, (B) somatic embryo, (C) artificial endosperms.

When plants are multiplied vegetatively- as distinguished from those grown from seeds- whether by tissue culture or by cuttings, all the offspring from a single plant can be classified as a clone. This means that the genetic make-up of each offspring is identical to that of all the other offspring and to that of the single parent. On the other hand, plants propagated by seed, resulting from sexual reproduction, are not clones because each seed (and the resultant plant) has a unique genetic make-up- a mixture from 2 parents, different from either parent and different from one seed to another. The term *cloning*, with respect to tissue culture, refers to the process of propagating in culture large numbers of selected plants with the same genotype (the same genes or hereditary factors) as their respective parent plant.

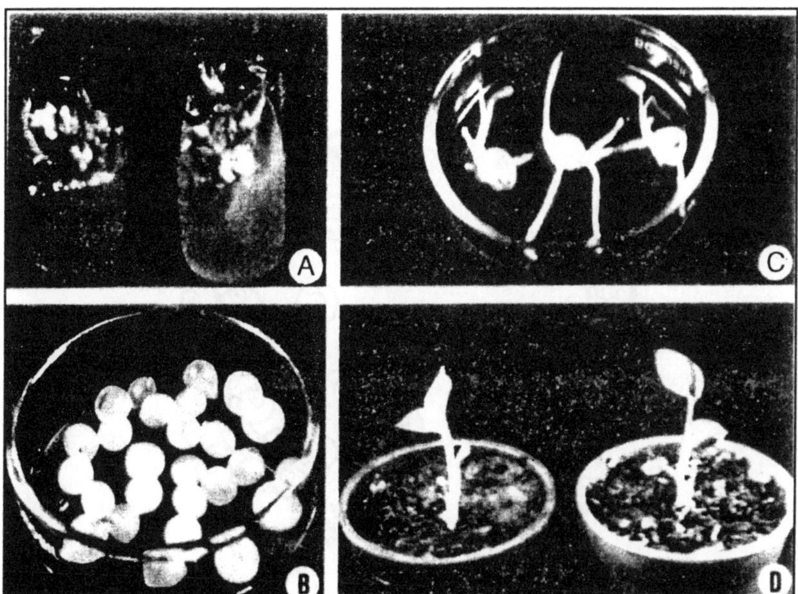

Fig. 6.2. Encapsulation of shoot tips of cardamom. A - multiple shoot cultures. B - Shoot tips encapsulated in 3% sodium alginate matrix, C - emerging shoot roots from the encapsulated shoot tips and D- plantlets derived from encapsulated shoot tips paper cups. (After Rao et. al, 1997).

The liquid cell suspension cultures have particular significance for the mass production of cells. One common source of cells for cell suspension is from friable callus, although specific cells, such as from leaf mesophyll (the thin, soft tissue between the upper and lower epidermis of the leaf), are also grown in suspension. Cells in suspension can form embryoids (somatic embryos) in the process of embryogenesis. Embryos may multiply and/or be induced to form plantlets in the process of morphogenesis. Many hybrid plants produce embryos that do not mature to viable seeds. These embryos can be rescued, removed from the seed at immature stage, and then grown in culture.

Suspension cultures have been enhanced by new methods, which can continuously introduce fresh medium into the suspension culture, thereby, enabling the production of thousands of cells or embryos in a single container with a minimum of manual transfer. This is one way that tissue culture can compete with the plentiful seed production in nature.

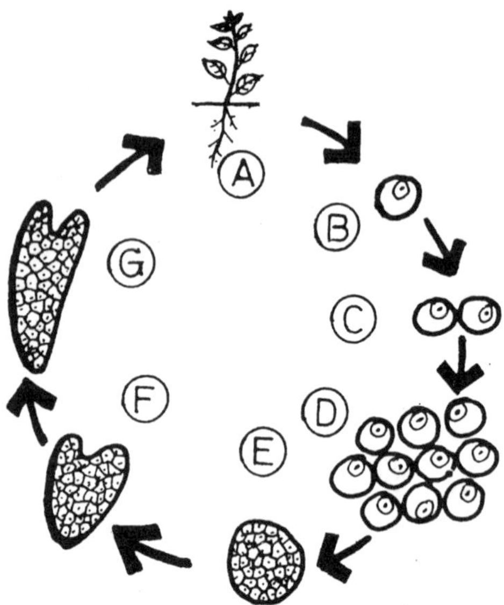

Fig. 6.3: Different stages of somatic embryogenesis. (A) a young plant, (B) isolated single cells from leaves , only one cell is shown. (C) doublet formation after first mitosis in culture. (D) cell colony (proembryo mass) following many mitosis. (E) Globular embryo. (F) heart shape embryo. (G) torpedo stage embryo.

Interest in anther and pollen culture- the tissue culturing of anthers or pollen to obtain haploid (cells with half the normal number of chromosomes of vegetative cells) clones- is spurred by the practical applications of such haploid cultures. Haploid (n) plants are sterile, but if the chromosomes duplicate, either spontaneously or by induction, the plants will be diploid (2n, which is normal for the vegetative state), and their progeny will be true to form. Considering the fact that it takes several generations of inbreeding to obtain a pure line by conventional means, it is little wonder that plant breeders are interested in anther culture.

DISCOVERY OF SYNTHETIC SEEDS

The origin of the idea of an artificial seed is difficult to determine. Certainly, those who first produced somatic embryos may have considered such an application (Steward, et al., 1958 and Reinert, 1958). The discovery of somatic embryogenesis in carrot in the year 1958 almost simultaneously by

F. C. Steward (USA) and J. Reinert (Germany). F. C. Steward, a renowned plant physiologist, at Cornell University in New York, was so impressed by the dramatic effects of coconut milk in carrot culture media that he set beside his other objectives in order to dedicate himself to the study of growth factors in this and other liquid endosperms. Among the active materials he extracted from the coconut milk were several ingredients that are now commonly included in purified form in many tissue culture media. Coconut milk is still used in some orchid culture media. They provided a new way of propagating the plant species.

Later, S. Guha and S.C. Maheshwari (University of Delhi, Delhi) in 1964 discovered the formation of pollen embryos from cultured anthers of wild *Datura innoxia.*

However, it was not until the early 1970's that the concept of using somatic embryos began to be presented as a potential propagation system for seed-sown crops. Toshio Murashige gave a number of seminars on tissue culture propagation where he concluded with this concept. For a period of time he conducted research in his laboratory that was focused on the developmental physiology of somatic embryos which he felt to be the limiting factor for large-scale propagation. He formally presented his ideas on artificial seeds at the Symposium on Tissue Culture for Horticultural Purposes in Ghent, Belgium, September 6-9, 1977. His terse comments in the proceedings, however, were to be applicable, *the cloning method must be extremely rapid, capable of generating several million plants daily, and competitive economically with the seed method* (Murashinge, 1977).

During the mid-1970's, two separate research groups began work on somatic embryogenesis for crop propagation. Keith Walker, then at Monsanto Company, directed a group of scientists that identified basic concepts of delivery of cloned, agricultural crops. Since the focus was to develop thrifty somatic embryo systems that would recapitulate zygotic embryogenesis, their choice was the advanced system developed for *Medicago sativa* L. (alfalfa) using a line (Regen S) identified by Bingham, et al. (1975). Soybean and vegetable crops were also of interest to them. Walker cited two reports that had a strong impact on their thinking about the use of somatic embryos for crop propagation. Early in 1980, Walker moved to Plant Genetics, Inc. where Redenbaugh, et al. (1984 and 1986) discovered that hydrogels such as sodium alginate which could be used to produce single-embryo artificial seeds. In a few experiments, the artificial seeds were planted in the greenhouse with plant production (7% for alfalfa and 10% for celerly).

Street (1977) advocated the problem of reliability in embryogenesis. According to him morphogenic competence is determined from the time of culture initiation, such that there is the need to have an initiation medium that will ensure that the competent cells are involved in callus formation. Sunderland (1977) demonstrated that the production of hundreds of morphologically uniform embryos from *Datura* and *Nicotiana* pollen.

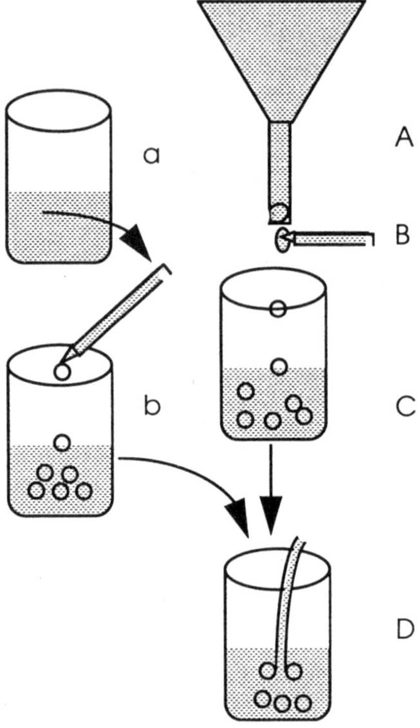

Fig. 6.4: Preparation of artificial seeds via encapsulation with calcium alginate beads using a droping method. Alginate (2%) is either mixed with somatic embryos (a) or poured in a separatory funnel (A), droping of alginate beads along with embryo into a bath of calcium nitrate (100 mM) solution (b) or a single embryo is inserted into alignate drop using a plastic pipet having 4mm inside diameter (B), falling of alginate beads into a bath of calcium nitrate (C) and washing of the capsules in water (D). Diagram modified from Redembaugh et. al., 1991.

Robert Lawrence (of Union Carbide) started to develop various methods for cloning forest trees. It was difficult for him to produce hybrids for crops such as celery and lettuce. This group focused on delivery of somatic embryos using fluid drilling technology (Lawrence, 1981) and using polyoxyethylene to form seed tapes or sheets. Lawrence and Walker's groups came together at a symposium workshop, Advances in Methods of In Vitro Cloning for Large Scale Propagation of Plants, held September 21-22, 1981 at the W. Alton Jones Cell Science Center in Lake Placid, New York. They discuss the various concepts like how low-cost, high-volume propagation system can be developed for vegetable and agronomic crops using somatic embryos and delivered by fluid drilling (in a seed tape, or as an artificial seed). The Lawrence group (then at Agrigenetics) subsequently began to use alginate for encapsulating carrot and celery somatic embryos (Lutz, et al. 1985). Because their encapsulation results with alginate were similar to those at Plant Genetics, they focused on the problems of somatic embryo physiology. In this way, they were successful in obtaining germination of carrot somatic embryos in vermiculite in a growth chamber.

Drew (1979) was active in developing methods to commercially propagate crops using somatic embryos. He suggested delivering carrot somatic embryos in a fluid drilling system, but was able to produce only three plants from carrot embryos on a carbohydrate-free medium. He could not get success in producing many plants through this system. He faced a crucial problem and found the very slow rate of development of plantlets derived from culture.

Kitto and Janick (1982) coated clumps of carrot embryos, roots, and callus with polyoxyethylene. Some embryos survived the coating process as well as a desiccation step (Kitto and Janick, 1985a and 1985b).

The early assessments of Murashige and Street (1977) on the difficulty of somatic embryogeny are still valid today. The quality and fidelity of somatic embryos are the limiting factors for development and scale-up of artificial seeds.

Interestingly, artificial seeds prepared from shoot buds can also be used for plant propagation, and this was reported by P.S. Rao's group from BARC, Bombay. Rice is the world's most important food crop and a primary food source for more than one third of the World's population. This crop has received considerable attention in biotechnological research programmes. Research on artificial seeds in rice is still in infancy and this technology through somatic embryogenesis would offer a great scope for large scale propagation of superior, elite hybrids (Brar and Khush, 1994). P. S. Rao and his associates have reported high frequency somatic embryogenesis from indica rice cultivars (Suprasanna *et al.,* 1995) and utilized this embryogenic system for the production of artificial seeds.

Table 6.1: Important crop plants in which artificial seed production and plant conversion has been demonstrated

In vitro propagules for encapsulation	Crop
Somatic embryos	Alfalfa, Celery, Brinjal, Carrot, *Brassica,* Lettuce, Sandalwood, Rice, Horseradish
Axillary buds/ Adventitious buds	Mulberry, Eucalyptus, Vitis
Shoot tips	Banana, Cardamom,*Carum carvi*

Since then the induction of somatic and/or pollen embryogenesis has been reported in a wide array of plants, including several crop plants such as rice, wheat, triticale, maize, pearl-millet, sorghum, sugarcane, potato, sweet potato, eggplant, lettuce, carrot, alfalfa, soybean, cucumber, *Brassica* species, asparagus, coffee, tobacco and cotton. However, the production of high quality and uniform embryos (which is very important for the preparation of artificial seeds) has been limited to only certain crops like carrot and alfalfa.

Somatic embryogenesis has, also been obtained from non-zygotic explant tissue (coffee leaf), suggesting that somatic embryogenesis may be obtained from sexually immature tree, zygotic embryo or ovule tissue.

Difficulties in developing somatic embryogenesis systems for tree species are similar to those for herbaceous species. However, tree species may also exhibit a unique set of tissue culture-dependent variability's.

USES AND LIMITATIONS OF ARTIFICIAL SEEDS

Plants are traditionally propagated either by seeds or the vegetative propagules like stem cutting. Now most plants can be multiplied through tissue culture techniques, particularly shoot tip culture (clonal micropropagation).

Micropropagation through artificial seeds may be commercially exploited on a large scale, generating millions of plants in a few days, and this may become a profitable multibillion rupees industry in near future. This technology would be feasible and even competitive economically with the seed method.

Artificial seeds offer the possibility of a low-cost, high-volume propagation system that will compete with true seeds and transplants. Most likely, the technology will first be used with hybrid vegetable crops such as celery or with high-value flower and ornamental species. Because of the relative ease of producing large numbers of somatic embryos, artificial seeds will be applicable for monoculture as well as mixed genotype methods of agriculture. The artificial seed coating also has the potential to hold and deliver beneficial adjuvants such as growth-promoting rhizobacteria, plant nutrients and growth control agents, and pesticides.

Potential Uses of Artificial Seeds

1. Delivery Systems.
2. Reduced costs of transplants.
3. Direct greenhouse and field delivery of elite, select genotypes, hand-pollinated hybrids, genetically engineered plants, sterile and unstable genotypes, large-scale monocultures, mixed-genotype plantations.
4. Carrier for adjuvants such as microorganisms, plant growth regulators, and pesticides Protection of meiotically-unstable, elite genotypes.
5. Analytical Tools.
6. Comparative aid for zygotic embryogeny.
7. Production of large numbers of identical embryos.
8. Determine role of endosperm in embryo development and germination.
9. Study of seed coat formation.
10. The synthetic seeds so developed are breed true.
11. There are potential advantages of artificial seed technology specially for tree genetic engineering.
12. The artificial production of seeds has already been obtained successfully in *Zea mays*, *Apium graveolens*, *Daucus carota*, *Lactuca sativa*, *Madicago sativa*, *Brassica* spp., *Gossypium hirsutum*, *Velerina* sp., *Santalum* spp., etc.
13. The encapsulation of somatic embryos (hydrated or desiccated) provides a potential method to combine the advantages of clonal.
14. Propagation with the low-cost, high-volume capabilities of seed propagation.
15. These seeds can be produced within a short time (one month) whereas natural seeds are the end product of complex reproductive process and breeders have to wait for a long time for development of new varieties.
16. Artificial seeds can be produced at any time and in any season of a year.

17. Dormancy is the common feature of natural seeds, but by means of artificial seeds the dormancy period can be reduced to a great extent, thereby shortning the life cycle of a plant.
18. They are useful in preserving germplasm.
19. Synthetic seeds are applicable for large scale monocultures as well as mixed genotype plantations.
20. Such seeds give the protection of meiotically unstable, elite genotypes.
21. The synthetic seeds provide us knowledge to understand the development, anatomical characteristics of endosperm and seed coat formation.

Both these propagation methods have **certain limitations** such as the need of intensive labour, rooting of regenerated shoots and transplantation, slow and small scale multiplication. Micropropagation has some additional problems like the need of acclimatization of tissue culture derived plants before they are transferred into field conditions (hardening) because of their tenderness due to the absence of lignification and low cuticle formation. By contrast, plant propagation via artificial seeds has several advantages over classical methods as well as micropropagation (with shoot tip culture). The advantages of this technology include the rapid and large-scale multiplication, minimal labour and low cost propagation. In addition, artificial seeds can be directly delivered to the field, thus eliminating transplantation and tissue hardening steps. They can also be provided with various kinds of adjuvants like plant growth regulators, useful microorganisms and pesticides to tailor a field-specific, plantable unit for a desired crop. However, the genetic uniformity is maintained in all these propagation methods.

Artificial seed technology can be very useful for the propagation of a variety of crop plants, especially crops for which true seeds are not used or readily available for multiplication (e.g. potato) or the true seeds are expensive (e.g. cucumber and geraniums), hybrid plants (e.g. hybrid rice) and many vegetatively propagated plants which are more prone to infections (e.g. day lily, garlic, potato, sugarcane, sweet potato, grape and mango).

This newly emerging technology would also be useful for multiplying genetically engineered plants (transgenic plants), somatic and cytoplasmic hybrids (obtained through protoplast fusion techniques), sterile and unstable genotypes. Besides, artificial seeds would be useful material for preservation of desirable elite genotypes (cryopreservation). They would also be valuable tools in experimental research to study the process of zygotic embryogenesis and understanding the role of endosperm in normal embryo development and germination.

Synthetic seed technology offers many useful advantages on a commercial scale. The resultant plant population from the synthetic seed will be uniform and the direct delivery of somatic embryos will save many subcultures to obtain plantlets from regenerated embryos. The encapsulated embryos could also be packed with pesticides, fertilizers, nitrogen fixing bacteria and even microscopic destroying worms.

At the Biotechnology Division of BARC, research on the development of protocols for synthetic seeds using somatic embryos, axillary buds and shoot tips is in progress in five economically important plants, sandalwood, rice, mulberry, banana and cardamom. The following pages describe the results obtained in this direction.

The artificial seed systems coupled with artificial intelligence and microcomputer systems like the most advanced robots which can mimic the motions and functions of a living being (i.e., automated encapsulation) would tremendously increase the efficiency of encapsulation and production of artificial seeds, and revolutionize the plant propagation method in the years ahead. This technology is gradually moving towards the commercial propagation of high value crops. However, there is a great need for refinement of this technology by the tackling of certain technical problems such as the need to produce high-quality and high-fidelity somatic embryos, and to avoid the genetic instability and variability of tissue culture derived plants (somaclonal variation, which is not preferred for crops where true-to type plants are important). These problems can be overcome if the process and regulation of somatic embryogenesis and origin of somaclonal variation are well understood. Intensive research is being carried out in several laboratories to address these vital issues. Further, the understanding about the storage, transport, handling, growth habit and harvest index of artificial seeds is essential. Similarly, the efforts to increase the output of embryos/plants per gram callus tissue of input (mass balance) and conversion frequency of artificial seeds are also needed. If all these problems are rectified with technical progress, no doubt this novel method can become valuable tool in agriculture to propagate crop species.

Although synthetic seed research is being carried out in many laboratories, the principle limitation for commercialization has been the somatic embryogenesis. Despite the fact that somatic embryogenesis has been achieved for many species, during the last decade for many of the important, high value genotypes, somatic embryogenesis has not been reported. Therefore there is a need to shift the focus on somatic embryogenesis in valuable crop plants. Research on encapsulation of axillary buds, shoot tips or any other propagules should be taken up. Equally important is the need to develop new synthetic seed coatings for encapsulating embryos and other vegetative propagules. Synthetic seed technology in years to come would certainly find its application in plant propagation and delivery of tissue cultured plants.

PRODUCTION OF SYNTHETIC SEEDS

Synthetic or artificial seeds are the living seed-like structure derived from somatic embryoids in vitro culture after encapsulation by a hydrogel. *The preserved embryoids are termed as synthetic seeds.* Somatic embryoids are identical with zygotic embryos and give rise to plants only under controlled laboratory conditions. Somatic embryoids are without seed coats. In vitro embryoid develops from callus tissue and their induction is initiated by somatic embryogenesis supplementing the medium with auxin and cytokinins in proper ratio.

Such seeds are contaminated with microbes and desiccate quickly when they are subjected to field conditions. Therefore, to get rid from this problem, they are encapsulated by a protective gel like **calcium alginate**. These encapsulated embryoids can resist unfavourable field conditions without desiccation. These seeds so developed behave like a true seed and are used as a substitute of natural seeds. They can also be sown directly in the greenhouse or in fields.

ESTABLISHMENT OF CALLUS CULTURE

Induction of Somatic Embryogenesis

⇓

Maturation of Somatic Embryos

⇓

Encapsulation of Somatic Embryos

⇓

Evaluation of Embryoid and Plant Conversion

⇓

Planting in Fields/green House

Fig. 6.5: Schematic presentation of steps of synthetic seed production

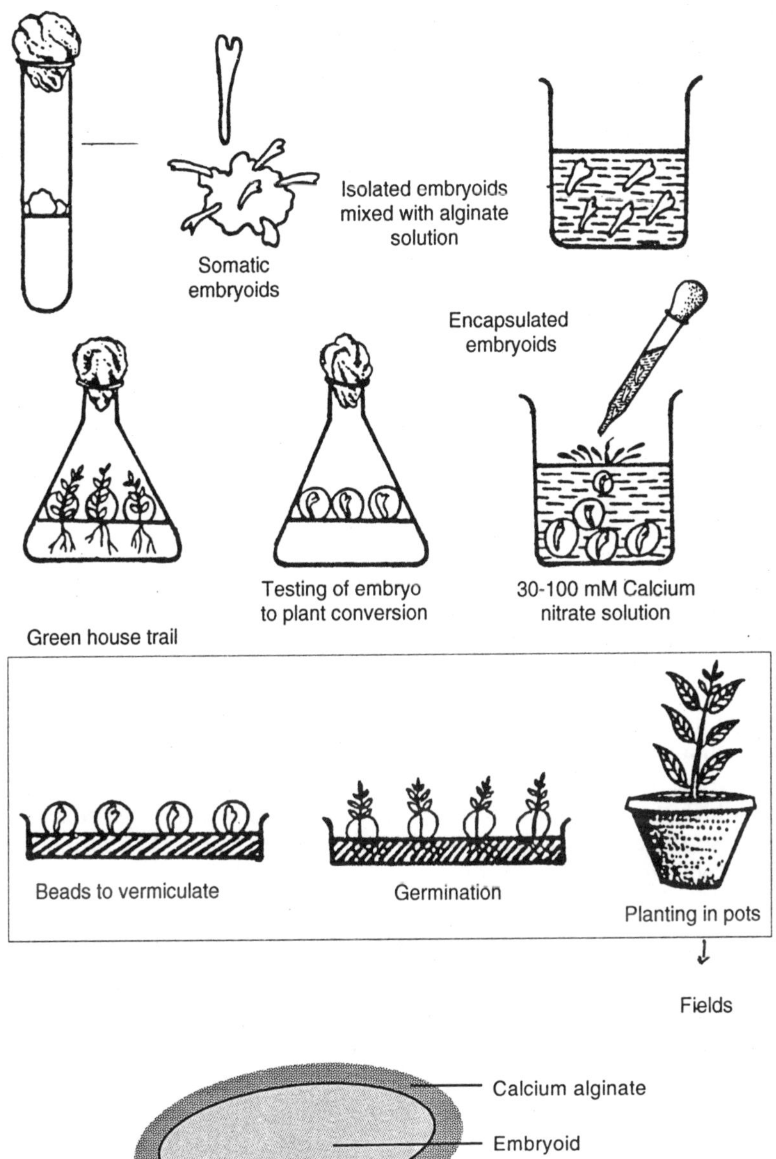

Fig. 6. 6: (A) Flow diagram presenting the procedure of synthetic seed production. (B) Cross section of synthetic seed.

ENCAPSULATION OR COATING OF SYNTHETIC SEED

Encapsulation is necessary to produce and to protect synthetic seeds. The encapsulation is done by various types of hydrogels which are water soluble. The gel has a complexing agent which is used in varied concentrations.

Table 6.2: Various type of hydrogels

Gel (Concentration) (% w/v)	Complexing Agent Concentration (μM)
Sodium alginate (0.5-5.0)	Calcium salts (30-100)
Sodium alginate (2.0) with Gelatin (5.0)	Calcium chloride(30-100)
Carragenan (0.2-0.8)	Potassium chloride
Locust Beam Gum (0.4-1.0)	Ammonium chloride (500)

The encapsulation or coating of carrot and celery somatic embryos is done by polyoxyethylene (Polyox) and dried the embryo in polyox mixture.

Two standard methods have been used for encapsulation of somatic embryos.

Gel Complexation via a Dropping Procedure

The most useful encapsulation system is to drip two per cent LF60 sodium alginate from a separatory funnel into a 100mM calcium nitrate solution. As the sodium alginate drops from at the tip of the funnel, the somatic embryos are inserted. The encapsulated embryos complex in calcium salt for 20 min, after which they are rinsed in water and then stored in a air tight container otherwise the capsule will dry out within 24 hour. It is a slow method of seeds production.

Automate Encapsulation Process

Automate encapsulation process is the recent and quick method of artificial seed production. An encapsulation machine can be used successfully to encapsulate somatic embryos, e.g., for alfalfa. Blank alginate capsules were planted in Speeding™ trays using a vacuum seeder. The blank capsules are planted in the field using a Stanhay planter. However, because for the rapid drying and the thickness of the alginate capsules, a hydrophobic coating is required for mechanical handling. For coating, an Elvax 4260 Copolymer™ (Dupont), is suitable for producing a slow-drying, non-tacky coating which allows embryo conversion Based on this system of production of artificial seeds the cost factor is not high for most of the cash crops.

Alginate or carrageenan is also used for artificial seed production, e.g., of carrot, asparagus, Norway spruce, etc. Mascarenhas (India) reported the encapsulation of *Eucalyptus* somatic embryos. He obtained 50 per cent germination from such seeds. Alginate artificial seeds are spherical and transparent. The alginate capsule is generally non-inhibitory.

MASS BALANCE CONCEPT

An additional concept that has greatly aided the improvement of artificial seed performance is *mass balance*. Mass balance considers the amount of tissue at the beginning of the experiment (or production run) and the number of high quality plants produced at the end of it. Simply emphasising the number of embryos per gram of fresh callus or the number of embryo-producing calli is not adequate. In fact, treatments that lead to higher numbers of embryos may actually produce fewer superior quality of embryos than another protocol. At this time, the artificial seed package, consisting of a calcium alginate bead coated with a hydrophobic Elvax polymer, appears to be sufficient.

STEPS OF COMMERCIAL ARTIFICIAL SEED PRODUCTION

The following steps are needed for commercial synthetic seed production:

1. Production of embryogenic tissue from transformed cells or tissue.
2. Large-scale production of synchronous somatic embryos.
3. Maturation of somatic embryos.
4. Non-toxic encapsulation/coating process.
5. Artificial endosperm/megagametophyte, depending on species.
6. Storage capability of artificial seeds.
7. High frequency, direct green house/nursery field conversion, depending on production requirements.
8. Low genetic and epigenetic variation.
9. Appropriate expression of engineered trait.
10. Artificial Seed Propagation
11. Artificial seed propagation could potentially reduce the time needed to insert a desirable gene into a productive forest, as compared to using seed as the propagation method. A considerable advantage would be to

eliminate or minimise the requirements for seed production using the following process:

12. Production of large-scale embryogenic tissue from genetically engineered cells;
13. Concurrent plant regeneration, confirmation of transformation, and *progeny* testing;
14. Cryogenic storage of potential superior lines;
15. Scale-up production and maturation of somatic embryos;
16. Encapsulation of somatic embryos as artificial seeds;
17. Either greenhouse/nursery establishment, growth, and transplanting into the field or direct seedling.
18. The final stage will be the evaluation of production plantations for increased yield/ performance due to engineered trait.

Unlike herbaceous species, the growth and maturity of trees far exceed the time required for tissue culture manipulations. If the engineered trait is expressed only in the mature tree and is not stable during meiosis, then mass clonal propagation would be accomplished only through tissue culture methods (or, by traditional relatively low-volume ramet production).

Cryopreservation of desirable genotypes is one of the key components for artificial seed propagation of tree species. It retains the genetic gains from genetically engineered tree species without having to establish clonal orchards. Once the progeny tests are completed, then embryogenic tissue corresponding to the superior clones can be thawed for rapid scale-up production via somatic embryogenesis.

Although the use of artificial seed technology should be extremely valuable for the rapid introduction of genetically engineered material into production of forests.

HYDROGEL ENCAPSULATION OF ARTIFICIAL SEEDS

Water soluble hydrogels have been found suitable for making artificial seeds. Two methods have been used to coat somatic embryos: gel complexation via a **dropping procedure** and **molding**.

Redenbaugh *et al.* (1986) mixed alfalfa somatic embryos with sodium alginate (2% w/v) and dropped them into a calcium nitrate solution (100 mM). Surface complexation began immediately and the drops were gelled completely in 30 minutes. Alternatively, the embryos could be mixed in a temperature-dependent gel such as Gel-rite™, placed in the well of a microtiter plate, and gelled as the temperature was lowered. Somatic embryos from several crops have been encapsulated in alginate with plants recovered *in vitro*.

Table 6.3: Useful gels for encapsulation of somatic embryos

Gel	Conc. % w/v	Complexing Agent	Concentration (mM)
Sodium Alginate	0.5-5.0	Calcium Salts	30-100
Sodium Alginate with Gelatin	2.0	Calcium Chloride	30-100
Carrageenan with	0.2-0.8	Potassium	
Locust Bean Gum	0.4-1.0	or Ammonium Chloride	500
Gelrite	0.25	Temperature lowered	

Table 6.4: Crops encapsulated in calcium alginate beads

Species	Common Name
Apium graveolens L.	Celery
Brassica species	Rapid-cycling Brassica
Daucus carota L.	Carrot
Gossypium hirsutum L.	Cotton
Lactuca sativa L.	Lettuce
Medicago sativa L.	Alfalfa
Zea mays L.	Corn

Initially, the effect of encapsulation was difficult to assess because of the overall poor quality of the somatic embryos. Although visually normal embryos were produced (i.e. bipolar, root and shoot axis, cotyledons), the germination and continued development of the embryos was very inconsistent. In fact, the use of germination (root elongation and emergence) as an efficacy assay was found to be unsuitable when shoot production and further growth was not observed concomitantly. Consequently, the approach for developing artificial seeds was shifted away from one focused on somatic embryogenesis (initiation of embryo formation) to one of somatic embryogeny (initiation, development, and maturation of embryos).

The concept of somatic embryogeny and the production of high quality embryos is not one that is widely followed by many researchers who have either focused on the production of a somatic embryogenesis system that results in some plant recovery or who have interest only in the study of the early stages of embryogenesis. This focus separated from an emphasis on producing mature, true-to-type, high quality embryos can possibly lead to conclusions based on abnormal somatic embryogenesis. To overcome this problem and to achieve high quality embryo production for scale-up of artificial seeds, measure the embryo response in terms of embryo-to-plant development or *conversion*. Essentially, embryo conversion frequency is the per cent of the somatic embryos that produce green plants having a normal phenotype. This assay has been critical for developing conditions and media that select for uniform plant production. Following are the events which are associated with the process of embryo-to-plant conversion.

1. Germination (radicle elongation)
2. Development of a vigorous root system
3. Growth and development of the shoot meristem
4. Production of true leaves
5. A direct shoot-to-root connection
6. Absence of hypertrophy of the hypocotyl
7. Minimization of callus growth in the hypocotyl
8. A green plant with a normal phenotype

SYNTHETIC SEED AND FOREST TREES

The use of biotechnological approaches in forestry may be greatly enhanced and considerable time could be saved by using artificial seed technology. Genetic engineering in forestry will be similar to that for field and horticultural crops. Desirable genes should be identified, cloned and inserted into the tissue (protoplasts, cells-pollen, zygotic embryos, needle tissue, etc.). The putatively transformed tissue will be regenerated to plants and tested for expression of the genes. With annuals, the transformed individual plants can then be backcrossed with the original population for large-scale production of transformed seed within one to few years. However, for most tree species, after adequate gene expression is confirmed, scale-up production of transformed seeds deviates significantly at this point from that of annual and biennial crops because of the very long generation time for trees, particularly conifers.

Plain steps in genetic engineering of tree species focus on transformation, followed by traditional forest seed production and tree evaluation are as under:

1. Regeneration of genetically engineered tissue;
2. Confirming presence and expression of engineered gene;
3. Bulking up propagules through vegetative propagation or seed production;
4. Greenhouse/nursery establishment and growth;
5. Seed orchard establishment with concurrent progeny testing, including evaluation of the engineered trait;
6. Seed orchard roguing;
7. Seed production in seed orchards and;
8. Either nursery production with transplanting into the field or direct seedling.
9. The final stage will be the evaluation of plantations for increased performance due to the engineered trait.

Somatic Embryogenesis and Synseeds

Somatic embryogenesis in many woody species is less successful than other methods of micropropagation. The earliest report of somatic embryogenesis in a coniferous tree is that for spruce *Picea abies*. Since then, somatic embryogenesis has been reported in several other *Picea* species. Mainly immature, and occasionally mature embryos, were used successfully in conifers, yielding somatic embryos and plantlets. Although the number of species that can be propagated through this method has been increasing, the percentage of recovery for plantlets remains relatively low. Studies of somatic embryo formation in pines show that they can arise either in the conventional manner, as for spruce, or by a process termed somatic polyembryogenesis, as found with sugar and loblolly pines. The resulting embryos elongated and formed numerous cotyledons in a repetitive and true-to-type polyembryonic process. In the loblolly pine, the somatic embryos thus produced developed transplantable plants. The rate of somatic embryo formation has been improved recently, making it commercially viable for some species. Procedures for induction of somatic embryogenesis and subsequent conversion of the embryos to artificial seeds have been developed in a limited number of spruce and pine genotypes, such as Norway spruce and loblolly pine, with treated embryos being stored for further use as artificial seeds.

7

Conservation of Fruit and Plantation Crops Through Biotechnology

Plantation crops meet human requirements for food, timber, medicines, spices and beverages. It is necessary to increase the output of plantation crops several-fold per unit area because of increasing human population associated with shrinking land area.

Plantation crops present certain unique problems for the plant breeder in terms of their improvement. Tissue culture technique is an important tool for rapid multiplication of several economically important plantation crops which can be of immense use in the multiplication of true-to-type high yielding plants on a large scale for planning. This technique has been successfully applied in banana, cardamom, eucalyptus, and teak.

In India, there is an urgent need to increase productivity of plantation crops in order to meet the growing demands for inland consumption and export. Plantation crops have a vital role in the Indian economy. About 41 per cent of tea, 50 per cent of coffee, 57 per cent of cardamom and 77 per cent of pepper produced in India and is exported to other countries (Muliyar, 1983).

The National Research Centre for Spices (NRCS) of the Indian Council of Agricultural Research has been conducting research on crop improvement, management, protection and post-harvest technology of spices.

Use of biotechnology in crop improvement programmes began in 1983, with the rapid clonal multiplication of cardamom at the Central Plantation Crops Research Institute. Tissue culture propagation in turmeric, ginger, black pepper and tree spices nutmeg, cinnamon, all spice and clove took off subsequently.

Increasing human population associate with shrinking land area to meet with the increasing demand of growing population it is necessary to increase the output of plantation crops several fold per unit area.

Table 7.1: Area and production of some plantation crops in India

Plant species	Area '000ha	Production per annum 000 tonnes
Comellia sinensis	400	645
Coffea arabica	220	190
Theobroma cacao	23	4
Elettaria cardamomum	100	3.5
Piper nigrum	110	30
Curcuma longa	88	200
Zingiber officinale	40	80
Anacardium occidenatale	460	150
Hevea brasihensis	350	235
Cocos nucifera	1100	5700*

* million nuts

Plantation crop present certain unique problems for the plant breeder in terms of their improvement. Therefore, tissue culture can be of immense help in the multiplication of true-to-type high yielding plants on a large-scale for planting. This technique has been successfully applied in banana, cardamom, eucalyptus, and teak.

Cashew and rubber are two important plantation crops in India. Indian cashew (*Anacardium occidentale* L.) account for 27 per cent in the world market. However, imports, about 300,000 tonnes of raw nuts from East African countries in order to meet the demand of the processing industry (Bavappa, 1982). Cashew has been traditionally neglected as a wasteland crop with a very low average yield of less than one ke per tree. There are tree which are known to give very high yields of 20 kg per tree. Seed progeny show marked variation in fruit and nut characters. If this can be doubled or trebled through somatic embryogenesis in a shorter period of time, the rate of multiplication of elite trees can be greatly accelerated.

To date no serious attempts have been made for in vitro embryos genesis of this crop. Apart from a brief abstract in cashew. There is ample scope to develop technology for rapid multiplication of this important earner of foreign exchange.

CITRUS

India ranks seventh in the production of citrus fruits in the world. They are grown in tropical and subtropical climates. The fruits belong to family Rutaceae and subfamily Aurantiodeae. Orange occupies the first place among the citrus fruits produced in the country. However, our oranges are different from the kind cultivated and consumed in the international market. The internationally accepted variety are Satsuma and Clementine. In Nagpur, mainly mandarins are being grown. Sweet orange, *santra*, *kinoo*, *lemon*, etc., are produced in the country. Citrus fruits occupy third position after mango and banana in the production of fruits in India. The per hectare yield in India, however, is only 40 per cent of that in Spain or Japan.

Oranges are cultivated in India in Maharashtra, Punjab, Tamil Nadu, Assam, Tripura, Rajasthan, Madhya Pradesh and West Bengal. In Maharashtra, oranges are grown in Vidharba region which covers districts like Amravati, Nagpur, Wardha and Yavatmal. The orange belt in this region spans Warud, Morshi (Amravati), Katol, Kondhali, Kalmeshwar, Narkhed, Pandhurna (M. P), Arvi, Yavatmal, Chandrapur and Wardha area. There are two crore oranges tree in this region which yield crop valued at Rs. 270 crore annually. In Madhya Pradesh, orange cultivation is largely confined to Chhindwara district. Besides India, other major orange growing countries in the world are USA. Brazil, Spain, Morocco and some African Countries. In the US, orange are mainly cultivated in California and Florida districts. While Brazil exports most of its produce to the US, Spain and Morocco export to France and Germany.

Tissue culture

Tissue culture research in *Citrus* has been mainly directed in three directions, viz., micropropagation, somatic hybridization using protoplast fusion and genetic transformation. A research on basic and applied aspects of tissue culture has been done extensively.

Micropropagation

Regeneration of superior fruit tree cultivars has been facilitated by organogenesis or somatic embryogenesis from leaf or nucellar explants. Modern genetic approaches helped the researchers and growers to propagate new cultivars. Protoplast fusion technique has been used to develop superior rootstocks for citrus. They are being developed by interspecific and intergeneic method. The recovery of haploid plants from anther culture of citrus, and triploid plants from citrus should aid crop improvement (Litz. 1990).

Navarro (1984) described following in vitro techniques for recovering virus-free citrus plants:

1. **Nucellus culture :** This technique is used to obtain nucellar plants of monoembryonic citrus varieties. It consists of culturing the nucellus, from immature seeds, to induce the formation of nucellar embryos that develop into plants. Many of the plants so formed have juvenile characteristics, and many are not true-to type.
2. **Ovule culture :** This technique is used to obtain nucellar plants of the polyembryonic seedless variety. The plants developed by this technique have similar characteristics to those obtained by nucellus culture.
3. **Shoot-tip grafting :** This technique is used to recover plants of all citrus species free of most viruses. A *small shoot-tip grafting* technique is used to recover plants of all citrus species free of most viruses. In this method a small shoot-tip is grafted onto a young seedling rootstock, and the resulting plant is non-juvenile and true-to-type.

Anther culture

This technique is useful to develop embryoids and plantlets from anther. Hidaka *et al.* (1979) studied in vitro differentiation of haploid plants using anthers of *P. trifoliata* on MS medium supplemented with 0.2 and/or 2.0 mg/L of 2,4-D, IAA, NAA and Kinetin. After three weeks of culture, heart-shaped or cotyledonary embryoids appeared from the anthers. Embryoids are then develop into callus in the medium containing 0. 2 mg/l IAA. The callus formation can occur on all media, however, it increases by the addition of 2, 4-D. Roots are induced by the transfer of the embryoids from the induction media containing growth regulators to the medium lacking growth regulators.

Hidaka et al. (1981) studied plantlet formation from anthers collected from immature flower buds of *C. aurantium* on MS with or without IAA, Kin and yeast extract at 28°C in darkness. Embryoid formation was generally evident 14 weeks after inoculation of anthers of sour orange and *Choshu-to.* Hidaka and Omura (1989) further studied the origin and development of embryoids from microspores in anther culture of *Poncirus trifoliata* and *C. aurantium* cultured on agar media. Ling et al. (1989) obtained diploid callus cells from the hypocotyl region of plantlets differentiated by anther culture of calamondin. The callus grew vigorously on MT medium containing 10 mg/l BA and differentiated embryoids on a medium with 5% lactose and no plant growth regulators. This embryogenic potential was transferred to protoplasts which regenerated plantlets.

Protoplast Culture

Citrus is an important fruit crop. Much experiments have been conducted involving protoplast for production of somatic hybrids.

Isolation and Culture

Oliveira *et al.* (1995) isolated protoplasts from the rootstocks Cravo lime (*Citrus limonia*) and Cleopatra mandarin (*Citrus reshni*) using enzyme mixtures to digest cell suspensions 7-10 times, following purification. Protoplast were cultured in droplets of semi-solid (0. 6% agarose) KM8P medium in the bark at 28°C following density adjustment to 2 x 10^5 protoplasts/mL. Isolation produced 0. 7 x 10^6 protoplasts/ml for Cleopatra and 0. 5 x 10^6 protoplasts/ml for Cravo, with a viability from 86 to 92% and 73 to 80% respectively.

Protoplast fusion

Grosser et al. (1988) reported that form *Severinia disticha* seedling epicotyls when grown in 1/10 strength MT medium with 0. 55 mg/L 2, 4-D, 0. 02 mg/L Kinetin, 20 ml/L coconut water and 50 g/l sucrose; protoplasts were prepared by treating 6-month-old callus with enzymes in BH3 protoplast culture medium at 28°C for 2-4 hours on a shaker.

Somatics hybrids

Protoplast fusion has been used to produce numerous citrus somatic hybrids for scion and rootstock improvement, including interspecific hybrids and intergeneric hybrids between sexually compatible or incompatible parents ((Grosser, 1993). Somatic hybrids in citrus were obtained with the help of PEG-induced chemical fusion of competent citrus protoplasts, isolated from nucellus-derived friable embryogenic callus or suspension cell cultures. The one parent was of protoplasts and the second parent was isolated from seedling leaves.

Somatic Embryogenesis

In citrus, embryos arise from nucellus or integument adventively. They are used for producing true-to-type plants. Nucellar embryony was reported first by Ohta and Fursato (1957) and Maheshwari and Rangaswamy (1958). Kobayashi (1978) studied the mechanism of formation of nucellar embryo and development of ovule in polyembryonic *Citrus sinesis* and monoembryonic Iyo (*Citrus iyo*). The nucellar cells containing dense cytoplasm and one large nucleus were found in the ovule of mature flowers of polyembryonic Trovita. Spiegel-Roy and Saad (1986) evaluated callus lines of orange *C. sinensis* for tolerance to NaCl stress to an unselected line. Beloualy (1991) regenerated complete plants from callus-derived embryo cultures of *C. aurantium*, *P. trifoliate* and Carrizao citrange (*C. sinensis* × *P. trigoliata*) via somatic embryogenesis and organogenesis. Callus cultures were grown on MT medium containing various combinations of BA, Kin, NAA, IAA and malt extract. Incubation was at 25°C under a 16-h photoperiod and with subcultures every 4 weeks. The use of undeveloped ovules from mature fruits allows the initiation of embryogenic cultures from a wide range of citrus species. Parthasarathy and Nagaraju (1992) using excised ovules (8-10 weeks old) of Khasi mandarin and Cleopatra mandarin revealed that MS medium alone or supplemented with male extract or IBA induced cotyledonary embryoids while NAA alone delayed further development of torpedo embryoids. Growth inhibitors like abscisic acid and paclobutrazol have also been tried in ovule culture. Nieves et al. (1995) found that supplementation of abscisic acid ($10x^{-6}$) delayed germination, and the presence of amino acids in the culture medium enhanced germination of Cleopatra somatic embryos. Encapsulated somatic embryos cultured on media containing various starch sources (potato, corn or rice) failed to germinate. Mishra et al. (1997) found that ovules of four citrus species, viz. , Khasi mandarin (*C. reticulata*), Cleopatra mandarin (*C. reshni*), Thorny mandarin (*C. reticulata*) and Malta (*C. sinensis*) culture on MS medium supplemented with malt extract (500 mg/L) and paclobutrazol showed differential response to different levels of paclobutrazol. Marrero et al. (1997) found stimulatory effect of 5 mg/l of chitosan hydrolysate in embryos of *C. aurantium*.

In vitro Conservation

Citrus collections are generally conserved in the field genebanks in many of the citrus growing countries, however, still there is a loss every year due to biotic and abiotic hazards. A few collections of selected cultivars are conserved as potted plants in the greenhouse. **Cryoconservation techniques** are considered one of the most promising alternatives to the traditional-type conservation of lives plants. Survival of citrus tissues is subject to freezing-thawing treatments e.g., in seeds (Mumford and Grout, 1979), ovules (Bajaj, 1984), embryos (Marin and Duran-Vila, 1988; Marin et al., 1993), embryonic axes (Radhamani and Chandel, 1992) and shoots (Chen and Wang, 1995) obtained from a few species and cultivars of citrus using cryoconservation technique.

Embryonic callus and cell cultures can be preserved using a number of cryoconservation treatments like, conventional cryoprotection with DMSO followed by slow cooling (Kobayashi *et al.* , 1990 ; Aguilar et *al.* , 1993), vitrification followed by fast cooling (Sakai *et al.* , 1990, 1991) and other simplified procedures (Engelmann et al., 1994). Cryopreservation of embryogenic cultures is one of the most valuable tools not only for the conservation of genetic resources of citrus which has a great economic importance but also to conserve plant materials in a form ready for further genetic manipulation.

Marin and Duran-Vila (1991) developed a tissue culture system using juvenile Pineapple sweet orange (*C. sinensis*) nodal stem segments. A cycle of secondary cultures was set up where nodal stem segments were excised from *in vitro*-grown plantlets and transferred to fresh medium. Plants were obtained from the secondary cultures by transferring the shoots to rooting and elongation media for genetic conservation of citrus, e. g. , juvenile tissues of 2 sweet orange varieties, trifoliate orange (*P. trifoliata*), Mexican lime (*C. aurantifolia*) and Eureka lemon (*C. limon*). Bhat *et al.* (1992) regenerated plantlets using 9, 12 and 36-month-old lime (*C. aurantifolia*) root cultures. The low frequency of plant propagation mitigates against using root cultures for germplasm conservation until the frequency of shoot regeneration can be improved. Oh *et al.* (1997) studied the pre-freezing and pro-protection on cryopreservation of somatic embryos. Somatic embryos were induced from the micropylar region of ovules of *C. junos* x *C. grandis* (*C. maxima*), *C. grandis* × *C. junos* and *C. platymamma* × *C. junos* on MT medium supplemented with zeatin. The highest rates of direct somatic embryogenesis were induced in the presence of 0. 01 or 1. 0 mg/L zeatin.

Perez *et al.* (1997) reported that nucellus-derived embryogenic callus cultures of Salustiana sweet orange subjected to cryoconservation assays. Cryoprotection with 10% (v/v) dimethylsulphoxide, freezing by slow cooling and thawing by fast warming was suitable to recover viable growing cultures and whole plants through embryogenesis. Chen and Wang (1995) stored germplasm using seedling of *C. madurensis*, and cultures were maintained without subculture for five years. Normah and Siti Dewi Serimala (1997) successfully cryopreserved *C. aurantifolia* seeds after desiccating them to a moisture content of 12. 93% (50% viability) while seeds of *C. halimii* exhibited only 25% viability after cryopreservation at a moisture content of 9. 5%. Seeds of *C. hystrix*, however, are sensitive to desiccation as they failed to germinate when the moisture content was reduced to 27%, and thus did not survive cryopreservation.

BANANA AND PLANTAIN

Bananas and plantains (*Musa* spp.) are the fourth most important global food commodity after rice, wheat and milk in terms of gross value of production (INIBAP, 1997). They are giant perennial herbs that thrive in the humid tropics and subtropics. Although their fruits are onc of the major commodities in international trade, they are far more important as starchy staple crops in local food economies (Stover and Simmonds, 1987). Cultivated banana and plantain are mostly triploid (2n=3x=33) and exhibit a marked degree of sterility. However, diploid and tetraploid cultivars also occur in much lower numbers. The crop is vegetatively propagated from suckers and fruits develop through parthenocarpy.

These are major fruit crops in the tropics and subtropics and make a vital contribution to the economies of a number of countries. During 1997 their annual world production was estimated to be around 88.4 million tones (FAO, 1997). They are very important in the nutrition of local populations as well as tradable commodities with large markets throughout the developed world.

India is the second largest producer of banana in the word, after Brazil. Though, Kerala and Tamil Nadu have a larger area under banana, but the production is highest in Maharshtra. The production of bananas in the country was estimated at about 6.2 million tonnes. The Jain group of Jalgaon is planning a banana revolution in India. They have already started off on home turf in Jalgaon district in Maharashtra. Jalgaon accounts for a lions's share of the all-India banana production.

Scientists at BARC have also developed banana plants using tissue culture technology. They have standardized the method of multiplication of three well-known varieties of banana in India, namely, Bhusawal, Rasthali and Srimanthi. The Assam State Industrial Development Corporation Ltd. is setting up a RS.4 crore, 100% EOU to manufacture banana puree at Goalpara. The state produces a sizeable quantity of banana in the plains as well as hilly areas throughout the year. In 1991-92, banana production in the state was 5,18,218 tonnes. The yield was 13,117 kg per hectare.

Knowledge of molecular and cell biology has provided a better understanding of plant regeneration, genetics, growth and development. The new approaches include genetic manipulation and faster regeneration of plants through cell or prevailing genetic diversity, the mechanism by which that variation is generated, and significance of that variation in adaptation and performance of the plants. These techniques also help to improve fruit quality, to enhance abiotic and biotic stress tolerances, as well as to transform the genomic constitution of the plant for better productivity. The present text will deals with the cryopreservation, cell or tissue culture, somatic embryogenesis and hybridization, synthetic seeds, molecular DNA markers and plant transformation techniques. Attempts have also been made to assess the current challenges of genetic improvement and disease or pest control through biotechnological means.

Tissue Culture

Procedures for micropropagation of Musa plantlets and the development of other tissue culture protocols have assisted germplasm handling and conventional plant breeding, and paved the way for nonconventional genetic improvement of *Musa* spp.

Micropropagation

The development of micropropagation techniques has been a major focus of Musa research during the past two decades and such techniques have now become well established (Cronauer and Krikorian, 1984; Banerjee and De Langhe, 1985; Vuylsteke, 1989; Israeli et al., 1995).

There are now many commercial companies producing over 30 minutes of *Musa* plantlets by micropropagation every year. It is possible to produce 2000 *in vitro* plantlets from one original explant within a year (Robinson, 1996). The most common method is to excise apical meristem and shoot-tips (meristem plus a few attached leaf primordia) and to culture these on suitable media. Several buds may be taken from a single mother plant as a source for explants. Viruses are not automatically eliminated by *in vitro* techniques unless this goal is built into the system. Only virus-free or pathogen-indexed plants are selected as a source of explants.

Micropropagation by shoot-tip culture is routinely and increasingly being used by researchers and nursery personnel in both the public and private sector. It is simple, easy, and applicable to a wide range of species and genotypes (Vuylsteke, 1989). Its advantages relative to conventional propagation include substantially higher rates of multiplication, production of clean or disease-free planting material, and the small amount of space required to multiply very large numbers of plants.

Application of micropropagation has greatly improved germplasm handling for the purposes of clonal propagation and production of uniform propagules. The capacity to derive many plantlets from the original germling of a single seed, and subsequently to generate quantities of propagules of selected clones has played a key role in the operation of Musa breeding programmes worldwide (Rowe and Rosales, 1996; Vuylsteke et al., 1997).

The micropropagated planting material is capable of growth and yield performance in the field which is equal or superior to conventional material (Smith and Drew, 1990a; Vuylsteke and Ortiz, 1996; Vuylsteke, 1998).

Establishment (Methodology)

1. First select the explant (shoot-tip, leaf bases, corm tissue, outer-leaf segment).
2. The selected explants are surface-sterilized with a surfactant like calcium or sodium hypochlorite. Continues the process under aseptic conditions.
3. Rinse at least three times in sterile water.
4. After rinsing the shoot-tips (or the explants) are trimmed to approximately 5 × 5 × 5 mm and transfer directly to the culture medium. A further surface-sterilization is optional (Israeli *et al.*, 1995).
5. There are many basal medium formulations to sustain growth and proliferation of the explants (Cronauer and Krikorian, 1986). The most widely used medium contains the Murashige and Skoog (1962) (MS) formulation with certain modifications. Smith and Murashige (1970) version with additional phosphate.

Table 7.2: Murashige and Skoog (1962) (MS)/ Smith and Murashige (1970) version with additional phosphate Medium

Supplements	Amount
Myoinositol	100 mg/L
L-tyrosine	200 mg/L
Thiamine-HCl	0.5 mg/L
Adenine sulphate	160 mg/L
BAP	5 mg/L
Indole-3-acetic acid (IAA)	2 mg/L
Sucrose	30 g/L
PH	5.8
Agar	7 g/L

Bekheet and Saker (1999) developed an efficient medium for in vitro propagation of banana cultivar William, Grande Naine and Maghraby for Shoot cultures. It is based on MS medium supplemented with 2 mg/l BA. When shoots were transferred to medium containing 0, 2, 4 or 6 mg/l BA for shoot multiplication, the highest number of proliferated shoots was recorded with 6 mg/l BA. Shoot length was greatest on hormone-free medium. Among different types of auxin used for rooting of banana shoots. NAA was more effective than IAA or IBA.

6. Test tubes (25 150 mm) with medium are autoclaved at 121°C and 103.4 kPa fro 15 min.
7. Maintain the cultures at 28 ± 2°C and 60-70% relative humidity in a 16-h light cycle with fluroscent light of 1000 – 3000 lux (Israeli *et al.*, 1995).
8. After a few days on the medium observe the swelling and greenness of the explant.
9. The shoots will appear 20\-3 weeks later.
10. Subculturing is done after 4-6 weeks or earlier if blackening occurs. Blackening occurs due to phenolic oxidation. It can be reduced by dipping the explant in antioxidants (cysteine, citric or ascorbic acid). Before being transferred to the medium; alternatively, antioxidants may be included itself (Banerjee *et al.*, 1986; Vuylsteke, 1989;Vuylsteke *et al.*, 1990). Antioxidants (ascorbic acid and citric acid) may be used in the preliminary washing may also be useful to prevent blackening.

Multiplication Stage

1. In this stage every explant gets expanded into a cluster of shoots which develop from a common expanded mass of basal, callus like tissue.
2. They are divided into separate propagules and then transplanted into a fresh culture medium. This procedure is repeated at 4-6 weeks intervals.
3. The medium for multiplication remains the same as used in Stage I (establishment). Sometimes, the cytokinin (BA) level is increased.
4. Thereafter, the propagules are removed from Stage I containers and placed into fresh medium in other containers. Proliferation and multiplication of propagules depend on genotype, culture medium composition, size of initial explant, its preparation procedure and age of culture (Israeli *et al.*, 1995).
5. The number of propagules produced for transfer varies from 5 to 10 or more from each explant. Multiplication stages are repeated several times to build up the supply of materials to a pre-determined level for subsequent rooting and transplanting.

Pre-transplant

In this stage the plantlets are transplanted from the artificial heterotrophic environment of the test to an autotrophic free-living existence in the greenhouse and onto their ultimate location. The important aspects of this process involves rooting and a change in the physiology of the plantlet (stimulation of photosynthesis, nutrient and water absorption through roots, and resistance to desiccation and to pathogens). This can be obtained by subculturing propagules to a medium with reduced or omitted cytokinin (BA), an increased auxin concentration, and often a reduced inorganic salt content. Israeli *et al.* (1995) suggested the use of MS medium supplemented with 2 mg/l IAA and 5 mg/l kin. Fitchet (1987), Gupta (1986) Vuylsteke (1989) quite often used NAA min the medium. Banerjee et al. (1986) Raut and Lokhande (1989) used IBA in the medium. It seems that the optimal hormone concentration of the medium and cytokinin/auxin ratios have to be adjusted for each clone and laboratory procedures.

Transplanting (Methodology)

1. Rooted or unrooted plantlets are removed from the culture vessel.
2. Wash the agar completely to avoid contamination.
3. The plantlets are transplanted into a standard pasteurized soil mix either in small pots or in a bigger tray at 45 cm apart.
4. Reduce the root length (up to 3 cm) because longer roots break and interfere with planting.
5. Prepare well aerated soil mixture for potting with high water holding capacity. They may be in following combination:

Table 7.3: Various soil mixture for acclimatization and further growth of the plantlets

Mixture	Reference
60% peal and 40% shredded Styrofoam	Israeli *et al.*, 1995
Peat and washed sand in 1:2 ratio	Daniells and Smith, 1991
Perlite	Novak *et al.*, 1990
Soil promix with vermiculite	Krikorian and Cronauer, 1984
Peat and perlite	Drew and Smith, 1990

6. Established the plantlets in the rooting medium then expose them to a higher relative humidity (>90%) and a higher light intensity. For hardening the plantlets the humidity of the rearing chamber should be decreased gradually.
7. Protect the plantlets from desiccation in a shaded, high humidity tent. Keep them under mist in a greenhouse or shade house.
8. The potting mixture (soil mix) should be wetted daily.

9. During the first two weeks of hardening light should be reduced by 80% with screens, decreasing to 60% after 2 weeks.
10. Under normal conditions new roots appear 4-5 days after the start of acclimatization, and new leaves start to grow within 8-10 days at the optimal temperature (25-32°C).
11. Ventilation in the shed or greenhouse should be proper. This will help to reduce heating.
12. Nutrients are applied about 10 days after the start of acclimatization on by foliar spray of compound nutrient solution.
13. The acclimatized plants are further transferred to bigger size pots (10-15 cm diameter) and reared up to 30-40 cm stature under controlled conditions gradually changing to the normal outdoor level. Then they are ready for field planting.

Performance of Micropropagated Plants

Cabrera et al. (1998) conducted two-year trial in Tenerife to test the viability of single-cycle cultivation using hardened in vitro-cultured Grade Naine plants (pseudostem height ≈ 4.5 cm), at a spacing of 5.0 × 2.5 m with 3 plants per hole (2400 plants/ha). He harvested the first crop 14 months after planting and the second 12 months later, thus effectively obtaining an annual crop. Yields were 73.1 and 77.7 t/ha for the 2 respective harvests, while > 90% of the fruits graded as extra. A comparative study of harvest date and yield was made between the 3 plants in each hole. Results shoed significant differences due to orientation although absolute differences were slight

Nandi *et al.* (1998) grew 3- to 4-month-old healthy sword suckers and ≈ 30 cm tall micropropagated plants of the banana cultivars Basrai and Shrimanti in the field. Growth and fruiting were observed until March 1995. In both cultivars, total height of micropropagated plants was significantly greater than that of the conventional, sucker-derived plants throughout the growth period. Flowering and harvesting in both cultivars occurred earlier in micropropagated plants than in conventional ones, and bunch weight, number of fingers, average fruit weight, fruit length and girth and total soluble solid contents were significantly higher.

Embryo Culture

Hybrid plant production of cultivated triploid Musa clones is hampered by low seed set and low seed germination rates. For example, only about 1% of hybrid plantain seed from 3x-2x crosses germinates when planted in soil. Embryo culture is useful in rescuing embryos of sexual crosses that fail because the hybrid embryo aborts. Embryo culture was the first tissue culture application in *Musa* spp., dating back nearly 40 years (Cox et al., 1960). In vitro embryo culture can increase rates of seed germination by a factor of 10 or more and thus is used routinely by breeding programmes (Vuylsteke et al., 1990). The presence of two incompatible genomes often results into a breakdown of the embryonic development cycle. As such plantain hybrid seeds germinate poorly in the soils at the rate of only 1% (Vuylsteke and Swennen, 1993). Therefore, the embryo culture technique is routinely applied to increase seed germination rates by a factor of 3 to 10 (Ortiz *et al.*, 1995).

Methodology

1. For embryo culture, the 0.7-1 mm diameter embryo is excised under a stereo-microscope from the hardened seed coat at 50 days after hand pollination.
2. Place these extracted embryo on half-concentration MS medium, modified with a range of hormones and inorganic salts (Vuylsteke *et al.*, 1990)
3. Maintained in the culture room for 4 – 6 weeks.
4. Culture conditions usually include a 2- to 4-week incubation in darkness at 20°C or less, followed by a short incubation above 20°C while exposed to light.
5. The developing embryos are transferred to a solid medium for enhancement of root and shoot growth.
6. Such plants (with roots and shoots) may be transferred to soil and kept at high humidity and moderate temperatures until established. Although with the available *in vitro* techniques germination rate has been enhanced to 20%, further investigation to refine the technique are under way.

Direct Adventitious Embryogeny

The embryogenic calluses and embryos are isolated and placed in a container (Alvard et al., 1993) and allowed for temporary immersion (for 1 min six times in 24 h) in a new medium containing picloram. These conditions develop highly intensive adventitious embryogenesis and existing embryos produce adventitious embryos. Such new embryos develop structure and an epidermis. They also enter the process of secondary embryogenesis (Escalant et al., 1994). A single inoculation of 200 mg (about 200 primary embryos) into one temporary immersion container results in nearly 50,000 embryos within 3 months. These embryos can be sampled regularly and transferred to liquid or semi-solid germination medium. The rate of conversion into plantlets is about 67% (Escalant *et al.*, 1994). This method seems to be applicable to all kinds of bananas which contain male flowers whether they are di, tri or tetraploid.

Somatic Embryogenesis

Somatic embryogenesis is the technique for the production of embryo-like structures (embryoid) from somatic cells. Somatic embryos can develop and germinate to form plants in a manner analogous to germination of zygotic embryos. It offers advantages over regeneration via organogenesis. It has following stages:

1. It begins with induction of the embryogenic state.
2. Followed by differentiation into an embryo.
3. And subsequent maturation, germination and conversion into plants.

Advantages

1. Through this technique one can get complete propagules.
2. Somatic embryos have both shoot and root meristems, separate shoot induction and root induction steps are not required (Parrott, 1993).
3. Somatic embryogenesis tends to be less prone to somaclonal variation than organogenesis. Percentage of off-types does not exceed 2% (Schoofs, 1998).
4. Useful during genetic transformation of a given genotype, in which case changes other than the engineered trait are undesirable.
5. This technology provides not only exciting alternatives for large scale clonal propagation and somatic hybridization but also a fast cell regeneration system for the genetic engineering.

An appropriate explant, culture media and environmental conditions are required for development of somatic embryos. Several plant tissue can serve as explant, with the caveat that stage of development or maturity of the tissue.

Production of somatic embryos from the selected explant may proceed either directly or indirectly after some form of callus culture. The procedure adopted earlier consisted in cultivating immature zygotic embryos of wild species on a semi-solid medium with picloram (4-amino 3,5, 6-trichloropicolinic acid) to obtain an embryogenic callus with somatic embryos. The somatic embryos were germinated and developed into plantlets (Cronauer and Krikorian, 1988; Escalant and Teisson, 1988, 1989). Later, Escalant *et al.* (1994a) described a method for deriving primary somatic embryos (or pro-embryos) from the explants and their subsequent development into somatic embryos by direct secondary embryogenesis or cell suspension cultures.

Table 7.4: Various sources of plant tissue (explant) of *Musa* for development of somatic embryo

Source of material	Reference
Slow growing cell suspensions from immature fruit-derived callus in *Musa*	Mohan Ram and Steward (1964)
Recovery of plants from somatic embryos obtained in cell suspension from rhizome tissue in diploid and triploid banana cultivars.	Novak *et al.* (1989)
Reported the embryogenic callus, induced from basal leaf sheaths and rhizome tissue from *AA*, *AAA* and *ABB* clones, when placed in liquid medium with zeatin	Novak (1992)
Meristematic 'scalps' of cooking bananas cv. Bluggoe as explants	Sannasgala) 1989, Dhed'a et al. (1991)
Calli from immature zygotic embryos and triploid somatic embryos from male flowers	Escalant and Teison (1989 and 1993)
Meristematic tissues from shoot-tips can also be induced to undergo somatic embryogenesis	Cronauer and Krikorian (1993)
Immature female and male flowers	Escalant et al. (1994), Cote et al. (1996), Grapin et al. (1998)

Procedure

1. Primary embryos are obtained from the explants (male flowers).
2. They are cultured on a semi-solid medium consisting of a mixture of auxins.
3. The first callus appear after one to two months and the first embryos after three to four months.

Callogenesis is the production of a compact nodular callus with a high starch content. Certain external parts of the callus is transformed succeedingly into translucent, flaky areas on which embryogenic cells with large protein reserves appear. Somatic embryos follow. succeeded in obtaining Somatic embryos from a variety of *Musa* genotypes (commercial *AAA* and *AAB* cultivars) were obtained by Escalant et al. (1994) by this method. Such The embryos are perfect and can regenerate whole plants after transfer to a germination medium.

Scalps, somatic derived from in vitro proliferating meristem cultures, can be generated from any given banana or plantain landrace (*Musa* spp.). A preculture of meristems on a the medium with 100 μM BA enhanced significantly competence for somatic embryogenesis. Out of 21 landraces covering all diploid and triploid genome groups and the most important banana types, 18 cultivars showed an embryogenic response. From 15 accessions an embryogenic cell suspension could be established (Schoofs *et al.*, 1998).

Embryogenic Cell Suspensions

Cell suspensions are established by transferring embryogenic callus to liquid medium. The development and composition of such cell suspensions (whether they originate from female or male flowers) are similar, irrespective of the genotype used (Grapin *et al.*, 1998). Cell suspension of *Musa* diploid species was obtain by culturing embryogenic callus initiated from immature zygotic embryos in liquid medium (Marroquin *et al.*, 1993). The somatic embryos are obtained after planting of aliquots (0.5 ml) of filtered (200 μm sieve) cell suspension. Plant regeneration was achieved by transferring the somatic embryos obtained on an MS-modified semi-solid medium with BA (0.22 μM) and IAA (1.14 μM).

Cell suspensions of *Musa* triploid cultivars (*Musa AAA* cv. Grande Naine, *Musa AAB* or French somber) were initiated by culturing embryogenic cultures from immature male flowers in liquid medium (Cote *et al.*, 1994). Cell suspensions were obtained which contained 8 million embryogenic clumps per litre, a multiplication rate (packed cell volume) of 2 every 3 weeks and a somatic embryo conversion rate of over 8%. Plant regeneration was obtained by transferring somatic embryos on an adequate semi-solid medium (Escalant *et al.*, 1994). Grapin *et al.* (1998) have demonstrated that the cell suspension technique can be very efficient for the cultivar French somber, planting 1 ml of packed cells led to the formation of 105 embryos of which 10 – 40% could be converted into plantlets.

A very high rate of somatic embryogenesis has also been achieved by cultivation in fermentors (also called bioreactors). The use of a bioreactor will increase several fold the multiplication capacity of one particular cell line. Also several units can be operated simultaneously for multiple lines. Bioreactors permit a high degree of culture control (maintenance of genetic integrity, synchronized embryo development, etc.). This process is amenable to robotics (Kozai et al., 1991) and cost-effective.

Somaclonal Variation

Thus, somaclonal variation has been observed in plants regenerated from shoot-tips, protoplasts, other cultured explants (immature embryo, flower, leaves, rhizome, etc.), micropores, anthers and ovaries, and also in plants regenerated from cultured tumorous tissues. Somaclonal variation has generated interest as a potential source of novel and useful variability, although it has so far had a limited direct contribution of Musa improvement. Somaclonal variation may be used as a secondary source of variability for genetic improvement, e.g., the utilization of a plantain variant with increased female fertility made a sterile plantain pool accessible to conventional breeding schemes (Vuylsteke et al., 1995;Vuylsteke, 1998). The extent of somaclonal variation depends on a number of factors, viz., the genotype, ploidy level, tissue culture procedure employed and time in culture, the source of explant cells and the culture media used (Withers, 1993; Israeli *et al.*, 1995).

This phenomenon has been studied extensively in bananas and plantains (Israeli *et al.*, 1991; Vuylsteke *et al.*, 1991, 1996; Novak, 1992; Cote *et al.*, 1993; Withers, 1993; Ortiz *et al.*, 1995; Israeli *et al.*, 1995) where variations in useful traits like plant height, growth habit, fruit shape and size, pseudostem colour, fertility status, disease and pest resistance have been described. Many somaclonal variants have been screened for enhanced fruit quality and resistance to diseases and pests and are being used in conventional breeding for developing superior banana or plantain hybrids (Withers, 1993; Ortiz *et al.*, 1995; Vuylsteke *et al.*, 1997).

Musa plants produced by shoot-tip culture show somaclonal variation rates of 0-70%. It depends on the genotype. (Israeli et al., 1995). This problem or risk remains always associated with the use of in vitro culture techniques for handling Musa germplasm. Israeli et al. (1995) have suggested following measures to reduce somaclonal variation

1. The source plant must completely match the known description of the cultivar.
2. It must have the desired agronomical characteristics and of a clone which is relatively stable in vitro propagation.
3. The number of plants (and obviously also the number of generations) produced from a primary explant must be reduced.

4. Many primary explants per batch must be used. It will minimize the risk having a high percentage of variants contributed by a single explant that had mutated in an early state of propagation.
5. Small undeveloped plantlets in each subculture should be discarded and not used for new generations.
6. Off-type plants should be screened out and discarded at all stages from regeneration to the final stage in the nursery.

Somaclonal variation is important for increasing genetic variability in the genus *Musa* particularly with respect to pest or disease resistance (Trujillo and Garcia, 1996). Somaclonal variation in breeding rests on the ability to recover new genetic variants with desirable characteristics at high frequency. Hence, it would be useful to devise methods to identify efficiently those rare but useful somaclonal variants among large populations of useless variants. Such large populations might be produce through cell culture procedures.

Anther Culture

The production of haploid plants is important in plant breeding practices because one can get the rapid production of homozygous lines following chromosome doubling to the original ploidy level. It also helps for the detection of the recessive mutants. There are several problems to obtain such lines in banana specially the conventional approach of using successive self-fertilization, the very strong female sterility of the clones, the large amount of space required by the plants, and the length of the successive growing season. To overcome these problems Bakry et al. (1993) and Escalant et al. (1994) have tried to establish a protocol to obtain homozygotic plants directly from immature pollen by the in vitro culture of banana anthers (androgenesis). Attempts have also been made to produce haploid plants from seeds fertilized by irradiated pollen to stimulate parthenogenetic development of the oosphere (parthenogenesis). It is important to note that genetic instability in anther culture is common.

Regeneration Procedure

1. The process involves cutting off the flowers at the optimum stage i.e., at uninucleate stage, 1-2 days before first mitotic division.
2. Anthers are surface-sterilized.
3. They are then excised and cultured on appropriate media containing relatively high levels of cytokinin and auxin.

4. After 3 months the microspores divide and can develop indirectly (via callus) into somatic embryos.

5. These can be later converted into intact haploid. The induction of androgenetic calluses originating from pollen of *Musa acuminata malaccensis* (Pahang clone) permitted the production of 47 different calli from 114 cultivated anthers. These calli have generated haploid banana plants (Escalant *et al.*, 1994).

6. All cells or plantlets produced in anther culture system are not always haploid or monoploid. Somatic (2n) cells surrounding the microspores and able to produce callus. The callus from the somatic cells may completely surround the haploid cells, making it extremely difficult to identify and propagate the haploid callus. Therefore, confirm the chromosome number of each culture or plant.

7. If longer cell line is cultured one can get great instability.

Table 7.5: Different stages of anther culture, material and culture conditions (Bakry *et al.*, 1993)

Stage	Material	Conditions of culture
I - Culture initiation	About 160 anthers of M. *acuminata burmannica* L. Tavoy containing microspores at uninucliate stage, 1-2 days before the first mitotic division	Medium BM + IAA, 27°C Obscurity duration : 35 days
II – Callus maintenance	Isolation and subculture of one anther-derived callus	Medium and conditions as in Stage I duration : 60 days
III - Shoot initiation	Callus growth, development of green shoots on the callus	Medium BM + BA + adenine hemisulfate conditions : 14-h photoperiod, 33 wm^{-2} of fluorescent light, 27°C, duration : 65 days
IV - Shoot culture	Subculture of individualized shoots rooting	Medium BM and conditions as in Stage III duration: 45-70 days

Stage	Material	Conditions of culture
V - Weaning in greenhouse	Plant growth, checking: chromosome counting electrophoresis	Culture in Jiffy-7, and then in 1 litre plastic bags containing a 1:1 (v/v ratio) river sand and sugar cane compose mixture

Germplasm Exchange and Conservation

In respect of banana and plantain landraces and also improved genotypes, shoot cultures, in combination with third-country quarantine (Vuylsteke et al., 1990a) or preferably with virus-indexing procedures (Diekmann and Putter, 1996), have been used as a vehicle for the safe exchange of Musa germplasm since the early 1980s. The advantages of this method include the reduce volume and weight of in vitro cultures and the improved phytosanitary status of germplasm. Viruses are not effectively eliminated by tissue culture (Drew et al., 1989), hence virus testing of germplasm now is recommended as a routine procedure to ensure its safe international distribution. Shoot-tip cultures, often maintained at lower temperatures (15-18°C) to reduce the need for subculturing, are also used for the slow-growth storage of Musa germplasm (Banerjee and De Langhe, 1985; Van den houwe et al., 1995). Recently, a simple technique was developed for the cryopreservation of banana meristem cultures, involving preculture on high-sucrose medium followed by rapid freezing (Panis et al., 1996). This should greatly facilitate the long-term conservation of Musa genetic resources.

Protoplast Culture

Protoplast are cells without cell wall removed either mechanically or enzymatically. Protoplasts are used for genetic engineering experiments (microinjection, transformation and somatic hybridization, etc.). Regeneration of plants from protoplasts have been achieved in *Musa* (Pains and Swennen, 1993; Pains et al., 1993; Haicour et al., 1993; Sagi et al., 1995b). **Embryogenic cell suspension** (ECS) cultures have been the source or protoplasts. Megia et al. (1992) were the first to describe protoplast isolation from banana ECS.

The cell wall is removed by treating the cells to hydrolytic enzymes that degrade cell wall materials. Once the protective cell wall is removed, protoplasts must be maintained under carefully controlled conditions. The yield up to 6 x 10^7 protoplasts per ml packed cells was easily obtained after an enzymatic treatment of a Bluggoe (*Musa ABB*) suspension for 16 hours (Pains and Swennen, 1993).

Protoplasts require culture media and other environmental conditions similar to those required for cell suspensions. To prevent rupture of the naked cells, careful attention should be given to osmotic potential of the medium.

Procedure

1. **Basic culture media :** Panis and Swennen (1993) described the composition and concentration of plant growth regulators (2,4-D), regeneration culture methods such as culture in liquid media, culture on semi-solid medium (medium solidified with agar, agarose or gelrite at different concentrations), culture of protoplasts embedded in semi-solid medium, hanging or sitting drop method, and feeder layer systems; protoplast inoculum density and age of suspension used.
2. Banana protoplasts has their low potential to divide, therefore, Haicour and Rossignol (1993). feeder cells for stimulation of protoplast division.
3. Matsumoto et al. (1998) isolated protoplasts from cultured embryogenic calli and suspension cells of banana cv. Maca and embedded in solid medium. They were floated on liquid medium E containing a rice cell suspension nurse culture. Best results were obtained when protoplasts at a density of 1 x 10^6/ml protoplasts were cultivated on a membrane isopore filter with nurse cells in the dark and transferred on new nurse cells after 10-14 days of incubation. Plantlets were regenerated on medium containing 2 μM of BA and IAA.

Cell Suspension and Protoplast Culture

Methods of plant regeneration from callus or by direct somatic embryogenesis in cell suspensions or protoplasts have been investigated for the recombinant DNA propagation and as a tool in transformation by recombinant DNA technology. Such regeneration has been achieved in cell and protoplast culture derived from in vitro meristems, immature zygotic embryos, and young male flower tissue. Limitations are that the procedures are genotype-specific and, in spite of attempts to refine them, remain difficult. An aspect of these several regeneration procedures (from both cell and protoplast culture, and using different explants) is that they could generate substantially more somaclonal variation than that obtained with shoot micropropagation. By such means the full potential of somaclonal variation to contribute to plantain and improvement could be explored.

Protoplast Fusion

Protoplasts amenable to various techniques of genetic manipulation, particularly those based on fusing protoplasts of different types together. These techniques are:

1. Combination of two complete genome.
2. Partial genome transfer from a donor to a recipient protoplast to produce partial asymmetric hybrids.
3. Transfer of organelles (cybrid production) notably chloroplasts and mitochondria in relation to transfer of properties such as herbicide resistance and cytoplasmic male sterility (Lindsey and Jones, 1989).

This technique has yet to be fully exploited for production of cybrids or nuclear hybrids in *Musa.* Haicour et al., (1993), Panis et al. (1993) and Sagi et al. (1995) made some attempts in this direction.

GRAPE

Grapes have in recent times emerged as an important item of exports from India. Our major customers of grapes are Persian Gulf countries like United Arab Emirates, Saudi Arabia, Bahrain, Kuwait, Oman and Bangladesh. Recently, however, the European market has also opened up for Indian grapes, for the first time in 1992 because of a bad season in Chile, the traditional supplier. Grapes grow wild in many temperate regions of Europe, Asia, Africa and America. The Present global production of Grapes is estimated at 100 million tonnes.

India's present production of grapes is estimated at about 6 lakh tonnes from an area under cultivation of 60,000 hectares. In terms of area and production Maharashtra leads the country followed by Karnataka, Andhra Pradesh and Tamil Nadu. The main grape growing regions in the country are Nasik, Aurangabad, Ahmednagar, Beed, Sholapur, Sangli and Pune in Maharashtra and Hyderabad in Andhra Pradesh. A small quantity of grape is also grown in states like Haryana, Rajasthan and U.P. In Karnataka, Grape cultivation was earlier confined to Bangalore and Dharwar districts only, but now it has spread to Belgaum, Bijapur and Bellary districts. In Tamil Nadu grape cultivation is concentrated in Coimbatore and Madurai districts.

Indo-French Biotech Enterprises Ltd.: Indo-French Biotech Enterprises has entered into an agreement with MMTC of India and Agrexco Export Co. of Israel for marketing Indian fruit and fruit products in markets abroad. Agrexco will also provide market forecasts, quality assurance and packaging assistance to Indo-French. Agrexco has started marketing grapes sourced from Indo-French in European market under a famous brand name `Caramel'. The company is setting up a vineyard in Nasik district for the production of Thompson seedless and ltalia variety of table grapes with technical assistance from Richter SA of France which has supplied the virus-free phyto sanitary root stalks of grapes. The unit will also manufacture raisins by drying grapes for exports.

Micro Plantae Ltd.: Phytotech Australia Pvt Ltd. has appointed Micro Plantae Ltd., Bombay as its exclusive licensee for the Indian sub-continent, to grow and market the premium black `Marroo' variety of seedless grapes. Marroo's advantages include above average output per acre, resistance to downy mildew, a widely prevalent vine infection and minimal consumption of giverallic acid, a plant growth regulator used by all growers. Because of its different but better taste, Marro grape commands a premium in the US market. Marroo is believed to be the world's largest sized black table grape, a cross between Carolina Blackrose and Ruby seedless. With judicious trimming and thinning, it is possible to obtain berries of average diameter of 21.5 mm. The grape also can be dried to obtain a raisin of very attractive colour. Micro Plantae Ltd., which has a tissue culture facility near Pune, proposes to set up a captive plantation of 100 to 150 acres and another 2,000 acres under contract for growing Marroo in India, primarily for export.

Nath Seeds Project: Nath seeds Ltd. is venturing into promotion of grape cultivation in Aurangabad in Maharashtra with a view to export them to Europe. There is a 70,000 tonnes annual market for grapes in Europe. The company will assist farmers in growing grapes on commercial scale in Aurangabed and will purchase their produce at a pre-determined price. The crop will be grown by the farmers, the pre-cooling and packing facilities will be provided by the company. The company has planned to export some 2,000 tonnes of grapes in late 1995 as the grape vines require 18 months to attain maturity for fruit yield.

Tissue culture

Micropropagation

The tissue culture technique is used for clonal propagation of grapes for the following reasons:

1. Development of virus-free planting stocks.
2. *In vitro* screening for resistance to biotic and abiotic stresses.

3. Long term conservation of germplasm under *in vitro* conditions including cryo-preservations.
4. Micrografting of scions to rootstook cultivars.
5. Propagation and exchange of virus-free plant material.
6. Rapid multiplication of plant material.

Much research work has been done on *in vitro* shoot multiplication techniques and the procedures have been summarized by Monette (1988) and Bouquet (1989). This has led to the development of new media for improved rooting of micropropagated shoots. Studies have also been conducted to improve *ex vitro* plant establishment following the transfer of plantlets to soil medium. Carbon dioxide levels enhance a six-fold increase in root dry weight on plants *ex vitro.* Paclobutrazol in medium along with a reduction in humidity improved resistance to wilting of *in vitro* plantlets of *Vitis vinifera* cv. Moscato Blanco (Smith *et al.*, 1992). *In vitro*-raised plants of Carignane and Grenache cultivars also showed morphological modification particularly in the form of leaf shape and pigmentation of shoots under field conditions. This variation turned out to be reversal modification in due course of time (Grenan, 1994). Orgenogenesis has been accomplished primarily from shoot apices, leave, buds tendrils and internode tissues.

Mur (1978) indicated that the buds of grapevine cv. Pinot Meunier cultured on modified MS medium produce tendrils, inflorescence and berries. They did not reach veraison stage. Adventitious buds are developed from the fragmented shoot apices of Cabernet Sauvignon and Sultana cvs. cultured on MS medium + BA (10μM).

Table 7.6: Micropropagation of rapes by using meristem and shoot-tip explants

Species and cultivars	Medium	Response	Explant(s)	Reference
V. vinifera cv. Cabernet Sauvignon	White Nitsch	Plantlets	Axillary buds	Mullins *et al.* (1979)
Marechal Foch; Cascade	MS	Plantlets	0.7-1 mm apex	Li and Eaton (1984)
Cabernet Sauvignon, French Colombard, Grenache, Tyompson	MS and Nitsch and Nitsch	Shoot organogenesis	Shoot	Stamp *et al.* (1990)

Species and cultivars	Medium	Response	Explant(s)	Reference
Seedless, White Riesling, *V. rupestris* cv. St. George, *V. vinifera* x *V. rupestris* cv. Ganzin 1				
Muscadine grapes 9 cvs.	MS and C_2D	Plantlets	Shoot apical meristem	Gray and Benton (1991)
V. vinifera cv. Thompson Seedless, Ribier and Black Seedless	MS	Shoot proliferation	Apical and Axillary buds	Botti *et al.* (1993)
V. vinifera cv. Banaty and Khalas	MS	Plantlets	Shoot-tips	Al-Maurri and Ghamdi (1995)
V. vinifera cv. Kyoho	MS	Plantlets	Immature and mature embryos	Yamashite *et al.* (1995)
Rizamat, Sekires				

Wafa *et al.* (1999) reported that MS medium supplemented with 0.2 mg BA, 30 g sucrose plus 7 g agar/L gave the best results. They produced uniform normal shoots from stem node cultures of Flame Seedless. Botti *et al.* (1993) found a significant interaction between genotype, culture date, size and type of shoot apices in *in vitro* shoot proliferation of *V. vinifera* cv. Thompson Seedless, Riber and Black Seedless. Hormones (cytokinin BA, gibberellin GA_3, and auxin IBA) *in vitro* also control morphogenesis of grapevine inflorescence primodia (Yahyaoui *et al.*, 1998).

Embryo Culture

The ovules of seed and seedless cvs. of *V. vinifera* excised from 0-101 days post anthesis and cultured on MS and ½ MS medium showed more survival. An addition of 0.1% activated charcoal to medium reduced browning of the tissue.

The ovules of *V. vinifera* cv. Thompson Seedless, when cultured after 68 days of anthesis on Cain's medium supplemented with 10 mM L-cysteine, enhanced the growth of embryos. (Emershad and Ramming, 1984). In Flame Seedless, the ovules cultures 49 days after anthesis gave higher frequency germination and viable plants. The ovules of 10 seedless grape clones and cv. Concord were cultures six weeks after full bloom on Cain's medium containing 10 mM L-cysteine. Burger and Goussard (1996) reported that NN or Bouquet and Davis (BMB) basic media supplemented with IAA and GA3, did not increase the percentage of Sultanina embryos compared to corresponding hormone-free media. The highest embryo percentage was obtained from eight-week-old ovules of Muscat Seedless, Festival Seedless grapes cultures on BMB + activated charcoal.

Anther Culture

Lebrun *et al.* (1985) reported that when anthers of *V. rupestris* cv. Rupestris St. George are cultured initially in the dark on Nitsch and Nitsch medium containing 2,4-D (5 μM) produced callus after one or more subcultures on proliferation medium (each of two weeks). The callus produced by the anthers disintegrated to produce suspensions of single cells and cell aggregates.

Salunkhe *et al.* (1999) cultured anthers of *V. latifolia* (wild grape) cultured on NN medium supplemented with 20 μM 2, 4-D and 9 μM BA produced callus after 4-6 weeks. One gram of callus produced more than 400 somatic embryos with 13.7% being converted to complete plantlets, which are subsequently established in soil.

Table 7.7: Culture of non-embryogenic callus or explants, with or without adventitious bud

Species and cultivars	Medium	Response	Explant (s)	Reference
FS-4-201-39 and Kober 5BB	Heller	Callus±	Stem segments	Alleweldt and Radler (1962)
Sultana	Heller	Callus	Berries	Radler (1964)
V. vinifera Riesling, Portugieser	MS	Callus	Shoot segments	Staudt *et al.* (1972)
V. vinifera (6 cvs.)	Gamborg and Eveleigh	Callus	Immature beries	Hawker *et al.* (1973)
27 genotypes	MS	Haploid callus from 3 genotypes, no organogenesis	Anthers	Gresshoff and Doy (1974)
V. thumbergii	Nitsch	Shoot differentiation from callus	Anthers	Hirabayashi *et al.* (1970)

Species and cultivars	Medium	Response	Explant (s)	Reference
V. riparia x *V. rupestris*	MS, Heller	Adventitious buds (1-2%)	Leaf blade from *in vitro* cultures	Favre (1977)
V. vinifera cv. Sylvaner	Schenk Hildebrandt Hawker *et al.*	Callus, no organogenesis	Shoot, petioles, tendrils	Jona and Webb (1978)
6 genotypes	Various	Callus	Various	Iordan *et al.* (1981)
17 genotypes	Nitsch and Nitsch	Adventitious buds from 4 genotypes, plantlets from 3 genotypes	Internode segments	Rajasekaran and Mullins (1981)
V. vinifera (5 cvs.)	MS	Callus, plantlets	Stem, fruit stalk, tendrils, internode	Wang *et al.* (1985)
Hybrid grapevine forms H_1, H_2	MS	Callus proliferation	Shoot segments, excised berries	Cholvadova (1989)

High yield of leaf protoplast from regenerated plants via somatic embryogenesis of cv. Chardonnay was obtained by incubation for 2-3 hours in a solution containing one per cent cellulase Onozuka R-10, 0.5 per cent macerozyme R-10 and 0.7 M mannitol, and most of the protoplasts obtained in this way regenerated cell wall within first few days. Cell division occurred after 10 days of culturing, and some cell divided tow or three but division ceased afterward (Barbier and Bessis, 1990). Reustle *et al.* (1994) developed a protoplast isolation method for successful regeneration of grapevines from protoplasts. Protoplasts were isolated from somatic embryos induced on leaf discs of grape cv. Seyval Blanc and were cultured in alginate layers.

Somatic Embryogenesis

Mullins and Srinivasan (1976) first reported somatic embryogenesis in *V. vinifera*. They cultured unfertilized ovules of *V. vinifera* cv. Cabernet Sauvignon on liquid Nitsch and White's medium containing 5-10 μM BA and 5-25 μM NAA formed callus which gave rise to somatic embryos.

Table 7.8: Protoplast isolation and culture response

Species and cultivars	Tissue utilized	Response	Yield	Reference
V. vinifera cv. Muscat Gordo Blanco	Peeled berry skin	Two sugar uptake processes evident	1.4 x 10^6/g	Brown and Coombe (1984)
V. vinifera (3 cvs.)	Leaves from cutting *in vitro* cultures	Some cell wall regeneration, no division	15-20 x 10^6/g	Bessis *et al.* (1985)
V. vinifera (3 cvs.)	Leaves	Protoplast 90% intact, respiration and photosynthesis	10^6 protoplasts/g	De Filippis and Ziegler (1985)
V. Vinifera cv. Grenache	Leaves from *in vitro* cultures	Cell wall regeneration, some division	30 x 10^6/g	Lebrun (1985)
V. vinifera cv. Seyval Blanc	Leaf discs	Conditions for greatest yield incompatible with further development	30 x 10^6/cm^2 leaf	Wright (1985)
V. vinifera cv. Chardonnay	Leaves	Regenerated cell walls	40 x 10^6/g	Barbier and Bessis (1990)
V. vinifera cv. Seyval Blanc	Leaf disc	Plantlets	-	Reustle *et al.* (1994)

Embryogenesis is well established in several grape species but response among cultivated genotype varies greatly. Embryogenic suspension of *Vitis vinifera* were developed from somatic embryos of Thompson Seedless and Chardonnay. The suspension consists of pro-embryonic masses (PEM). It proliferated without differentiation in medium containing 2,4-D. Chardonnay somatic embryos and developed fully from PEMs following subculture in medium without 2,4-D. The development of such embryo did not advance beyond the heart stage in Thompson Seedless variety. Jayasankar *et al.* (1999) also developed somatic embryos in large numbers from suspension-derived PEMs of both cultivars on semi-solid medium lacking 2,4-D.

Table 7.9: Somatic embryogenesis from ovules, anthers and other explants

Species and cultivars	Medium	Response	Explant(s)	Reference
V. vinifera cvs. Cabernet Sauvignon, Grenache and *V. vinifera* *V. rupestris*	Nitsch *et al.*	Somatic embryogenesis	Unfertilized ovules	Srivnivassan and Mullins (1980)
V. vinifera	MS	Haploid plantlets	Anthers	Zou and Li (1981)
23 genotypes	Nitsch and Nitsch	Plantlets from 12 genotypes	Anthers	Bouquet *et al.* (1982)
28 genotypes	Nitsch	Somatic embryogenesis, plantlets from 13 genotypes	Anthers	Hirabayashi and Akihama (1982)
V. vinifera cvs. Pinot noir, Chardonnay, Grmay	Nitsch	Somatic embryogenesis plantlets	Ovules	Bessis and Labroche (1985)
V. rupestris and hybrids	MS	Somatic embryogenesis, plantlets	Anthers	Bouquet *et al.* (1985)
V. vinifera cv. Cabernet Sauvignon	MS	Somatic embryogenesis, Plantlets	Anthers	Fallot *et al.* (1985)
26 genotypes	Nitsch and Nitsch	Somatic embryogenesis from 6 genotypes	6 mm leaf-discs	Hirabayashi (1985)
V. rupestris	Nitsch and Nitsch MS	Somatic embryogenesis, no plantlets	Anthers	Rajasekaran and Mullins (1985)
V. vinifera cv. Cabernet Sauvignon	Nitsch *et al.*	Somatic embryogenesis, plantlets	Ovules	Rajasekaran and Mullins (1985)
V. vinifera cv. Cabernet Sauvignon	MS	Somatic embryogenesis, plantlets	Anthers	Mauro *et al.* (1968)
V. rupestris *V. vinifera* cvs. Cabernet Sauvignon, Grenache, Thompson Seedless, White Riesling	Nitsch and Nitsch	Somatic embryogenesis	Anthers	Stamp and Meredith (1988)

Species and cultivars	Medium	Response	Explant(s)	Reference
V. longii, *V. vinifera* cvs. Chardonnay, French Colombard, Grenache, White Riesling	Nitsch and Nitsch	Somatic embryogenesis	Zygotic embryos	Stamp and Meredith (1988)
V. longii, *V. vinifera* cv. Thompson Seedless, *Vitis* hybrids	MS	Somatic embryogenesis	Anthers, unfertilized ovaries, leaves	Gray (1989)
V. vinifera Koshusanjaku	Nitsch and Nitsch	Somatic embryogenesis	Leaves	Matsuta and Hirabayashi (1989)
V. vinifera cvs. Krymskaya, Zhemchuzhina Podarok, Moldovy, Magerach 100-74-1-5, (*V. vinifera* x *V. rotundifolia*)	MS	Somatic embryogenesis	Stem, leaves	Marchenko (1991)
V. rotundifolia Muscadine grapes	MS	Somatic embryogenesis, plant regeneration	Immature zygotic embryo	Gray (1992)
V. vinifera cvs. Bianca, Poderok, Magaracha, etc. Kober 5BB, 101-14 *V. rotundifolia*	Cl liquid medium	Plantlets	Petiole	Zlenko and Troshin (1993)
Muscadine grapes cv. Regale, Fry	Nitsch and Nitsch	Somatic embryogenesis	Immature leaf lamina, petioles	Robacker (1993)
V. rotundifolia cultivars	MS	Somatic embryogenesis	Ovules	Gray and Hanger (1993)
Interspecific hybrids, Chancellor	MS	Somatic embryogenesis, plantlets	Embryogenic callus	Hebert *et al.*(1995)

Somaclonal Variation

Somaclonal variation is used as a mechanism to increase variability within existing grape cultivars. It is random in nature. Krul and Mowbray (1984) compared somatic embryo-derived vines of *V. vinifera* cv. Seyval Blanc with normally propagated vines and found that vines from somatic embryos had darker leaves, reddish vs green canes and cylindrical rather than conical fruit clusters. Furthermore, cuttings from regenerated vines rooted and grew more rapidly.

Possibilities exist to screen for a variety of traits *in vitro*, such as resistance to the toxin *eutypa* produced by *E. lata* (Mauro *et al.*, 1988; Tey-Rulf *et al.*, 1991; Phillippe *et al.*, 1992) or resistance to a possible toxin associated with Pierce's disease (Gray and Meredith, 1992). It helps in targeted improvement in important cultivars. To obtain disease tolerant variety a rapid *in vitro* propagation of 5 grapevine cultivars was carried out and microapical plantlets were obtained through microapical culture and heat treatments. Grapevine fan leaf nepovirus (GFLV) was detected by PAS-ELISA and GFLV coat protein (*CP*) gene cDNA probes, labeled by long arm photosensitive biotin. Microapical plantlets showed that the heat treatment of 38°C (night) was more effective to release GFLV, and long-term tissue culture of >20 generation resulted in releasing GFLV. GFLV-free plants of cvs. Teng Ren, Red Globe and 9307 were obtained for the first time (Zhang *et al.*, 1988).

MANGOES

Mangifera indica (mango) belongs to family Anacardiaceae. It is the most important tropical fruit. The plant is believed to have been cultivated for more than 4,000 years. It is originated in the Indo-Burma region. It is the third largest cultivated fruit in the world in terms of both area and production after bananas and citrus fruits. India is the largest producer of mango in the world with an annual output of about 10 million tonnes and a share of 60 per cent in the total world production. Apart from India, the other major producers of mango in the world are Brazil, Pakistan, Mexico, Indonesia, the Philippines, Haiti, China and Tanzania. Mango is also grown to a small extent in Egypt, madagascar, Zaire, Venezuela and Sri Lanka.

Mango is believed to be an allopolyploid, and is the result of an interspecific hybridization followed by chromosome doubling (Mukherjee, 1950). The chromosome number of mango is $2n = 4x = 40$. Conventional breeding in mango is very difficult because of the long generational cycle, i.e., 5-10 years, pollination problem (mostly self-incompatible), alternate bearing habit, etc. The mango improvement work by *in vitro* systems was mainly initiated in the early 1980s, and a significant progress in somatic embryogenesis, *in vitro* selection, transformations, imparting disease resistance, control of ripening, etc., were made in the last 20 years. The origin of mango lies in eastern India and Burma (Singh, 1960). Most of the monoembryonic. Indian mango cultivars are usually propagated asexually by grafting (approach, veneer, stone) or by budding (chip), while the seed-propagated cultivars that originated in the humid tropics of South-east Asia and the Philippines are polyembryonic; each monoembryonic mango seed contains a single zygotic embryo, while the polyembryonic mango seeds showed one or more embryo, while the polyembryonic mango seeds showed one or more embryos of which one is expected (not always) to be zygotic. Seedling populations of monoembryonic cultivars usually loose some valuable characters, while non-variability is normally expected in seed-propagated polyembryonic cultivars.

In India, more than 100 varieties of mango are cultivated, each differing in size, shape, colour, texture and taste. The prized varieties among them are Alphonso and Pairi of Maharashtra and Gujarat, the Dasheri and langra of Uttar Pradesh and Punjab, Malda of West Bengal, Romani of Tamil Nadu and the Banganapalli and Magoba of Andhra Pradesh. The other important varieties include Vathua, Banesham, Bangalore, Fajli, Jahangir, Totapuri, Neelum, Imampasand, Mondapa, Chausa, Sapheda and Rataul. The share of Dasheri in the total production of mangoes in India is about 8 per cent, followed by Banganapalli 7 per cent, Langra 4 per cent, Alphonse 2 per cent and Pairi 1.5 per cent.

In India, Andhra Pradesh, Bihar, Goa, Gujarat, Haryana, J&K, Karnataka, Kerala, Madhya Pradesh, Maharashtra, Orissia, Punjab, Rajasthan, Tamil Nadu, Tripura, Uttar Pradesh, West Bengal are the major States which produce mango. Apart from these, Arunachal Pradesh, Assam, Manipur, Mizoram, Nagaland, Sikkim, A&N Islands, Chandigarh, Daman & Diu, Delhi and Pondichrry also produce mango.

Mango is propagated through seed or by vegetative means through inarching, budding, grafting and gootee. The planting is usually carried out during monsoon. Some farmers plant trees of Shisham, jamun or casuarina around the periphery of their orchard to prevent the falling of fruits by strong winds. A liberal dose of manure and irrigation is required for vigorous growth of the plant. The plant starts bearing flowers and fruits in 5 to 7 years and reaches its peak production in about 17 years. The life span of the tree is about 50 years.

At present, mango comprises about 23 per cent of the total fresh fruit exports from India. Indian mango is fast emerging as important foreign exchange earners in the fruits group. alphonso, Neela, Totapuri, Chausa, etc are the varieties in great demand abroad. The major markets for fresh Indian mango are in the West Asian countries like United Arab Emirates, Saudi Arabia, Kuwait, Bahrain, Qatar and European countries like U.k., France, Holland

and Germany. India exports dried mango slices to UK, Hong Kong, Bangladesh, Kuwait and Canada; puree and paste to Saudi Arabia, erstwhile USSR, the Netherlands and Nigeria; mango juice to erstwhile USSR, UK, Yemen, Ethiopia, USA and UAE; mango slice in brine to UK, lraq, USA, erstwhile USSR and Germany; and mango squash to USA and Canada.

Mango tissue seems to be relatively amenable to regeneration in vitro. Considerable success has been achieved in somatic embryogenesis. The prospects for genetic manipulation in mango seems quite good. The *Agrobacterium*-based transfer system may be useful to extend storage life, resistance to pests and diseases. The techniques of in vitro conservation of mango germplasm should be a potential target. Following are the reports available in context to work done in the field of tissue culture and genetic engineering of mango.

Table 7.10: Tissue culture (Micropropagation) and genetic engineering researches conducted on mango

References	Work conducted
Somatic embryogenesis	
Litz *et al.* (1982)	Obtained somatic embryos from nucellar explants of polyembryonic mango cultivars Chino, Heart, Ono and Turpentine on modified MS medium.
Litz *et al.* (1984)	Demonstrated that more efficient recovery of somatic embryos occurred when the nucelli of polyembryonic mangoes were excised and cultured on medium with 1-2 mg/1 2,4-D.
(Litz (1987) and (Litz and Schaffer, 1987)	Induced embryogenic cultures of monoembryonic cvs. Lippens and Keitt.
DeWald *et al.* (1989a)	Demonstrated that the entire tissues can produce embryogenic cultures on a modified liquid B5 (Gamborg *et al.*, 1968) basal medium. They suggested subculture from liquid to solid medium when the somatic embryos reach to the heart stage of development.
DeWald *et al.* (1989b)	Successfully achieved multiple somatic embryos by sequential transfer to a medium containing 20% (v/v) coconut water and reduced sucrose concentration, maturation of embryos and germination were obtained.
Monsalud *et al.* (1995)	Indicated that hyperhydricity of immature somatic embryos has been a limiting factor for the development of highly embryogenic suspension cultures of many important mango cultivars. They studied the reversion of hyperhydricity in mango somatic embryos derived

References	Work conducted
	from cultured ovules and achieved.
(Pliego-Alfaro *et al.*, 1996a)	Studied inhibition of mango somatic embryos growth from ovule halves (collected 40-50 days after pollination). They induced in vitro by treatments for 4 or more weeks with abscisic acid (ABA) (0-100μM) with and without high osmolarity provided by mannitol (0-10%).
Litz *et al.* (1998)	Studied the efficacy of different genotypes on induction of embryogenic cultures.
Litz and Yurgalevitch (1997)	Compared Effects of aminoethoxyvinylglycine (AVG), 1-aminocyclopropane-1-carboxylic acid (ACC), dicyclohexylammonium sulfate (dicyclohexamine) (DCHA) and methylglyoxal bis (guanylhydrazone) (MGBG) on induction of embryogenic competence from nucellus of *Mangifera indica* cultivars Tutehau (TT; polyembryonic) and Tommy Atkins (TA; monoembryogenic).
Thomas (1999)	Achieved high frequency somatic embryogenesis from nucellar explants in the monoembryonic mango cv. Arka Anmol.
Organogenesis	
Rao *et al.* (1981)	Initiated induction of adventitious roots from callus initiated form cotyledons was obtained by using MS basal medium supplemented with Kin and NAA.
Raghuvashi and Srivastava (1995)	Observed that the exudation of phenolics from the cut-ends of mango explants. It hinders the their regenerative ability in any *in vitro* growth medium. Pre-treatment of explants using liquid shaken culture helped overcome this problem.
Somaclonal variations	
Litz *et al.* (1991) and Mathews and Litz (1992)	Demonstrated the potentiality of in vitro selection of mango showing enhanced resistance to *Colletotrichum gloesporioides* (it causes Anthracnose) which produce a phytotoxin, identified as aspergillo-marasmin A (lycomarasmic acid).
Jayasankar and Litz (1998)	Selected embryogenic nucellar cultures of two polyembryonic cultivars (Hindi and Carabao) for resistance to the culture filtrate and phytotoxin of a virulent strain of *Colletotrichum gloesporioides*. It was isolated from leaves.

References	Work conducted
Biochemical and molecular markers	
Gan *et al.* (1981)	First demonstrated the feasibility of using isozymes as biochemical markers in mango.
Degani *et al.* (1991)	Characterized 41 mango cultivars derived from self-pollinated and open-pollinated trees using isozymes. They identified 6 loci with 17 allelomorphs and determine the outcross origin of some mango cultivars.
Schnell and Knight (1992)	Successfully differentiate zygotic from nucellar seedlings using IDH, LAP, PGI, PG and TPI.
Degani *et al.* (1992)	Demonstrated that there are two distinct zones of PGI activity in mango (PGI1 and PG12) and suggested that four alleles control the PG12 banding.
Truscott *et al.* (1993)	Characterized 88 mango cultivars on the basic of isocitrate dehydrogenase, cytosol aminopeptidase, glucose 6-phosphate isomerase, phosphoglucomutase and triose-phosphate isomerase.
***In vitro* conservation**	
Wang and Fu Jiarui (1994); Pliego Alfaro *et al.* (1996b)	Indicated that the seeds of mango are recalcitrant; hence long term storage under ordinary condition is not possible. They observed that the excised embryonic axes of mango were more tolerant to desiccation than whole seeds. They explored the possibilities to induce developmental arrest in recalcitrant mango embryos with high concentration of ABA, mannitol or low temperature (7.5°C) treatment.

PAPAYA

Papaya, a native of Southern Mexico and Costa Rico. It grows all over India, year round. However, Andhra Pradesh, Karnataka and Tamil Nadu provide ideal climatic conditions for its cultivation. At one time, India was a leading producer of papaya in the world. Today, it is because in other countries its production has gone up dramatically while in India it has risen at a much smaller pace. The major producers of papaya in the world are Zaire, Uganda, Tazania, Kenya, Taiwan, the Philippines, India and Sri Lanka. The importers of papaya are USA Australia, Argentina, UK and other European countries.

Papaya is table fruit in most of the tropical countries and is consumed as a part of breakfast, dessert and as fruit salad. It is used for making soft drinks jams, ice-cream flavouing, crystallized fruit and also sold as canned cubes and juice.

The annual production of papaya in the country is estimated at about 3.8 lakh tonnes from the area under cultivation of 11,000 hectares. The varieties production in Karnataka, Tamil Nadu, Pusa (Bihar), and Madhya Pradesh are most popular.

In 1992-93, India exported 339,55 tonnes of papaya valued at Rs.37.98 lakh. Of this 121.44 valued at Rs.8.26 lakh was exported to United Arab Emirates and 81.18 tonnes valued at Rs.17.50 lakh to Japan.

Papain: Papain and carpain alkaloids obtained from latex of papaya are of great commercial importance. Papain is used in meat processing industry as a tenderizer, in breweries to remove cloudy appearance of bear, in malt food industry to remove enzyme and save malt consumption in woollen textile industry to reduce shrinkage and in leather industry for batting hides and skins. It is also used in the manufacture of chewing gums.

Research Priorities for Papaya

1. To avoid the musky flavour and odour in fruits which are not readily accepted by people unaccustomed to them
2. To develop dwarf lines with long peduncles which would reduce the risk of harvesting
3. To develop homozygous diploids from haploids obtained through anther culture which would be of greater benefit to papaya breeders.
4. To develop lines resistant or tolerant to anthracnose that might resist shocks in transit
5. To develop lines resistant or tolerant to virus and fungal diseases
6. To develop lines with increased latex content.

Table 7.11: Tissue Culture (Micropropagation) researches on Papaya

References	Work conducted
Axillary bud/shoot-tip culture	
Litz and Conover (1981) and Pandey *et al.* (1986)	Observed that explants from staminate plants respond well to *in vitro* establishment and proliferation when compared with their respective pistillate plants.
Purnima and Bisht (1988)	Observed a varietal differences in establishment and proliferation of papaya.
Miller and Drew (1990);	Indicated that explant type have direct effect on proliferation.
Reuveni and	Obtained a high rate of establishment in primary culture

References	Work conducted
Shlesinger (1990)	by using axillary bud explants collected from lateral shoots of rooted cuttings grown in greenhouse.
Mondal *et al.* (1990)	Found that explants taken from fruit bearing trees showed higher per cent of contamination while those from saplings lesser per cent of contamination.
(Saraswathi, 1996)	Cultured shoot-tips of the papaya line CO3 and Sun Rise Solo and observed good amount of callus on MS medium supplemented with 2.0 mg/L IBA and 0.5 mg/L B.
Litz and Conover (1978a)	Indicated that when explants from matured field-grown trees showed 95 per cent contamination in primary cultures.
Shlesinger *et al.* (1987)	Sterilized the axillary buds collected from three-year-old field-grown papayas using rifampicin and sodium hypochlorite solution.
Embryo Culture	
Yang (1986)	Successfully rescued the hybrid embryos resulting from the interspecific crosses. In 30 days.
Khuspe *et al.* (1980)	Successfully rescued the hybrid embryos resulting from the interspecific crosses. In 60 days
Rojkind *et al.*(1982	Successfully rescued the hybrid embryos resulting from the interspecific crosses. In 90 days
Saraswathi (1996) Magdalita *et al.* (1996).	Successfully rescused the 90 days old selfed embryos on MS (half-strength) medium.
Farinas *et al.* (1990) Jayalatha (1994) Shanthi (1996) Saraswathi, 1996).	Indicated beneficial effects of organic additives (coconut water or White's vitamin) to MS medium on germination of both mature and immature hybrid embryos have by various researchers
Ovule Culture	
Rojas and Kitto (1988 and 1991)	Successfully initiated callus in ovules *No reports are available so far with respect to recovery of homozygous diploid.*
Anther Culture	
Litz and Conover (1978)	Recovered low frequency of haploids by culturing anthers excised for 16-20 mm flower buds in liquid MS medium supplemented with 87.6 mM sucrose, 10 g/L activated charcoal, 8.8 μM BA and 2.7 μM NAA.

References	Work conducted
Protoplast Culture	
Chen and Chen (1992)	Isolated protoplasts from the immature hybrid embryos of cross *C. papaya* × *C. cauliflora* cultured in KM8PS medium for 2 weeks and then plated in the same medium with 1% agarose. It resulted in somatic embryos. Protoplast-derived somatic embryos attain maturity on MS medium.
Suspension Culture	
Jordan and Velozo (1996)	Successfully recovered somatic embryos through cell culture of axillary bud derived from one-month-old papaya calli.
Somatic Embryogenesis	
Chen *et al.* (1987) De Winnaar (1987)	Obtained induce callus from shoot-tip and stem explants from fro leaves, cotyledons and roots
Indirect Somatic Embryogenesis	
De Bruijne *et al.* (1974)	Demonstrated the technique of three-step procedure to induce somatic embryos in petiolar segments of papaya.
Yie and Liaw (1977)	Devised a two-step procedure to obtain somatic embryos from internodal segments of papaya seedlings by inducing callus on MS medium containing 5.4 μM NAA and 0.5 μM Kin followed by somatic embryo formation on MS medium with 0.0-0.25 μM IAA and 5.0-10.0 μM Kin.
Castillo *et al.* (1998).	Cultures embryogenic calli of papaya in a liquid phase maturation medium resulted in a significantly higher frequency of somatic embryogenesis and regeneration than parallel treatments on agar-solidified medium. An average of 455 somatic embryos/hypocotyl explant were produced in a liquid system after 112 days, compared with 174 somatic embryos/explant on solidified medium.
Yang and Ye (1992)	Adopted three-step procedure to induce somatic embryos in petiolar segments, using different growth regulators.
Detection of Pathogens of Resistance	
Guthrie *et al.* (1988)	Detected phytoplasma DNA using PCR with primers specific for phytoplasms in general and for the stolbur group of phytoplasmas
Gibb *et al.* (1988)	Studied the genetic relatedness of phytoplasmas associated with dieback (PDB), yellow crinkle (PYC) and mosaic (PM) diseases in papaya be restriction fragment length polymorphism (RFLP) analysis of the *16S rRNA* gene and indicated that PYC and PM photoplasmas. They

References	Work conducted
	were identical and most closely related to members of the faba bean phyllody strain cluster.
Genetic Transformation	
Pang and Sanford (1988	First to made successful transformed the papaya leaf-disc calli
Fitch *et al.* (1990)	Made first stable transformationusing microprojectile bombardment method. Embryogenic tissues following bombardment were cultured on MS (half-strength) medium
Ponce *et al.* (1995)	Produced a herbicide-resistant transgenic papaya plants using same particle bombardment method and phosphinothricin (*bar*) and kanamycin (*nptII*) resistant genes as selectable markers and the *gus* gene as a reporter gene,
Yang *et al.* (1996)	Produced transgenic papaya plants resistant to PRSV with the help of vectors like C58 strain of *Agrobacterium*
Ponce *et al.* (1996)	Reported the *Agrobacterium rhizogenes*-mediated transformation using somatic embryos as the initial explant. They used coat protein (*cp*) gene of a Taiwan strain of PRSV and efficiently transferred to papaya by *Agrobacterium tumefaciens*-mediated transformation
Yeh *et al.* (1998)	Developed about 38 transgenic lines and micropropagated by tissue culture
Cai *et al.* (1999)	Obtained papaya lines with transgenic resistance to PRSV, a reproducible and effective biolistic method.
***In vitro* Conservation**	
Drew (1992)	Developed improved technique for *in vitro* propagation and germplasm storage.
Suksa-Ard *et al.* (1997).	Conserve *C. papaya* shoots *in vitro* by lowering the temperature, applying growth retardants and increasing the osmotic potential of the culture medium

Table 7.12: For Hardening (mixtures used for establishment of in vitro plantlets)

Potting mixtures	Reference
Peat-perlite mixture	Litz and Conover (1978)
Soil-compost mixture	Litz and Conover (1980)
Soilless potting mixture	Litz and Conover (1982)
Sand-soil-FYM mixture	Rajeevan and Pandey (1983),
Peat-perlite-polystyrene mixture	Drew (1988

Potting mixtures	Reference
Soil-FYM-sphagnum mixture	Singh and Pandey (1988)
Sand-soil-vermiculite mixture	Saraswathi (1996)

GUAVA

Table 7.13: Tissue Culture (Micropropagation) work on Guava

References	Work conducted
Amin (1987) and Jaiswal and Amin (1987)	First time reported successful plantlet formation from somatic tissue of mature guava plant. The apical portion (5-7 cm) of shoots from new growth on mature branches and from basal sprouts of 15-years-old guava plant was cultured.
Amin and Jaiswal (1989)	Studied the effect of sucrose, agar and pH on the growth and proliferation of guava shoots with a view to optimize the cultural conditions for consistent high production of shoots using MS medium with 1 mg/1 BA). They also reported the best growth and multiplication rate of guava shoots at pH 5.0).
Fitchet-Purnell (1989)	Able to establish shoot-tip and nodal sections of the dv. Fan Retif in *in vitro*. The oxidative browning which caused the explants death was successfully prevented by placing the rinsed explants in petri dishes to dry in an airflow for 30-45 minutes, because the surface was moist when the explants were incubated. The stem and buds remained green and all buds subsequently developed. Broodrijk (1989) also reported the method for in vitro propagation of cv. Fan Retif.
Khattak *et al.* (1990	Used shoot-tips from cv. Safeda seedling and from plastic wrapped or unwrapped branches of 10-years-old trees. They were surface-sterilized with $HgCl_2$, $Ca(OC1)_2$, NaOC1 or ethanol at various concentrations for different durations. The best results were obtained by treating the seedling explants with 70% ethanol for 1 minute followed by 5% NaOC1 for 5 minutes and treatment of shoots from plastic-wrapped branches with 0.05% HgC12 for 5 minutes.
Prakash and Tiwari (1993 and 1996)	Reported in vitro clonal propagation of guava.
Siddiqui and Farooq (1997)	Used nodal explants of the glasshouse-grown (2 to 3-month-old) seedlings of guava plants for culture.

References	Work conducted
Embryo culture	
Ramirez-Villalobos and Salazar Yamarte (1988)	Cultured immature embryos taken from adult field-grown guava plants on MS medium supplemented with zeatin, ribozeatin, 2ip, BA or Kin. Zeatin and ribozeatin were each used at 0.1, 0.5 or 1 mg/1 while 2iP, BA and Kin were used at 1,5 or 10 mg/1. High rates of callus formation were obtained in the ribozeatin, zeatin and control treatments, as well as in the 1 mg/1 2iP treatment.

LITCHI

Litchi, first originated in China some 3000 years ago. was introduced in India in the seventeenth century. It belongs to the family Sapindaceae and subfamily Nepheleae. It is grown in temperate regions of Australia, South Africa, Thailand, Mauritius and Hong Kong. At present, India is the second largest producer of litchi in the world, after China. The genus Litchi has three species, *i.e., chinensis* (the commercial litchi), *philippinensis* (developed in the Philippines and Papua New Guinea at high altitude) and *avenesis* from the Malay Peninsula and Indonesia (Menzel, 1991). The other members of horticultural importance of the family are the longan (*Dimocarpus longan*), rambutan (*Nephelium lappaceum*), pulasan (*N mutabile*) and Spanish lime. Although about 100 varieties of litchi are identified the world over, only 15 of them are reported to be commercially cultivated in India. The more popular varieties of litchis grown in India are *Shahi*, *China* and *Rose Scented* that all have good domestic demand and export market provided they are properly handled and packed for export.

The chromosome number of L. chinensis is $2n = 2x = 30$ (Smith, 1976; Lu *et al.*, 1987) and that of rambutan and longan are $2n = 2x = 22$ and $2n = 2x = 20$, respectively.

The trees are evergreen, recurrent flushing, round topped, medium to large, reaching up to 10 metres in height. The compound leaves are arranged alternately consisting of 4-7 oblong leaflets. The flowers are small, apetalous, male, hermaphrodite and pseudohermaphrodite (functional male) types. Heterostyly is common in litchi. The edible part (aril) is an outgrowth of the outer integument of the seed coat.

(Ray (1999) studied the cultivars with varying period of maturity, viz., Bai-Tang-Ying (early) and Liquili (late) and indicated that they may provide useful gene for breeding work. Means to inhibit the PPO activity may provide a solution to the pericarp browning problem and extend the shelf-life. The potentiality of hybrids of *N. mutabile* and other wild species with rambutan as a rootstock may be explored, which can be stronger resistant to root disease than rambutan.

There are several constrains in commercial litchi cultivation, they are: (1) irregular bearing habit, (2) lack of suitable variety, (3) poor fruit set and severe fruit drop, (4) short period of a availability and shelf-life and pericarp browning, (5) sun-burning and skin-cracking and, (6) trees are exacting in soil and climatic requirements. To overcome these constrains, researchers are looking forward to use various biotechnological approaches to improve the litchi varieties. These reports are as follows:

Table 7.14: Tissue culture (Micropropagation) and genetic engineering researches conducted on litchi

References	Work conducted
Yu (1991)	Cultured shoot segments of cv. Liudong 1 in half-strength MS medium with various concentrations of benzyladenine (BA), gibberellic acid (GA_3) and IAA and cultured at 25-30°C and 2000 lux for 10 h daily.
Murray and Murray (1994)	Indicated that *Litchi chinensis* has great genetic diversity, with severe graft incompatibilities.
Cello and Espanto (1997)	Reported that when fruitlets of litchi (0.6 – 2.2 cm long) were cultured in modified B5 + 1 mg Kin + 400 mg glutamine + 6 g sucrose + 2g gelgro, the nucellus produced calli and these were transformed into somatic embryos.
Das *et al.* (1999)	Multiple shoot induction in *L. chinensis* has been achieved.
Embryo culture	
Kantharajah *et al.* (1992)	Litchi embryos (3 nm) could be cultured on MS medium with 2% sucrose supplemented with 150 ml coconut water/1 and subsequently grown into plants. Embryos were treated to induce adventitious buds from adventitious shoots as a means of achieving multiplication.
Protoplast culture	
Yu *et al.* (1996)	Friable and granular embryogenic calli, initiated from young litchi cv. Xiaofanzhi embryos. They were subcultured on STS (silver thiosulphate) containing medium for more than 2 years. Protoplasts were obtained from embryogenic cell suspensions with a mixed enzyme solution containing CPW salts, 11% mannitol, 0.8% cellulase R-10, and 0.4% macerozyme R-10.

References	Work conducted
Biochemical markers	
Cilliers and Visser (1995)	Studied isoenzyme banding patterns produced by 19 litchi cultivars and 6 vegetative selections revealed sufficient differences to distinguish between most cultivars.
Jankowski and Chojnacki (1995)	Indicated the presence of $C_{60-, 70-}$ polyphenols n leaves is a chemotaxonomic marker for the majority of members of Sapindaceae.

AVOCADO

Persea (Avocado) was originated from African-Laurasian and its subgenus *Eriodaphne* was originated from Africa. Avocado (*Persea americanan*) fruit is exclusively used freshly in Europe and the United States as a luxury item, but in Central America it is a staple fruit. Avocado seeds are genetically different from one another and highly variable, and hence each exerts a different influence on the variety budded to it (Storey, 1955). Avocado like other subtropical woody species has a high rate of fruit abscission. It is propagated from rootstocks using desirable nursery practice. There have been many reports of experimental procedures to induce rooting, but few of these are widely practiced for large-scale commercial propagation. The most important diseases of avocado are caused by the root-fungus *Phytophthora cinnamoni* and the avocado sunblotch viroid (ASV). These pathogen are of specific importance because they are very widespread, and chemical control measures are either unavailable for ASV or extremely expensive for *P. cinnamoni*. In addition, ASV restricts the international exchange of germplasm, e.g., budwood imported into Australia is routinely indexed for ASV, and all positive infections are destroyed.

Tissue culture

Micropropagation

Shoot cultures is used for an efficient proliferation of clonal varieties of avocado and other *Persea* species (Barringer *et al.*, 1996). Shoot proliferation could be easily induced in vitro from shoot-tips and single node cuttings of *P. indica* seedlings. Shoot regeneration could not be achieved (Nel *et al.*, 1983). Cooper (1987), Gonzales-Rosas *et al.* (1985), Harty (1985) clonally micropropgated *P. schiedena* and *P. americana* cv. Duke 7 as rootstocks. They were resistant to the root-rot disease. Zirai and Lionakis (1994) studied the effect of cultivar, explant type, etiolation pre-treatment and age of the plant material on the *in vitro* regeneration ability of avocado.

The proliferation and survival of avocado nodal cultures of juvenile origin were affected by the form and concentration of nitrogen. Optimum growth was achieved on modified MS medium containing 67% KNO_3 and 33% NH_4NO_3 with total N of 40 nM supplemented with 100 mg/l myo-inositol, 1 mg/l thiamine, HC1, 30 mg/l sucrose and 4.44 mM with a 16-h photoperiod .(120-150 mmol m^{-2} s^{-1}). Witjaksono *et al.* (1999) obtained good shoot growth and greater biomass accumulation in a CO_2 enriched environment than under ambient CO_2 conditions.

Involvement of Mycorrhiza on Regenerated Plantlet

Micropropagated plantlets of avocado exhibit a very slow rate of growth during the acclimatization phase. This may be due to absence of mycorrhiza. When the plantlets were inoculated with the vesicular-arbuscular mycorrhizal (fungus of *Glomus fasciculatum*), an improvement in the formation of well developed root system, converted into a mycorrizal system was observed. Introduction of the mycorrhizal fungus at the time when plantlets were transferred from axenic conditions to *ex vitro* conditions following changes were observed (Vidal *et al.*, 1992):

1. Improved shoot and root growth
2. Enhanced the shoot-root ratio
3. Increased the concentration and/or content of N, P and K in plant tissues and helped to tolerate environmental stress at transplanting.

The above findings show that mycorrhiza formation seems to be the key factor for subsequent growth and development of micropropagated of avocado.

Production of Virus-free Plantlet

The rapid *in vitro* propagation is used for elimination of viroid-like organisms causing sunblotch disease. Earlier, a usual non-tissue culture technique for virus elimination from vegetatively propagated woody species. It is high temperature treatment (thermotherapy) of plants in pots,. This method is not successful with material infected avocado sunblotch viroid or the combination of avocado viroid, AV2, AV3 and AV4 (Nel and Kotze, 1984). With reference to *in vitro* methods for virus elimination in avocado, there is little published work on the use of meristem tip culture beyond a preliminary report by Hendry and VanStaden (1982). Skene and Barlass (1988) used an alternative method for virus elimination. It is shoot-tip grafting. It has been extensively studied, however, till date no significant progress could be made.

Embryo Culture

Avocado has a high rate of fruit abscission. It is further exaggerated on small potted trees grown in a glasshouse. Skene and Barlass (1983) rescued immature embryos older than 6 weeks at abscission by culturing them in a liquid half-strength MS basal medium supplemented with 0.5 mg/l benzyladenine. The embryonic tissue explants can develop large number of multiple adventitious shoots in MS medium containing 30 g sucrose supplemented with 1.0 mg BA/1. Ahmed *et al.* (1998) could obtained an average of 10 adventitious multiple shoots from a single piece of embryonic tissue in a shoot multiplication medium. WPM medium supplemented with 1.0 mg/l IBA and 10-4 M putrescine plus 0.2% activated charcoal was optimal for root development

Protoplast Culture

Blickle *et al.* (1986) successfully isolated protoplasts from callus derived from stem explants. The callus cells were plasmolyzed with 0.7 M mannitol and 0.4% PEG (6000). Plasmolyzed cells were then treated with 1% Onozuka cellulase RS and 0.1% pectolyase Y2; which yielded 2-3 $\times$ 10^6 protoplasts/g of callus.

Somatic Embryogenesis

Callus can be readily initiated on virtually all organs of avocado placed into on *in vitro* culture (Schroeder, 1973). The ability of pericarp tissue from mature fruit to proliferate in vitro was first demonstrated in avocado. Regeneration of roots in callus derived from avocado pericarp also appears to be the first indication of partial totipotency in cells from organs other than those of vegetative origin (Schroeder, 1962). Twenty years later, embryoids were observed on callus derived from immature embryos (Pliego-Alfaro, 1982; Skene and Barlass, 1983), but in each case continued development was limited. Callus production from avocado and morphogenesis were carried out by using MS basal medium, supplemented with sucrose, vitamins and combinations of cytokinins and auxins (Barringer *et al.*, 1996). WPM (woody plant medium), Anderson mineral salts and vitamins and Dixon and Fuller medium have also been used for *in vitro* callusing and micropagation from *Persea*. Embryogenic avocado callus developed from immature zygotic embryos on MS medium containing the auxin, picloram (Pliego-Alfaro and Murashige, 1988). However, the ability of in vitro regeneration of many woody species including avocado, is restricted and seems to be limited to juvenile plant material. A procedure of recurrent somatic embryogenesis from cell suspension cultures of avocado was used for genetic transformation (Curz Hernandez *et al.*, 1988).

DATE PALM

Date palm (*Phoenix dactylifera* L.) is the oldest amongst the cultivated fruit trees. Date palm cultivation is believed to be established as early as 4000 B.C. in Mesopotomia (South Iraq). In Egypt, it was under cultivation since the pre- historic days. History of date culture is very ancient in India. It is believed that date palm was introduced by the soldiers of Alexander in the 4th century B.C. in the Indus Valley. Different types of palms including wild date palm (*Phoenix sylvestris* (L.) Roxb). have been seen in sculptures from Sanchi. The Arab gardeners working in the palaces of the former rulers of Kutch might have brought date seeds and offshoots from Arab countries. In early sixteenth century, Babar mentioned in his memoir about the date palm cultivation in India. The date palm cultivation by planting regular orchard in Rajasthan was first started by the then ruler of erstwhile Bikaner State, Maharaja Ganga Singh. He planted suckers of Halawy, Khadrawy and Zahidi at Sriganga-nagar.

Date palm fruits are eaten as raw dates (fresh fruits), dry dates (*chuhhara*) and soft dates (*pind khajoor*).These are used in religious ceremonies and are considered auspicious. Different products such as sugar, starch, vinegar, juice, toffees, wine, chutney, jam, pickles etc. are prepared from date fruits. The date palm fruits have been recognised for their highly nutritive, well mineralized, highly flavoured and high calorific food value. It is a good source of iron, potassium, phos-phorus and calcium.

Date palm is very specific to climatic conditions required for its successful cultivation. As per an Arabic saying *date palm should be grown with its feet in running water and its head in the fire*. It requires prolonged hot dry summers, moderate winter and almost rain free period at the time of fruit ripening (July- August). Its ability to produce fruits abundantly at extremely low humidity is possible only if the supply of ground water/canal irrigation is sufficient. The mean temperature required at the time of flowering and fruit ripening is 25°C and 40°C, respectively. The pattern of maximum temperatures at the time of fruit maturity determines the rate of maturity and fruit quality. Western Rajasthan, comprising Jaisalmer, Barmer, Bikaner and Jodhpur, districts is ideally suited for date palm cultivation and has vast potential.

Table 7.15: Date palm germplasm maintained and established at Bikaner

Name of the variety	Source of Receipt
Abdul Rahman	Abohar, Punjab
Agolani	Egypt
Amri	Egypt
Barhee	USA
Barshi	Oman
Bikaner	Abohar, Punjab
Bint-Aisha	Abohar, Punjab
Gizaz	Abohar, Punjab
Halawy	USA
Hamra	Abohar, Punjab
Hatemi	Saudi Arabia
Hayani	Abohar, Punjab
Khadrawy	Iraq
Khalas	USA
Khashab	Oman & Saudi Arabia
Khuneizi	Oman
Medini	Abohar, Punjab
Medjool	USA
Migraf	Abohar, Punjab
Muscat	Abohar, Punjab
Nagal	Oman
Nagal Hilali	Abohar, Punjab
Ruziz	Saudi Arabia
Sakloti	Egypt
Sayar	USA
Sedami	Abohar, Punjab
Sewi	Abohar, Punjab
Shamran	Abohar, Punjab
Sriganganagar	Abohar, Punjab
Suria	Abohar, Punjab
Tayar	Saudi Arabia
Umshok	Abohar, Punjab
Zagloul	Abohar, Punjab
Zahidi	USA

Tissue Culture

The degree of integration of biotechnology into the system of improvement of date palm has been reviewed by el-Hadrami *et al.* (1998). Aaouine (1998) suggested that biotechnology has played a key role in safeguarding the Moroccan date palm industry. Tissue culture has been used for large-scale propagation of endangered, sensitive cultivar selections, combining disease tolerance and fruit quality. Genetic transformation to develop resistant materials for Bayoud disease (*Fusarium oxysporum* f. sp. *albedinis*) would be an important approach for the future.

Table 7.16: Tissue culture work done on Date palm

References	Work conducted
Micropropagation	
Oppenheimer and Reuveni (1972) Reuveni *et al* (1972) Reuveni and Lilyen-Kipnis (1974)	Successfully propagated using tissue culture methods. Efforts were made to produce high yielding disease resistant clones for large number of plantations in Israel.
Tisserat (1979, 1984) Bouguedoura *et al.* (1990) Raj Bhansali and Kaul (1991)	Reported refinement in cultural conditions, type of media, addition to charcoal and antioxidizing chemicals, type of media compositions have improved the methods for mass propagation of date palm through tissue culture. Induction of vegetative bud development from shoot-tip or axillary bud explants in culture.
Khan *et al.* (1983)	Reported regeneration of plantlets by reculturing explants consisting of shoot apex, primordial leaves and axillary buds on RM-1965 medium containing NAA (1mg/l) and cytokinin (0.1 mg/l).
Loutfi and Chlyah (1998).	Histological analysis of cultured explants revealed that buds originated from petal primordia.
Suspension Culture	
Sharma *et al.* (1986) Bhaskaran and Smith (1992).	Tried to establish suspension culture in date palm friable callus for rapid embryogenesis
Somatic embryogenesis	
Ammer and Benbadis (1977) Reynold and Murashige (1979)	Obtained asexual embryogenesis from excised zygotic embryos
Tisserat et al. (1979 Mater (1986)	Obtained asexual embryogenesis from somatic tissue

References	Work conducted
Sharma et al. (1984, 1986) Raj Bhansali et al. (1990) Dass et al., (1989) Raj Bhansali and Kaul (1991) Bhaskaran and Smith (1992) Sudhersan et al. (1993)	
Zaid (1987) Raj Bhansali et al. (1988).	*During in vitro* culture,date palm explants release excessive browning substances. It occure at the time of establishment of embryogenic callus
Maier and Metzlier (1965) Zaid (1987) Raj Bhansali and Singh (1982)	Indicated that the Browning of explant tissue and tissue culture medium is assumed to be due to the polyphenols and formation of quinines which are highly reactive and toxic to the tissue. They further indicated that the inhibitory effects may result from the bonding of phenols with proteins and their subsequent oxidation to the quinones.
Raj Bhansali *et al.* (1988	Successfully established Tissue culture-raised plants the field. The repetitive somatic embryogenesis (RSE) process developed is highly efficient in raising date palm tissue culture in certain varieties
Omar (1988)	Investigated the response of explants of *Phoenix dactylifera* seedlings, offshoots and adult trees. He used medium containing various combinations of growth regulators.
Saker et al. (1998).	Studied various morphogenic responses (callus, root and shoot formation) initiated from immature embryo explants.

BER

Jujubes have been in cultivation since ancient times both in India and China. Indian archeological and literary records suggest fruits of ber during 1500 BC to 300 AD. The Indian scriptures, *Katyayana Grhya Sutra* (XV. 10 ,11) and *Astadhyayi* of *Panini* (V.2.24) in *Sutra* period (800-300 BC) ,mention thee varieties of ber, i.e., *Badara, Kuvala and Karkandu.* Another variety called *Sauvira* is mentioned in *Charaka Samhita* (Mehra, 1967). There are at least 200 varieties of ber in cultivation in India. At least 400 varieties of

Chinese jujube have been in cultivation for the least 400 years in China (Hayes, 1945). The *ber* belongs to the genus *Ziziphus* of the family Rhamnaceace having more than 600 species (Bailey, 1947). Of these, only a few, such as Z. *jujube* Lam., Z *sativa Gaetn.* , and Z. *lotus* Lam., yield edible fruits. Cultivated varieties referred as Z. *jujube* Mill. , are supposed to be of Chinese origin while Z. *vulgaris* Lam. represents those of Indian origin (Watt, 1908) .Chatterjee and Randhawa (1952)

The perpetuation of the heterozygosity has been maintained since *ber* is highly cross pollinated. Owing to cross pollination among the existing heterozygous population, hybridity is a rule in genus *Ziziphus*. Such hybrid seeds grow in nature and throw segregating population from which some desire types develop as new cultivars. Ber cultivars, in general, are polyploids. Segmental allopolyploidy also plays an important role in creating variability.

Tree growth habit ranges from spreading and vine-like as in the cultivars of Z. *mauritiana* Lam. to erect as in those of Z. *jujuba* Mill. (Chatterjee and Randhawa, 1952) and in Z. *mauritina* var. *rotundifolia* Lam. (Indian ber), variations in tree growth habit such as semi-tall, semi-erect (Vashishtha and Pareek, 1989) have been described. The cultivars depict wide variability in leaf shape, size, length, form of base and apex, pubscence on undersurface and in petiole surface (Singh et al., 1971; Vashishtha and Pareek, 1989), perhaps owing to such differences in the species from which these cultivars have evolved.

The important varieties of ber are: *Khatti, Gobindgarh Selection, Kaithali, Lam, Nimaj, Tikadi, Sanaur – 1, SUA, Thornless, Narikeli, Thornless, Narikeli, Narma, Chhuhara, Ilayachi, Banarasi pebandi, Dandan, Bahadurgarh, Mundia, Jogia, Pathan, Chonchal, JS II, Triloki, Banarasi Kadaka, Umran, Katha, Chencho, Madhuri, Deedwana, ZG – 3, Jullundhari, Vikas-2, Sanaur-2, Kala Gola, Noki, Maharwali, Wilayati, Mirchia, Pebandi Alwar, Gola, Gola Gudgason, Popular Gola, Kakrola Gola, Gola Gudgaon-1, Ponda, Akrota, Gola Gudgaon, Kheera, Kali, Katha Phal, Aligani, Laddu, Jhajjar Selection, Safed Rohtak, Meharun, Bagwadi, Rewa Selection, Sanaur-3.*

It is a hardy cultivated fruit tree in North Indian plains. It is distributed all over the warm arid and semi-arid regions due to the drought and heat tolerant abilities. It is now becoming more and more popular due to low price and having high nutritional value. It is rich in Vitamin C, A and B complex (Chandra *et al.*, 1994). Ber leaves are rich source of proteins and mineral matter, which provide a nutritious fodder for the animals. The major ber growing states are Haryana, Punjab, Uttar Pradesh, Rajasthan, Gujarat, Madhya Pradesh, Bihar, Maharastra, Andhra Pradesh and Tamil Nadu.

Seedling ber trees are found extensively growing widely in arid and semi-arid areas (Pareek, 1997).

Table 7.17: Tissue culture (Micropropagation) and Genetic Engineering work for ber improvement

References	Work conducted
Goyal and Arya (1985) Raj Bhansali (1998) Rathore et al. (1992) Kabir et al. (1994) Mathur et al. (1995) David and Vasikar, (1980)	Developed athe protocols for clonal propagation of *Z. nummularia* and *Z. mauritiana* and *Z. xylophyra* through tissue culture. They used explants from nodal segments, shoot-tip, seedling and root suckers for clonal propagation.
Somatic embryogenesis	
Raj Bhansali, (1989)	Induced asexual embryogenesis in *Z. mauritiana* in seedling explant on 2.5 mg/l 2,4-D, 0.5 mg/l BAP and 0.25 mg/l Kin on incubation in dark condition for initially 2-3 subcultures at 26°C.
Sung and Song (1992)	Reported the induction of somatic embryogenesis in Chinese jujube (*Zyzyphus jujuba*) from immature cotyledons and seedlings parts on modified MS medium.
Mitrofanova *et al.* (1997)	Induced somatic embryogenesis and developed complete plant regeneration in *Z. jujuba* (Chinese date) on half-strength MS containing 1 mg/l 1,2-D through direct somatic embryogenesis.

BAEL

Aegle marmelos (bael) is an important fruit crop of Indian subcontinent and has potentiality to be grown in tropical regions belongs to family Rutaceae. It is a hardy fruit tree and grows in arid climate. The tree is deciduous, medium sized with trifoliate leaves (divided into three leaflets). The plant can be grow in poor soil (acidic, alkaline or stony soils having pH range of 5 to 10). There are no standard bael cultivars, however, they are known by name of place such as *Kaghji Gonda, Kaghji Etawah, Deoria large, Mirzapuri, Rampuri, Basti No. 1*, etc. The plant is propagated through budding (T-budding, patch budding, chip budding), grafting and air-layering). Commonly the plant are developed through seed germination. The seeds are sown in the month of June. The seedlings become ready for transplantation after an year. The seedling trees are not true-to-type. Plants can also be propagated by means of is most popular for raising saplings for plantation in new orchards. The tree mythological and medicinal value. It is used for stomach aliments, diarrhoea and dysentery. The ripen fruits are used as tonic, laxative and good for heart and brain. The fruit contains riboflavin, carotene, niacin and vitamin C.

Table 7.18: Tissue culture (micropropagation) work on bael

Work	Reference
Regeneration from seedling from seed explant tissue such as hypocotyls	Arya *et al.*, 1981
Regeneration from nucellus	Hossain *et al.* (1993)
Regeneration from cotyledons	Hossain *et al.* (1994
Regeneration from leaf	Islam *et al.* (1992)
Regeneration from zygotic embryos	Islam *et al.* (1994, 1996a)
High frequency plant regeneration from radicle tissue	Arumugam and Rao (1996)
Multiplication of shoots by using microshoots	Hazarike *et al.* (1996)
Stem explants produced meristemoids and multiple shoots on MS medium containing BAP, Kin and NAA	Varghese *et al.* (1993)
Biochemical changes, particularly the composition of membrane lipids in relation to differentiation in callus cultures	Bhardwaj *et al.* (1995)
Somatic Embryogenesis	
Conducted studies on asexual embryogenesis and regeneration of plantlets from seeds obtained from unripe fruits of 20-year-old tree.	Islam *et al.* (1996b)

FIG

Ficus caria (edible fig) belongs to the family Moraceae and botanically it is known as syconium (multiple fruit). The fruits are rich in vitamin A and C and minerals. They are consumed as fresh, dried or after processing. The plant (*Ficus caria*) a chromosome number of 2n=26, while *F. elastica* cv.

Decora is triploid in nature. Most of the present day cultivars grown commercially have been selected either from wild seedling trees or from chance seedlings and are being maintained by clonal propagation. *Ficus caria* is probably native of the southern part of Arabian peninsula, Italy, the Balkan peninsula and the USSR.

A very little biotechnological research specially, on genetic transformation has been conducted, however, some of the reports are available on tissue culture, in spite of its economic importance. The variability available in this crop with wide distribution in many parts of the world could be well documented using molecular markers to study the phylogenetic relationships.

Table 7.19: Tissue culture (micropropagation) and genetic engineering researches on *Ficus*

Reference	Work conducted
Micropropagation	
Pontiis and Melas (1986)	Able to produce 1 to 2 new shoots from shoot-tips of cv. Kalamon within 4 weeks on MS medium supplemented with 0.5 mM phlorogucinol (PG) without BA. After 12 weeks, *in vitro* cultures they produced 12 to 20 shoots having 2 to 4 cm length. No shoot proliferation was observed in cultures without PG.
Pontikis and Melas (1986)	Shoot-tip for establishment of *in vitro* micropropagation
Bardosa *et al.* (1992)	Meristem explant for establishment of *in vitro* micropropagation
Barbosa *et al.* (1992)	Induced shoot elongation and proliferation and plantlet development on MS medium supplemented with thiamine, nicotinic acid, myoinositol and growth regulators at pH 6.3. To avoid browning of the tissue and medium, addition of antioxidants like activated charcoal was suggested.
Hu and Guo (1994)	Studied factors affecting Four factors adventitious shoot formation on explants, viz., cultivar source of explant, composition of culture media and pre-treatment of explant.
Nobre and Ramano (1998)	Cultured apical shoot-tips of the fig cultivars Berbera and Lampa on Muriithi or Jonard culture medium supplemented with the antioxidants polyvinylpyrrolidone (PVP-40) (0.025, 0.05 or 1.0% v/v) or ascorbic acid (56.8 or 11.3 mM). Growth and development were best on Muriithi medium supplemented with 0.05% PVP. Single node explants of shoots obtained in the establishment phase were cultured on basal MS medium supplemented with 3% sucrose and BA (0.7, 1.1 or 2.2 mM) alone or in combination with NAA (1.0

Reference	Work conducted
	μmM). After 3 weeks, the highest multiplication rate (5.3 shoots/culture) was obtained in the medium supplemented with 2.2 mM BA without NAA
Demiralay *et al.* (1998)	Cultured Meristems of fig cv. Bursa Siyathi taken in spring or autumn on MS supplemented with 0.5 mg/l BA 0.1 mg/l IBA and 0.1 mg/l GA with or without 89 mg/l phloroglucinol (PG) or 2 g/l active charcoal.
Kumar *et al.* (1998)	Developed a procedure for multiple shoot induction and plantlet regeneration with apical buds collected from 7 – to 8-year-old trees of *F. carica* using MS medium supplemented with 2.0 mg/l BA. An average multiplication rate of four per subculture was established with 90% success
Virus elimination through tissue culture	
Muriithi *et al.* (1982)	Indicated the possibility of eliminating mosaic virus in fig by meristem tip culture.
Haelterman and Docampo (1994)	Standardized a technique for *in vitro* propagation of mosaic-free fig cultivars using thermotherapy and shoot-tip cultures.
Biochemical and molecular markers and Characterization of germplasm	
Elisiario *et al.* (1988)	Used isoenzyme and RAPD markers to distinguish among 55 traditional varieties of *F. carica. They identified* six isoenzyme systems were revealed after starch gel electorphoresis of leaf extracts Glucose-6-phosphate isomerase (PGI), phosphoglucomutase (PGM), isocitrate dehydrogenase (IDH), malate dehydrogenase (MDH), aspartate aminotransferase (GOT) and cytosol aminopeptidase (LAP).
Khadari *et al.* (1995)	Identified 21 fig accessions (*Ficus carica*) representing different varieties was performed by using the RAPD technique.
Chessa *et al.* (1998)	Characterization of fig isoenzyme using electrophoresis. They analyzed seventeen enzyme systems. Acid phosphate (AcPH), diaphorase (DIA), fumarase (FUM), glutamate oxalacetate transaminase (GOT), malate dehydrogenase (MDH), peroxidase (PRX) and phosphoglucoisomerase (glucose 6-phosphate isomerase) (PGI) have shown high resolution of the bands and reproducibility of the analysis.

CARAMBOLA

Averrhoa carambola or carambola (family Oxalidaceae) believed to have originated in South Asia in Indonesia. Chandler (1958) reported that India and Moluccas might have been the original home of this genus. It s an evergreen fruit tree where it has been reported to occur in the wild state. The plants are grown primarily as a dooryard tree throughout the tropics and the subtropics.

Fruits from most seedling trees are sour and acidic. They contain oxalic acid, vitamin A, B and C. Such fruits are unsuitable for fresh consumption. They are used for preservation of beverages and for cleaning brass and other tarnished metals. The fruits of sweet and subacidic section are consumed as fresh fruits. The plants are usually seed-propagated. For large scale planting, they are propagated vegetatively by veneer grafting, chip budding and air-layering is a less reliable propagation method. The fruits are economically important, therefore, researchers have tried to use biotechnological tools for their improvement.

Table 7.20: Researches on *Averrhoa carambola* using tissue culture techniques

Reference	Work conducted
Litz and Conover (1980)	Reported regeneration of shoots from callus derived from young seedling leaf explants.
Litz and Conover (1980)	Young green leaves life Arkin carambola were used as explants after surface-sterilization in a 10% solution of commercial bleach with two to three drops of Tween-20 for 10-12 min, followed by three to four rinses with sterile distilled water. The callus developed from leaf explants of Arkin carambola on MS media containing a range of 2,4-D and 2iP together with 30 g/l sucrose.
Rao *et al.* (1982)	Demonstrated the differentiation of both roots and shoots from callus derived from cotyledon explants of carambola. It was observed that organogenesis could only occur if the cotyledons were cultured together with part of the embryo axis. It was not mentioned whether organogenesis did, in fact, occur either from the cotyledons or from the embryo axis.
Griffis and Litz (1993)	Callus from the youngest fully expanded leaflets of grafted carambola cv. Arkin plants was initiated on MS medium supplemented with 1 mg 1-cysteine, 30 g sucrose, 3.5 g agar, 1 g gelrite, 0.5 mg 2,4-D and 0.2 mg IAA/l and various concentrations of BA and/or 2iP. The shoots developed directly from callus initiated on medium supplemented with 5 g

Reference	Work conducted
	2, 4-D and 5 mg 2iP/l and transferred to a medium supplemented with 5 mg/l 2iP alone. However, neither rooting *in vitro* nor rooting of microcutting in the greenhouse was successful
Amin and Razzaque (1993)	Hypocotyl and cotyledon explants of aseptically grown seedling produced callus and subsequently adventitious shoots on MS medium with half-strength major salts, and 0.1-0.2 mg/l of NAA, along with 0.5-1.0 mg/l of BA.

LONGAN

Euphoria longan commonly know as longan (family Sapindaceae) is more closely related to litchi than any other species of the family, however, it differs morphologically. It is indigenous to subtropical China or possibly in the region between Myanmar and India. The chromosome number of longan is 2n = 2 x = 20. Ripe fruits may be are used freshly, dried or quick frozen. The canned longans are also being used.

Research work, specially biotechnological approache, is still needed to improve the longan for commercial purposes. Tissue culture techniques involving shoot-tip culture, leaf culture, and embryo culture have been attempted with some degree of success. Chen *et al.* (1997) indicated that the survival rates of *in vitro*-cultured plantlets are low. This may be due to microbial contamination and browning. Lai *et al.* (1997) suggested to use embryo culture techniques to breed large-fruit varieties and aborted seeds Use of micrografting techniques should be used to study the graft incompatibility problems in longan. There is no report on the application of biochemical and molecular markers.

Table 7.21: Tissue culture (Micropropagation) and genetic engineering researches conducted on longan

References	Work conducted
Litz, (1988)	Indicated that embryogenic callus could be induced from leaflets of new vegetative flushes of mature longan tree on growth medium (Gamborg's B5 with 400 mg/l glutamine, 60 g/l sucrose, 200 mg/l casein hydrolysate and 1.7 g/l gelrite) containing 0.25-2.0 mg/l of Kin and 2, 4-D in every combination. Somatic embryos developed to maturity on B5 medium containing glutamine, casein hydrolysate, 20 g/l sucrose, 10% (v/v) coconut water.and 1.7 g/l gelrite. Callus initiation and somatic embryos development occurred in darkness at 25°C. Mature somatic embryos germinated in the

	light.
Chen *et al.* (1999)	Developed the procedures micropropagation of longan using lateral and terminal shoots from seedlings and 6-year-old grafted trees.
Embryo culture	
Yang and Chen (1987)	Indicated that the wrinkled seed failed to germinate, but *in vitro* culture of embryos was successful. Large embryos (> 3.0 mm) produced plantlets in 20 days of culture, while embryoid usually formed from small embryos in 3-5 months.
Fu *et al.* (1990)	Indicated that about 80% of the dehydrated excised axes from longan seeds (moisture content of about 10%) could retain their viability when thawed at 10-30°C after storage in liquid N for 12 hours.

PASSIONFRUIT

Genus *Passiflora* (passionfruit) belongs to the family Passifloraceae. Two species (*Passiflora edulis* f. *flavicarpa* or yellow passionfruit and *P. edulis* or purple passionfruit) are mainly grown commercially. The plant is bisexual, colourful, fragrant flowers are born solitary at leaf axils of new growth. The fruit is round or oval and greenish yellow or purple in colour when ripe. Smooth seeds are black and enclosed in a yellow orange aromatic juicy pulp. Various species are probably native to South American rainforests in the Amazon region of Brazil and in Paraguay and northern Argentina. The tropical American species (*P. quadrangularis, P. ligularis, P. laurifolia, P. maliformis, P. mollissima, P. antiquiensis, P. caerulea)* are grown as fruiting plants or used as sources of cold or disease tolerance in breeding programmes. Menzel *et al.* (1990) listed a few minor tropical species (*P. alata, P. coccinea, P. mixta, P. popenovii, P. semanii, P. serratodigitata*) are cultivated as edible fruits. The basic chromosome number is normally 3, 6, 9 or 12, mostly 9.

Biotechnological Approaches for Conservation

The successful regeneration of passionfruit plant from tissue culture has several implications. It may be possible to use *in vitro* selection techniques to select passionfruit cell lines and plants regenerated from these cells that are resistant to disease toxins and environmental stress. The breeders are trying to breed dwarf cultivars with disease resistance and dense fruiting.

Table 7.22: Tissue culture (micropropagation) and genetic engineering research work conducted on passionfruit

References	Work conducted
Kawata *et al.* (1995)	Described a method for long-term micropropagation of passsionfruit from subcultured multiple shoot primordia.
Drew (1991).	Propagated cv. E 23, a F hybrid between *P. edulis* and *P. edulis* f. *flavicarpa* was by shoot-tip grafting
Micrografting	
(Biricolti and Chiari, 1994).	Used meristem culture as a potential regeneration technique for obtaining virus-indexed stock plants for mass propagation of *P. edulis* f. *edulis*
Protoplast culture	
D' Utra Vaz et al. 1993 Otoni et al. (1995 a, b)	Indicated that new plantlets can be regenerated and genetically improved by somatic cell technique through protoplast culture
Otoni et al. (1995a)	Indicated that protoplast yield and viability varied according to the enzyme mixture used for isolation.
D'Utra Vaz et al (1993)	Isolated protoplast from newly expanded leaves by enzymatic digestion exhibite maximum frequency when plated at a density of 1.5 x 10^5 protoplast/ml in agarose-solidified droplets of KM8P medium containing 250 mg cefotaxime/ml.
Otoni *et al* 1995b	Regenerated plants from dark-green compact callus derived from leaf mesophyll protoplasts after 3 subcultures at 30-day intervals on MS medium supplemented with either 1.0 mg BA + 0.001 mg NAA or 1.0 mg zeatin/l, in both cases containing 3.0% (w/v) sucrose and solidified with 0.7% (w/v) agar.
Otoni *et al.* (1995a)	They supplemented the medium with antibiotic (cefotaxime) which proved essential for inducing and sustaining cell division in isolated protoplasts.
Biochemical and molecular markers	
Genetic linkage and identification of genotypes	
Do *et al.* (1992)	Compared chloroplast DNA (ctDNA) of *P. edulis* (including f. *flavicarpa*), *P. alata, P. coccinea, P. suberosa* and intraspecific and interspecific hybrids by RFLP analysis using 6 restriction enzymes.
Fajardo *et al.* (1998)	Carried out genetic analysis based on RAPD on 52 accession representing 14 species of the genus *Passiflora* using 50 random 10 mer primers.

References	Work conducted
Genetic transformation	
Lin (1988)	Smeared *Agrobacterium rhizogenes* MAFF03-01724 over a stem segment of *P. edulis* inoculated upside-down on MS solid medium.
Mander *et al.* (1994)	Cocultivated leaf and stem explants with a disarmed strain of *A. tumefaciens* harbouring the co-integrate vector pMON200.

LOQUAT

Eriobotrya japonica (loquat) is the member of family Rosaceae and subfamily Pomoideae. The genus is native to the hills and moist regions of central Eastern China. It is a subtropical fruit. The other species under the genus *Eriobotrya* are *E. petiolata, E. longifolia, E. hookeriana, E. dubia, E. bengalensis, E. angustissima, E. elliptica* and *E. japonica.* The fruits are used for dessert purposes as well as for preparing jam, jelly and canning. It contain triterpenoids and sesquiterpene glycosides which have hypoglycaemic activity (Tommasi *et al.*, 1992). Following research work has been conducted with respect to embryo and protoplast culture:

Table 7.23: Tissue culture research work conducted on loquat

References	Work conducted
Lin *et al.* (1995)	Examined the effects of different carbohydrate sources (D-sorbitol, sucrose, D-fructose, D-glucose and D-galactose) embryo and protoplast culture. They indicated that the use of D-sorbitol in the culture medium was beneficial.
He *et al.* (1995)	Investigated the semi-thin section method. They revealed that there are transfer cells in both the stigma and style. Some cells in the inner integument and nucellus had outstanding wall ingrowtins. Exo-layer cells of the endosperm were transfer cell-type cells, and the embryo sac and central cell had wall in growths and haustorial structures.

MANGOSTEEN

Mangosteen (*Garcinia mangostana*) is a tropical plant and belongs to the family Guttiferae (Clusiaceae) that includes about 9 genera with 86 species of fruit trees. It does not grow in the wild state. *Garcinia mangostana* is native to the Malay Archipelago and the Molucca and Sunda islands. The fruits are consumed fresh. They are also used as topping for ice-cream. It is propagated by seeds which are not zygotic but develop from nucellar tissue of the carpel walls of fruit (apomictic). Plants are uniform because of parthenocarpy and parthenogenetic nature.

Mangosteen shows very low genetic diversity. Richards (1990) suggested that for improvement in propagation and establishment, efforts should be made to graft and hybridize mangosteen with its presumptive parents. Use of somatic embryogenesis to produce somaclones and genetic transformation may aid in developing certain genetic base.

Table 7.24: Tissue culture research work conducted on mangosteen

References	Work conducted
Goh et al. (1988)	Indicated that proliferating shoots could be obtained from cotyledon segments cultured on MS medium containing BA.
Teo (1992)	Indicated that *in vitro* culture of apomictic seeds have the presence of an embryonic axis in the seed.
Normah *et al.*, 1992	Indicated that short day (8h) can promote shoot formation in comparison with 12-h photoperiod in presence of BA (40 or 50 μM) and 0 or 2.5 μM NAA. The shoots were rooted in MS medium supplemented with 20-30 μM IBA. The plantlets formed were successfully transplanted into a vermiculite and sand mixture.
Goh et al. (1994).	Regenerated a high frequency direct shoot-bud from excised leaves
Lakshmanan et al. (1997).	Indicated that the presence of IAA adversely affects the state of competence and the process of caulogenic determination in the tissue. It has inhibitory effect on shoot-bud differentiation. They are insensitive to ethylene or its precursor ACC. Shoot-bud differentiation was greatly enhanced by BA, but selectively delayed by ethylene

APPLE

Malus domestica (apple) belongs to the family Rosaceae, subfamily Pomoideae and genus *Malus*. About 25 species are under this genus (Rehder, 1954). *Malus domestica is most common.* It appears that over centuries, several *Malus* species have contributed genes into the cultivated apple via sexual hybridization and introgression. As a consequence, there is some controversy regarding its correct botanical name. Korban and Skirivin (1985) after analyzing several aspects recommended *Malus domestica* which is the most accepted name.

Apple is a native of Eastern Europe and Western Asia. It has been grown for over 3,000 years. At present, North America, Western Europe, Australia, New Zealand and South Africa are the main producers of apple in the world. Apple is believed to have originated in western Asia (Watkins, 1976). The species available at present is the result of selection carried out over thousands of years as is evident from ancient Roman and Greek literature. Apple cultivation spread to various parts of the world having temperate climate, India is one of them. Global annual production of apple is about 40.2 million tones which is next to only the grapes and nut crops and fourth in production among all fruitcrops (FAO, 1990).

The commercial cultivation of apple began in India about 60 years ago. It has registered rapid increase in production only during the last 35 years. Because of inadequate infrastructure for post-harvest handling and an improper marketing network which gives little returns to the farmers despite a high price being by the consumers.

Rising of inputs, spread of diseases and production costs have resulted in meager return to the growers. In the developed countries lines in the West, apple cultivation is carried out on highly scientific lines which results in yields of 30 tonnes per heatare as against a meager 8 tonnes per hectare obtained in India.

In India, apple is cultivated in the northern states of Jammu and Kashmir, Himachal Pradesh and hilly regions of North Uttar Pradesh. The important regions in which it is cultivation are Srinagar, Kulgam and Uttarmmachipura in Kashmir; Udhampur and Kistwar in Jammu; Almora, Nainital and Garhwal, Chamba-Mussorie belt; and Kullu valley and Shimala hills. Apple is also cultivated to a small extent in Bangalore (Karnataka) and the Nilgiri Hills.

The annual production of apple in the country is about 1.75 million tonnes. Himachal Pradesh leads the country with an area of 1.51 lakh hectares under apple cultivation followed by Jammu and Kashmir with 75,000 hectares and Uttar Pradesh with 55,000 hectares. The annual production in these states is approximately 7,56,965 fruits vary in size, colour, flavour and quality.

In 1994, due to adverse weather conditions, the apple output in Himachal Pradesh was exceptionally low. The Himachal Pradesh Horticulture Produce Marketing and Processing Corporation (HPMC) for the first time in 1994, undertook the purchase of apple on a commercial basis from growers in different parts of apple growing areas in the districts of Shimla, Kullu and Kinnaur to free the growers from exploitation by middlemen. In 1994-95 season, it has handled and marketed more then 70.000 boxes of fruit worth Rs.1.5 core.

In India, the traditional market for Himachal apples exist in Madras, Delhi Bangalore, Thiruvanthapuram, Hyderabad, Bombay, Pune and Calcutta. In recent years, new market for Himachal apples have been explored in Patna, Cuttack and Nagpur.

Biotechnological Approaches for Apple Conservation

During the last few years, biotechnological advancements have made significant contributions for production and post-harvest management aspects of apple. A few of these techniques are as follows:

Hanke *et al.* (1994) suggested the applications of biotechnological methods in fruit breeding specially for efficient recovery of viable plants from *in vitro* cultures. The main emphasis was on the regeneration process and the genetic base and cultural conditions *in vitro*. Hanke *et al.* (1994) also suggested the importance of micropropagation system, the regeneration system from complex leaf explants and the protoplast system for improvement.

Micropropagation

Apple rootstock is the important material for micropropagation studies because of its specific attributes and profound influence on the cultivars. Clonal rootstocks of apple were the first to be micropropagated on commercial scale and have become a common practice in various countries of the world (Zimmerman, 1991). The trees of Cox's Orange Pippin and Bramley's seedling were grown vigorously on micropropagated rootstock MM106 than the controls using buds from non-micropropagated plants. For high density plantation, the conventional methods of propagation of dwarfing rootstocks have proved insufficient, therefore, micropropagation proved as a better alternative for commercial production of disease-free and uniform plant throughout the year. These plants are independent of the seasonal changes.

Anther Culture

The induction of embryos from anther culture of Topred and Starking on MS medium supplemented with 2.2 µM NAA or 0.5 – 2.3 µM 2,4-D was reported by Zhang *et al.* (1987). *In vitro* androgenesis is cultivar-dependent, and pre-treatment of anthers with cold temperature (6°C for 5 weeks) prior to culture enhanced embryogenesis. The irradiated apple pollen increase the rate of somatic embryogenesis from ovule somatic cells (James *et al.* 1985). Hofer *et al.* (1999) suggested the application of a developed protocol for anther cultures and induction experiments of *in situ* parthenogenesis followed by embryo culture to apple cv. Remo resulted in the production of two lines. These lines were characterized as diploid homozygous and resistant to scab and mildew Ploidy level. This determination was conducted by using flow cytometry and homozygosity of the shoots was analyzed by isoelectirc focusing and PAGE using 3 isozymes. At rooting stage, the plants were transferred to greenhouse for evaluation.

Protoplast Culture

Much information are not available about the use of protoplast technology in fruit trees (Hidano and Nizeki, 1988). Patat-Ochatt *et al.* (1989) conducted initial studies on the isolation of protoplasts from tissues of temperature fruit trees consistently resulted in low yield of protoplasts. The aim of these studies was to develop efficient methods for isolating viable protoplasts in large quantities. Varied factors were used by various workers to achieve this goal, like, the explants source, age, growth conditions of the mother plants, enzyme concentrations, incubation conditions, etc. However, no general strategy could be evolved for obtaining consistently high protoplast yields from different temperate fruit trees (Hidano and Niizeki, 1988). Plants were recovered from mesophyll protoplast of MM106 rootstock by transferring intact regeneration calli from regeneration medium to standard micropropagation medium, the latter with an increased level of group B vitamins (Patat-Ochatt and Power, 1990). On the ground of these studies, certain guidelines have been suggested for isolation of viable protoplasts from apple (Daughty and Power, 1988; Kouider *et al.*, 1984 a, b; Patat-Ochatt *et al.*, 1988; Belaizi, 1991). Saito and Suzuki (1999) reported regeneration from meristem-derived callus protoplasts of scion cultivars of apple that have been difficult to regenerate from leaf protoplasts. Calli were induced from the meristem of apple cultivars of *M. domestica* and *M. prunifolia* on MS medium (2 mg/l, 2,4-D, 1 mg/l BA, 0.8% agar) and subcultured in a liquid medium. The ability to regenerate plants from suspension calli was studied and two combinations, one with 0.1 mg/l IAA, 0.1 mg/l ABA and 2.0 mg/l TDZ and another with 0.1 mg/l IAA, 1.0 mg/l ABA and 2.0 mg/l TDZ, were effective for plant regeneration.

Somatic Embryogenesis

The induction of somatic embryogenesis from nucellus tissue and embryos of Golden Delicious MS medium supplemented with 0.1 mg/l thiamine HCl, 0.5 mg/l nicotinic acid, 0.1 mg/l pyridoxine, 2.0 mg/l glycine and 100 mg/l myoinositol was reported by Eichholtz *et al.*, (1979). Somatic embryo-like structure from protoplast culture isolated from callus and suspension cultures of Jonathan apple was obtained by Kouider *et al.* (1984a). These structures developed root but failed to developed root but failed to develop shoots.

Somatic embryo-like structures were obtained from leaf tissue of various apple cultivars, namely McIntosh, Wijick, Akeru and Graven Stem *in vitro* on N6 medium supplemented with 22 μM BA and 1 μM NAA (Welander, 1988). Globular and heart-shaped somatic embryogenesis has been induced in calli derived from different parts of *in vitro*-grown apple seedlings, flower buds and petals (Mehra and Sachdeva, 1984). Calli induced on Nitsch and Nitsch medium supplemented with 8.9 μM BA and 10.8 μM NAA gave the best result. Rahman *et al.* (1988) assessed the potential for somatic embryogenesis in cotyledon cultures of 4 apomictic *Malus* species. The frequency of somatic embryogenesis did not differ between species. The formation of adventitious roots, shoots and callus was also associated with somatic embryogenesis. Conversion of the embryoids into plantlets was achieved after 60 days of cold treatment on half-strength MS medium. Somatic embryos from cotyledons lacked clonal fidelity.

Somaclonal Variation

Phytophthors cactorum (a fungus) causes collar rot of several fruit trees and is one of the most important crown and root diseases of apple (Jeffers and Aldwinkle, 1988). Reaction of apple rootstocks to the disease varies in different regions which may be attributed to various pathological strains of *P. cactorum* and/or to different environmental conditions. M4, M9 and M26 are commercially propagated clonal rootstocks. They are less susceptible than MM106 clonal rootstock (Utkhede and Quamme, 1988). *In vitro* selection of apple rootstocks M26 and MM106 in culture filtrate of *P. cactorum* was carried out by Rosati *et al.* (1990) using *in vitro* plants as the source of explants. The selected plants were regenerated in both the clonal rootstocks and out of 40 somaclones, one was found to be superior in growth on medium containing growth regulator on culture filtrate.

The leaf-discs of apple can also be used as source of plant regenerated (Donovan *et al.*, 1994). About 270 glasshouse-grown somaclones with fire blight pathogen, *Erwinia amylovora*; 33 per cent of the somaclones showed an increase in resistance to *E. amylovora* compared to parental material. In contrast, after inoculation of plantlets maintained *in vitro,* only 21 per cent produced less severe symptoms than the parental stock. The 16 somaclones, which showed the highest levels of resistance, varied among themselves for the level of resistance. This study has clearly established that somaclonal variation can be utilized to increase the level of resistance to pathogens. These somaclones showed variation for other characters, such as rooting ability, root number and proliferation, etc. Such variation in somaclones was also reflected in the pattern of isoenzymes as reported by Marteli *et al.* (1993). Four somaclonal variants regenerated from adventitious buds of the apple variety Greensleeves were pre-selected on the basis of their reduced fire blight (*Erwinia amylovora*) susceptibility (Chevreau *et al.*, 1988). By using various inoculation techniques on *in vitro* leaves and microcuttings of greenhouse and field-grown plants the tests indicated that one clone (R 46/3) was clearly less susceptible than the control. This clone was also characterized as a spur variant with a reduced growth which can explain its limited susceptibility to fire blight.

Genetic improvement in apple in Angers, France, more than 20 well characterized isoenzyme loci have also been identified. Induction of haploid plants through anther culture or parthenogenesis produced 10 haploid plants. Evaluations of somaclonal variations after adventitious bud regeneration in apple produced improved colour and disease-resistant clones (Chevreau *et al.*, 1988).

Biochemical and Molecular Markers

Molecular marker plays a very significant role in apple genome mapping, characterization and identification of individual genotypes including cultivars and rootstock and for the detection of diseases.

Detection of Pathogens and Genes of Resistance

Viral Pathogens: Firrao *et al.* (1994) and Firrao *et al.* (1995) studied the sensitivity of PCR-based methods for the detection of mycroplasma-like organism (MLO) in apple by using as little as 300 mg of diseased tissue Avinent and Llacer (1994 and 1995) detected apple proliferation (AP) phytoplasma. They used *16S rRNA* gene as the template for DNA amplification through PCR. This method with the help of suitable primers and *16S rRNA* gene as the template for PCR amplification is a suitable approach for routine detection of apple proliferation phytoplasma-infected apple tissues (Malisano *et al.*, 1996; Minucci *et al.*, 1996). PCR with modifications has further been used for the detection of apple chlorotic leaf spot virus (ACLSV) and apple stem grooving virus (ASGV) with high accuracy (Kinard *et al.*, 1996; Nemchinov *et al.*, 1995; Candresse *et al.*, 1995).

PCR amplification of viral 3 terminal region of apple stem pitting (ASP) and pear vein yellow (PVY) viruses is useful in detecting these viruses following modifications in the method, e.g., making use to Immuno Capture (IC) PCR, IC-RT-PCR, etc. Further confirmations with cloning of the DNA fragments of interest and Southern hybridization have made these molecular techniques very reliable (Chen and Zhang, 1997; Jelkmann and Keim-Konrad, 1997; James *et al.*, 1997; Marinho *et al.*, 1998; El-Dougdoug, 1998).

Fungal and Bacterial Pathogens: Molecular markers techniques are also being used in differentiating disease resistant and susceptible varieties. Most of the information has been generated with respect to pathogens, such as *Venturia,* a fungal pathogen causing apple scab, one of the most severe diseases, *Erwinia amylovora* (causing fire blight) and *Podosphaera reucotricha* causing powdery mildew. The studies on Apple scab were conducted by Yang and Kruger (1994). They compared RAPD patterns of scab-resistant *Malus floribunda* clone B 21 and a commercial apple variety and showed scab-resistance gene (*Vf*) linked with OPD 20/600 fragment with a recombinational value of about 0.20 –0.25. It was claimed to be first DNA marker reported for scab-resistance. Yang and Korban (1996) identified a RAPD marker OPD 20/600 in *M. floribunda* linked with *Vf* gene. This marker was cloned and sequenced, and specific primers based on the marker were developed to screen apple germplasm for resistance/susceptibility to scab. Tartarini (1996) also reported 5 RAPD markers associated with *Vf* genefo *M. floribunda* clone 821. Out of these five, two markers, OPAM 192200 and OPAL wer closely linked with *Vf* locus.

The linkage mapping for resistance to apple scab was performed in a population derived from a cross between apple (*Malus pumila*) cultivars Prima (resistant) × Fiesta (susceptible) screened in replicated field and glasshouse trials throughout Europe.

The *Vf* gene is responsible for scab-resistance. It was constructed using 24 molecular markers linked to the resistance gene. One isozyme marker (*Pgml*), 6 RFLP markers and 17 RAPD markers formed a linkage group with the consensus measure of resistance to scab (King *et al.*, 1998). Almost 200 random sequence decamer primers were used by Yang *et al.* (1997) to screen a pair of bulked samples of apple (*Malus* x *domestica*) DNA and that of the donor parent *M. floribunda* clone 821 for molecular markers linked to the *Vf* gene conferring resistance to apple scab (*Venturia inaequalis*). They identified a single primer that generated a PCR fragment, OPAR4/1400 from the donor parent *M. floribunda* clone 821 and the scab-resistant selections/cultivars bulk, but not from the scab-susceptible recurrent parent bulk. The expression of *Vf* gene itself may be influenced by several facors which may limit the sole use of molecular markers for screening scab-resistance in germplasm. However, under usual conditions, these molecular markers are likely to platy major role in apple breeding programmes.

Apple canker : *Cylindrocarpon heteronema* is the pathogen of apple canker. A synthetic oligonucleotide primer specific to apple canker pathogen was developed by Brown *et al.* (1993). The causal organism could be detected in PCR products. This could also be confirmed with Southern hybridization.

Further heterogeneity in this pathogen was shown by using its mDNA as the probes (Brown *et al.*, 1994).

Fire blight : *E. amylovora* causes fire blight in apple. McManus and Jones (1995) and McManus *et al.* (1996) succeeded in specific amplification of a DNA fragment from *E. amylovora* infecting apple and nested PCR, PCR-dot-blot and reverse blot hybridizations were used. These approaches have been suggested to detect the pathogen in the leaf tissues of apple and have been compared among themselves for their accuracy. Based on PCR amplification of *Erwinia amylovora*, Berger *et al.* (1996) suggested that fire blight caused by this pathogen can be monitored in the orchards. It was concluded that blossom monitoring could be considered as an efficient means to control the pathogen. Norelli *et al.* (1999) tried to transform apple rootstock and scion cultivars with genes encoding proteins that lyse bacterial cells and to select transgenic linens that have increased resistance to *Erwinia amylovora*. They used *Agrobacterium tumefaciens*-mediated leaf piece transformation system. Over 250 tránsgenic lines of the rootstock M7 and the scion Royal Gala have been selected that contain genes encoding the lytic proteins attacin E, cecropin SB-37, cecropin shiva-1, and hen egg white lysozyme. In addition, lines transformed with pBI 121 vector plasmid not containing lytic protein genes were selected. Transformation of Royal Gala M7 with the gene encoding attacin E has resulted in a significant increase in resistance to fire blight.

Powdery mildew : This disease is caused by *Podosphaera leucotricha* Markussen *et al.* (1995) using RAPD-PCR based DNA amplification with about 850 deca primers, succeeded in identifying a linkage group containing gene *Pll* responsible for resistance to powdery mildew. The gene was introgressed from *Malus robusta* to cultivated variety. Seven RAPD markers representing the introgressed DNA sequences from *M. robusta* were identified and arranged with *Pll* locus in a common linkage group. Two RAPD markers OPAT 20450 and OPD 21000 were tightly with *Pll* gene with 4.5 and 5.0 cM distance respectively. These markers are suggested for marker-assisted selection for breeding powdery mildew-resistant apple varieties. Gianfranceschi *et al.* (1996) successfully applied DNA markers to prove that the resistance gene present in the cultivar Nova Easygro (said to be *Vr*) is tightly linked to *Vf* and is most probably *Vf* itself.

In vitro Conservation

In apple, cryopreservation of shoot-buds is done in liquid nitrogen for its subsequent survival. Tyler and Stushnoff (1988a) exposed shoot buds of 15 cultivars of apple (with 2.5 cm stem attached to them) to -30°C or -40°C for 5 minutes or 24 hours before immersion in liquid nitrogen. The buds were subsequently retrieved and could be successfully patch-budded to greenhouse-grown dwarfing rootstocks for further propagation or production of flowering shoots. Dehydration improved survival rate which in some cases was as high as 100 per cent. They observed variable response of the genotypes for their survival after their immersion in liquid nitrogen. The success not only depended on genotype of apple and other treatments but also on the season which has a profound effect. Success was reported to vary from zero to 100 per cent depending upon season (Suzuki *et al.*, 1989). Niino *et al.* (1992) incubated *in vitro* shoots of a few cultivars of apple at 5°C for 3 weeks and partially dehydrated the shoots by culturing them for one day on a medium containing 0.7 M sucrose. Glycerol, ethylene glycol and dimethyl sulphoxide were used as the cryoprotectants. This approach of making use of dehydration and cryoprotectants as well s exposure to low temperature has been used for a large number of apple cultivars and their rootstocks (Niino and Sakai, 1992; Suzuki, 1993). Similarly, meristems have also been successfully cryopreserved (Mi, 1993). Dereuddre *et al.* (1994) attempted cryopreservation of somatic embryos and shoot-tips and reported that recovery after immersion in liquid nitrogen depended on sucrose and DMSO concentrations in pre-culture media. Capuano *et al.* (1998) suggested that encapsulated buds excised from micro-propagated shoots of the clonal apple rootstock M 26 can be considered as non-embryogenic synthetic seeds.

COFFEE

Coffee (*Coffea arabica* L.) is their second important commercial crop. India accounts for only about 2-2.5 percent of world production and export about 50 per cent of its produce worth about Rs. 3 billion annually. Both *robusta* and *arabica* types of coffee are cultivated in almost equal quantities, in India. *C. conephora* var. *Robusta* shows great variability, since it is a cross-pollinating species. *C. arabica* is a tetraploid, self-pollinating species which can also be multiplied on a mass-scale through tissue culture. In coffee embryogenesis through callus was reported in the early 1970s. Callus cultures of several different varieties of coffee have been established without difficulty. Callus has been obtained from both seedling and mature leaf explants. High frequency somatic embryogenesis could be achieved and plantlets should obtained.

Table 7.25: Tissue culture of coffee spp.

Explant source	Response
C. arabica	
Stem	C
Endosperm	C
Various plant parts	R,E
Var. Bourbon	SE,
Leaf	plantlets
C. conephora var. Robusta	
Stem	C,SE
Leaf	C,SE
Anther culture	C
C. arabica	C,SE
C. canephora	C,SE
C. excelsa	Haploid
C. liberica & *C. arabica*	C, pro-embryos

Abbreviations: C - Callus; E - Embryoid; R - Roots; SE - Somatic embryos

Shoot apices of *Coffea arabica* seedlings were cultured to produce multiple shootlets and plants. Undifferentiated haploid callus tissue was also obtained in *C. arabica*. Therefore, tissue culture appears to have the scope for large-scale multiplication and would be helpful in yield stabilisation. A combined effort involving the use of new varieties (e.g., var. *Cauvery*), rapid clonal multiplication, mycorrhiza for better phosphorus utilisation and integrated pest management will be helpful in increasing productivity.

CASHEW

Cashew (*Anacardium occidentatlae*) is the member of family Anacardiaceae. It comprises 60 genera and 400 species of tropical and sub-tropical trees and shrubs. Cashew is believed to be native of South Eastern Brazil, from where it was introduced to India in the sixteenth century. It appears probable that it reached East African countries also be about the same time. It is grown in tropical and sub-tropical regions of the globe. However, it is cultivated on a large scale in Brazil, East Africa and the Southern States of India. Cashew was first introduced into India to cover bare hills and for soil conservation. Cashew is cultivated on a wide variety of soils and lands in India. It is suitable for fairly steep slopes with shallow top soils. In India cashew is grown mainly in a laterite, red and coastal sands in the states of Kerala, Maharashtra, Goa, Karnataka, Tamil Nadu, Andhra Pradesh, Orissa and

West Bengal. In India, Kerala, followed by Tamil Nadu have more area under Cashewnut. Cashew is growing is 1.56 Lakh ha in Kerala with a annual production of 1.49 lakh HT of raw nuts.

Chromosome number of cashew is 2n=42. The cashew tree is a low spreading evergreen tree with a number of primary and secondary branches. The number of primary branches in four year old trees vary from 9 to 30 and secondary branches from 246 to 412. The root system of mature cashew tree, when grown from the seed, consists of a very prominent tap root and a well developed and extensive network of lateral and sinker roots.

Nearly 50 species under 22 genera are represented in the India sub-continent. Cashew grows wild with little or no care in many countries of the world which are geographically located between a latitudes 30 degrees North and South of Equator. Cashew is a sun-loving tree.

Constraints on Cashew yield improvement

The following are said to be the main constraints on Cashew yield improvement.

1. Grown mostly in marginal lands;
2. Insufficient number of progeny orchards;
3. Lack of sufficient good quality seed nuts;
4. Non-adoption of inputs like fertilizers and insecticides, breeding procedure may be systematically dealt with as follows:
5. Collection, Conservation, cataloging and Evaluation of the germplasm.
6. Selection of Parents – promising characters of selected lines.
7. Clonal and pedigree method of selection.
8. Improvement by hybridization.

Germplasm Collection, Conservation, Cataloguing and Evolution

As cashew is a highly cross-pollinated heterozygous polyploid, there is a great variability in vigour, productivity, sex-ratio and also fruiting behaviour and hence, an in-depth study of germplasm collection is utmost essential. The term *Germplasm* is used in its widest sense to include all genotypes to be found in local population, and in collection of introduced material.

The main objective of germplasm collection is to preserve and document the genetic variation in representative samples of primitive varieties/land races of cultivated as well as wild relatives. Those which are endangered by genetic erosion, are also to be conserved because of the basic variation of the crop species. Hence, though the variability in cashew is limited, Collection, Conservation, Cataloguing and Evaluation are the basic need for evolving varieties by selection and hybridization.

Collection

The germplasm collection serves as the initial population out of which parents are to be selected. It is much easier to collect seeds than the collection of vegetative parts of the plant from various parts of the country or form abroad. Seednuts are fairly abundant and can easily be packed for transportation; whereas great care is required to keep vegetative collection alive, during transmission to the introduction station (Gene bank). The cuttings will easily dry out or rot, if they are not properly wrapped in water-proof material. Since seednuts incorporated a wide range of genetic diversity, the genetic integrity of a particular clone or genotype can be preserved through vegetative propagules only.

The efforts to collect and conserve germplasm of cashew in India started with the establishment of pioneering centers in the early fifties. There had been considerable exchange of materials between these centers. Since the exchanging of material were through seedling progenies in most part, the heterozygosity of the parent trees and the out-crossing behaviour render the materials so disseminated as a constantly evolving dynamic system. Even though most of the collections available are of indigenous sources, they are not strictly so, since the nucleus of material that spread widely and naturalized in India's habitat, have come entirely through introductions only. However, the variability with respect to several trials in cashew has been too large for a material originated from a limited gene pool. Hence, there is a good scope for introduction of germplasm from far and near. The exotic types are from Brazil, Malaysia, Sri Lanka, Tanzania, Nigeria, Mozambique, Kenya Singapore, and Panama Republic.

Conservation

Genetic resources are indispensable basic material for plant breeding and to carry forward the past generation to the future. These invaluable living material are preserved in gene banks for future use. Conservation is done for critical analysis and understanding of the available genetic resources. In this way of preserving materials, present and future needs of genetic variation can be met.

There are two ways of preserving living materials:

(i) In situ preservation and

(ii) Ex situ preservation.

In situ preservation is pertaining to protection of the plants in its original habit, after marking the elite plants in its original habitat, after marking the elite plants in a commercial plantation from where the required quantity of material are collected and utilized when necessary.

Ex situ preservation is the removal of seeds or vegetative material from their original habitat and transferred to the Gene bank where they are systematically planted and compared their relative performance in a compact block.

Cataloguing of Germplasm

The information on each genetic resource is a pre-requisite for efficient utilization of the living material and such information gathered must be so structured to enable not only for recording and retrieval of data; but also enable an analysis of connections and correlations.

Table 7.26: Cashew research Station in India and Germplasm assemblage upto 1989

State and Station	Station Start	Year of start	Indigenom	Exotic
Maharashtra				
1. Vengurla	1951	153	4	157
Andhra Pradesh				
1. Bapatla	1955	174	5	179
Kerala				
1. Anakkayam				90
2. Madakkathara	1963	86	4	93
Tamil Nadu				
1. Vridhachalam	1963	169	8	177
Karnataka				
1. Vittal/				
2. Shanthigodu	1972	283	9	292*
3. Puttur	1986	123	2	
4. Chintamani	1988	75		75
5. Ullal	1953	111	8	
Orissa				
1. Bhubaneshwar	1977	47		47
West Bengal				
1. Jhargram	1988	81		81

Germplasm Evaluation

Identification of superior genotypes is only the first step in the development of cashew nut production. In germplasm evaluation, in addition to yield, other desirable characters such as compact canopy, short flowering phase, bold nuts, cluster-bearing habits etc. are identified which are used in the breeding programme. A precise knowledge on the extent of variability present, among the genotypes and the association between yield and other component characters is important for a proper selection of characters. Exploitation of the genetic resources is possible only through proper characterization of collections already made. Evaluation is a difficult task for the perennial nature of crop and its requirement of large area for planning. Often the desired phenotypic characters are polygenically inherited and highly influenced by the environment; and hence the data obtained by such evaluation are not easily understood at the genic level. However, at present the characterization of genotypes and their selection procedure are based on phenotypic characters. Critical evaluation of germplasm after assembling is essential for identifying desirable types which are to breeding value.

The evaluation of germplasm has led to the identification and release of 21 selections by different cashew research centers. The variety names (source in parenthesis) are: three selections from NRCC, Puttur; Sel-1 (A 18/4), Sel-2 (2/9 Dicharla), Sel-3(3/8 Simhachalam); four Bapatla; BPP-3 (3/3 Sch), BPP-4 (9/8 EPM), BBP-5(T.No.1), BPP-6 (T.No.56); 4 from Madakkathara: Ana-1(BLA 139-1), NDR 2-1, K 22-1, BLA 39-4, two from Vengurla: V-1 (Ansur-1), V-2 (WBDC-VI; two from Virdhachalam VR-1 (M10/4), VR-2 (M44/3); two from Ullal; Ull-1 (8/46), VLL-2 (3/67); two from Bhubaneshwar; Tr No. 40, Ven. 36/3; two from Jhargram; Jhar-1 (BLA 39-4), Jhar-2 (T.No. 40). All these types gave nut yield more than 10 kg/tree/year.

Selection

If the phenotype is a correct identification of genotype, the selection is effective, superiority of selection can be further confirmed by a progeny test. The selection can be resorted and useful in cashew, where the population is heterogeneous. Phenotypically, good looking plant is selected, confirmed to their performance by other tests. The selection would be to obtain highest possible yield of nuts; however, the variation in the nut yield per tree is useful criterion in selection, particularly under uniform conditions of climate and plant density, and to judge yielding ability of any particular tree, at least yield records should be available for 5 successive years. Since there is a high degree of heterogeneity in seed progenies of cashew, there is a great scope for identifying superior genetic stock by proper selection from existing population or by a planned programme of selection and hybridization.

In a planned programme for the selection, the hedge-row system of planting can be advocated in cashew for screening test. In this system, the trees will suppress each other in the rows; but will develop strong brunches towards the open space between the rows which will be helpful for rigorous field selection of a better quality mother tree.

Selection of Mother Trees

The starting point of cashew breeding programme should be the choice of mother tree from the best available cultivars.

1. compact canopy,
2. intensive branching,
3. high lateral to leader ratio (4:1);
4. short to medium duration of flowering (45-60) days);
5. early, and cluster bearing (4-6 nuts/panicle), medium to big size nuts (6 to 10 g), high shelling percentage (30-35%) and resistance/tolerance to tea mosquito bug.

Direct selection of superior elite trees from the existing population is the first line of improvement of cashew crop.

After a most crucial visual observation, number of elite mother trees are selected from the initial material which should have as many good characters as possible.

Promising Characters of Selected Lines

While selecting superior lines, a few good characters have been identified which may be suitably utilized for further improvement in the breeding programme.

Among the released varieties, VR-2 (M 44/3) has proved to be prepotent (which could transmit its high yielding potential even to seedling progenies). Therefore, available seeds of M 44/3 may be utilized for planting purposes.

In case of other selected lines, the desirable characteristics may not be inherited by the seed progeny. As to their relative importance, vegetative propagation will allow fixation of existing genotype as a clone.

Clonal Selection

From the seedling progeny, the best performing individuals (elites) are selected and multiplied vegetatively as a clone. Clonal propagation results in immediate fixation of such characters observed in a plant. After testing of clones in a replicated trial for 5 to 6 years, the best one is released as a variety.

Pedigree Method of Selection/Multiplication Testing

Owing to intensive effect of environmental conditions on cashew yield, it is preferable to have a larger number of progenies with 12-15 plants each than to have fewer selected mother trees. It is also highly desirable to test them simultaneously at various localities. An average result thus can be obtained which will enable the selection of progenies well-adapted to an array of different ecological conditions.

It has been ascertained that one of the excellent progenies. VR-2 (M.44/3) so far isolated from comparative yield trial and widely adopted, bear the desirable characteristic of broad ecological adaptability. Such testing should be done in comparative yield trials including one established strain as check.

Standardization of Multiplication Technique

In breeding cashew, two objective may be aimed:

(i) The production of improved of improved seedling strains and

(ii) The development of superior clones.

All the improved varieties of cashew have so far originated from seedlings which have to be multiplied vegetatively to obtain true-to-type planting material. The potentialities of the selected trees can be maintained through vegetative multiplication.

The selected seedlings of F1 generation may be multiplied vegetatively. In scion material, nut size can be a critical factor, too small a nut or very large nut can be a disadvantage, however, early fruiting forms might be of special significance.

OILPALM

Another success story is that of the Oil Palm. The area under oil palm cultivation has increased tremendously over the past 20 years to meet the edible oil requirement of the developing countries. Oil palm boasts the highest productivity (4.6 tonnes/ha) among edible oil yielding crops and owing to the heterozygosity of the species, individual production may vary between 21 kg and 77 kg/ plant. The DBT has undertaken three Oil Palm Demonstration (OPD) projects in Maharashtra, Karnataka and Andhra Pradesh over an area of 1000 ha. Oil processing units will form an integral part of the OPDs. The Bhaba Atomic Research Centre (BARC) at Bombay and CPCRI at Kasargod are already engaged in developing protocols and micropropagating this palm. Malaysia ranks as the leading exporter of palm oil where vast plantations use vitro-plants. However, great caution should be exercised before planting callus derived clones to avoid the risk of encountering undesirable somaclonal variants.

8

Conservation of Vegetable Crops Through Biotechnology

An overview of the present developments and future prospects of biotechnology in the world becomes necessary before describing commercial application of biotechnology in vegetable crops. The genesis of biotechnology is about 25 years old, since then there have been fast developments in its application in plants, animals and micro-organisms. The new technology is able to access genes from any species or organism through transgenics in which a gene or genetic construct has been incorporated by molecular techniques, unlike the restricted gene pool utilized by the conventional plant breeder.

The scientists all over the world have a consensus that the conventional technology alone will not be able to meet the challenge of doubling the global food production by the year 2025. However, the target can be achieved by using biotechnology along with conventional technology.

Biotechnology research is targeted to benefit farmers, food processors and consumers. Presently, the emphasis has been on resistance to biotic stress, such as, herbicide tolerance and resistance to diseases and insects with a view to increasing crop productivity. About two-thirds of the genes incorporated in transgenic crops are for resistance to biotic stresses (James et al., 1991; James, 1996). Crop production in the world is estimated to get reduced by about 36% due to biotic stresses – insects (14%), diseases (12%) and weeds (10%). Besides, there is a considerable reduction in the use of pesticides resulting in reduced cost of labour and agrochemicals. It also helps overcome the problem of residual toxicity and protection of environment, the major constraints in vegetable production.

The next advancement in biotechnology research in 2000-2005 is expected to be in incorporating genes for resistance to abiotic stresses, like temperature, salinity, drought, frost and metal toxicity/phytoremediation to decontaminate soils and improved quality of starch, protein and oil. Experiments on these aspects are already underway. Later in 2005-2010, biotechnology is likely to be directed the production of pharmaceuticals, vaccines and therapeutic proteins in crop plants serving as a biological renewable resource. The future products of the first decade of the next century (2010) and onward will be specialty chemicals to be produced in crops (Fraley, 1994). The production of biodegradable plastics, monomers and polymers in transgenic plants – the biological renewable resources is considered to be better than the petroleum – based products because the non-renewable resource petroleum is becoming depleted.

POTATO

Potato (*Solanum tuberosum*), a native of the High Andes in South America,was first introduced in India some time around the end of the 16th or the introduction of the 17th century. Its commercial cultivation and consumption in largest quantities in India, however began in 1932 it became popular in the West, not only because of its nutritive valued but also because of its production potential and certain cultivation advantages over cereals.

Its maturity period is short and is adaptable to a wider variations in regional and climatic conditions and sowing and harvesting periods. It shows market response to fertilizers. About 150 different species of potatoes are found in the America continent which can be grown on land form the sea level to an altitude of 4,000 metres.

In India, potato is being grown on 9,38,400 hectares of land accounting for 0. 67 per cent of the total cropped area. It is grown in almost all the states under diverse climatic conditions although not all the states are self-sufficient. Over the year, Uttar Pradesh and Punjab in the north and Karnataka in the south have developed the crop to achieve sizeable marketable surplus uses which are supplied all over the country. In addition these states have also established and excellent network of cold storages which ensure both adequate supplies throughout the year as well as fair price to the farmers. Catching up with Uttar Pradesh, Gujarat which has started producing about 10 lakh tonnes of potatoes in Nadiad and Mehsana districts. In Maharastra, the crop is harvested in the motnhs of February in the Nasik district and in August in Mahabaleshwar and Satara districts. In Talegaon district in Maharshtra two crops are harvested in a year, even after putting together the four districts are not able to meet the entire needs of the state.

Nearly, 82 per cent of potatoes are grown in the plains during the shorter winter period form October to March and 8 per cent in the plateau regions of south-eastern, central and peninsular India, generally as a rainfed crop during rainy season (July to October) and as an irrigated crop during the winter (October to March) in irrigated areas. In Tamil Nadu potato is grown round the year both as a rainfed crop and irrigated crop. On the basis of soil and climatic conditions the various potato growing zones in India comprise the western Himalayas, very high hills situated between 2,500 metres to 3,000 metres above the sea level high Himalayan hills situated at 1,800 to 2,500 metres, mid hills at 1,000 to 1,800 metres, plains and northeastern plains situated at an altitude of about 300 metres.

Biotechnological Approaches for Conservation

Potato (*Solanum tuberosum*) is a vegetatively propagated heterozygous polyploid crop. Potato is an out-breeding species with tetrasomic inheritance. Potato is particularly difficult to manipulate by conventional breeding, and particularly well adapted to direct genetic manipulation by transformation. Its genetic nature imposes many limitations on seed multiplication, conservation of genetic resources and genetic improvement. Development of pest and disease resistance varieties are the main objectives of potato improvement for research institutions. Potato is perhaps the premier example of a crop plant to which biotechnology has been most extensively applied in all aspects of production, improvement and germplasm handling. Steward and Caplin (1951) made the first successful establishment of tissue culture from potato tubers, and subsequently *in vitro* potato cultures were developed from different plant parts such as leaves, petioles, ovaries, anthers, stems, roots and shoot-tips (Bajaj, 1987). the First biotechnological approach was made through meristem culture. It was successfully employed to obtain virus-free potato clones. Micropropagation of disease-free potato clones combined with conventional multiplication methods has become an integral part of seed production in many countries (Shekhwat et al., 1997a).

Biotechnology applications deriving from cell biology, molecular biology, and molecular genetics differ among crops because of the plant species' relative amenability to cell manipulation and by the nature of their reproduction systems. The techniques like, cellular selections, somaclonal variations, somatic hybridization and genetic transformation are now used for potato improvement. These techniques have generated novel genetic variability for synthesis of future potato. Recent advances in molecular breeding vis-à-vis marker-assisted selection have offered enormous opportunity to the breeders to tap desired genetic variability more efficiently and to use them across trans-specific or generic barriers.

Genetic Engineering

The potato is highly amenable to genetic engineering through the use of *Agrobacterium tumefaciens.* Potato was the first food crop to be genetically transformed (An et al., 1986; De Block, 1988), and it has long maintained its position as a leader among transgenic crops.

Direct Gene Transfer for Development of Transgenic Potato

Direct gene transfer is predicated in existing potato varieties, to add specific traits for release with no additional crosses. Gene transfer into potato via agro-infection is efficient, easy, and less expensive. The agroinfection protocol is therefore well adapted for use in developing-country laboratories, which do not always have the capacity to purchase sophisticated and costly equipment. Proper patterns of gene expression should be designed for using specific sequence for spatial and temporal control. Sequences driving organ (leaf, stem, tuber) and tissue-specific expression are available in the potato crop (Belknap et al., 1994) as well as for almost all intracellular compartments. The level of expression of a transgene often varies greatly among transgenic lines because of the effect of the chromosomal location (position effect) or either co-suppression or gene silencing effects. These problems are usually circumvented for potato by generating a large number of transgenic lines and subsequently selecting those with good level of expression. Variation in transgene (usually one to three), minimizing the co-suppression effect. Clonal propagation can also maintain the original transgenic lines with appropriate levels of expression in potato.

Gene Silencing

Gene silencing will grow upon transgene pyramiding, particularly as many gene constructs use the 35S promoter and the *nptII* selectable market gene. Studies on engineering disease resistance in plant revealed that the combination of several transgenes in a single genotype leads to a synergistic effect on disease resistance (Lorito et al., 1996). This may oblige a minimum number of crosses to be made. Therefore, it will favor methylation of homologous sequences, such as the 35S sequences. This may be expected to reduce overall expression of resistances genes (Thierry and Vaucheret, 1996). New promoters and selectable markers are required to co-transform plants and to build up quantitative traits by genetic-engineering means.

Genomic Analysis in Potato

Developments in comparative mapping, new types of molecular markers, and coordinated international efforts to map **quantitative-trait loci** (QTLs) have helped in the understanding of the genetic basis of agronomic traits of importance. Traits of particular interest in developing countries include broad-spectrum pest and disease resistance, tuberization response to daylength, and the accumulation of toxic compounds that may depend on environmental conditions.

Mapping efforts have established amplified fragment length polymorphisms (AFLPs) and **microsatellites** as valuable markers for potato genetics. AFLP markers were shown to be portable between crosses (to map at the same locus when generated by the same primer combination and with an identical fragment size. Thus, genetic maps generated by AFLP markers can be aligned to existing potato genetic maps (van der Voort et al., 1997). One hundred microsatellite markers have recently been generated, and over 70 of these have been mapped in two potato mapping populations.

Genes for virus, nematode, and fungus resistance have been mapped in potato, as have several genes that control important morphological characteristics (Kreike et al., 1996). Natural resistance genes from tuber-bearing *Solanum* germplasm mapped and currently under molecular characterization. The important genes are : *R1, R2, R3, R6, R7*, RX_{adg}, Rx_{acl},, Ry_{sto}, Ry_{adg}, *Grol, H1, GroVI*, *Gpa.*

Studies on genomic analysis for insect resistance, tuberization response to short days, and a long dormant period and its insect resistance, cumulative mapping information, preparation QTL genetic maps, tags approaches, etc. are also on the way to increase potato crop production.

TOMATO

Tomato, *Lycopersicon esculentum* Mill ($2n = 2 \times = 24$), is an important solanaceous fruit vegetable crop of widespread culture and popularity. All the *Lycoerpersicon* species are native to the Andean region of South America encompassed by parts of Chile, Columbia, Ecuador Bolivia and Peru. The evolution of modern tomato cultivars followed from hypothetical phylogeny precursor to *L. esculentum* var. *cerasiforme* to domestication in Mexico to the present large fruited cultivars. Modern tomato cultivars, while not necessarily arising directly in the center of origin, have their root beginnings on the west coast of Latin America. It is grown worldwide across all the continent covering an area of 2,891,000 hectares with a production of 78,282,000 tonnes (1995 FAO estimates). The leading tomato growing countries are; the USA, USSR, Italy, China, India, Turkey, Egypt, Spain, Greece, Brazil and Mexico.

Table 8.1: Biotic Stress Resistance varieties of Tomato

Variety	Source	Salient features
Arka Abha	IIHR, Banglore	Resistant to bacterial wilt disease.
Arka Alok	IIHR, Banglore	Resistant to bacterial wilt disease.
Sakthi	KAV, Trichur	Resistant to bacteria wilt, developed through single seed descent method of selection from LE-79. It is resistant to 3 isolates viz., K 60, 126408-1 Tifton 80-1 of *P. solanacearum.*
Sonali	KKV, Dapoli	Developed from exotic collection. Resistant to bacterial wilt disease.
Pant Bahar	GBPUAT, Pantnagar	Field resistant to *Fusarium* wilt and *verticillium* wilt, and suitable for processing.
Punjab kesri	PAU, Ludhiana	Moderately resistant to fruit borer and late blight.

Source: Tomato, P. K. Kalia, In: Planbt Breeding, Theory and Techniques, S. K. Gupta (Ed.), 2000, Agrobios (India), Jodhpur, pp 127-142.

Table 8.2: Multiple Resistant Varieties of Tomato

Variety	Characteristics
Kalianpur No. 1	Resistant to late blight and early blight, moderately resistant to *Fusarium* wilt.
Ohio 8243 and Ozark Pink	Resistant to race 1 or *F. oxysporium F. sp. lycopersici* and race 1 of *Verticillum dahlia* as well as radial and concentric fruit cracking.
Ohio 8556	It is an F_6 generation derived from five crosses. Fruit firm, jointless pedicel (j^{-2}), suitable for processing and machine harvesting. Resistant to *fusarium* wilt race 1, field tolerance to early blight and anthracnose fruit rot, resistant to radial and concentric fruit cracking.
LHT 24	It has been developed from a cross between 'Fresh Market 9' and 'Tamu Saladette' by pedigree method upto F_6 and then massed. Tolerant to heat, resistant to *Verticillium* wilt, *Fusarium* wilt race 1, root knot nematode and A*lternaria* stem canker.
Neptune	Determinate, heat tolerant, bacterial wilt tolerant, and crack resistant. Resistant to *Fusarium* wilt race 1 & 2, *Verticillium* wilt race 1 and gray leaf spot.
Micro-Gold	It is resistant to *Fusarium* wilt race 1 & gray leaf spot. Fruits have high level of resistance to radial and concentric cracking blossom end rot & ripening disorders such as blotchy ripenning

Variety	Characteristics
	and gray wall.
Praiarie schooner	It is F_{11} selection from a cross between Droplet x Ontario 7710. Determinate, resistant to *Verticillium* wilt, *Fusarium* wilt and bacterial speck.
Prairie Dawn	It is F_{10} selection from a cross Campbell 28 x Duke (F_1). Determinate. Resistant to race 1 of *Verticillium* wilt and *Fusarium* wilt

Source: Tomato, P. K. Kalia, In: Planbt Breeding, Theory and Techniques, S. K. Gupta (Ed.), 2000, Agrobios (India), Jodhpur, pp 127-142.

Table 8.3: Sources of Pest Resistance varieties of Tomato

L. pimpinellifolium	A major source of resistance to *Fusarium oxysporium* race 1 and 2 and also of buck eye fruit rot and leaf curl gemini virus.
L. hirsutum	Resistance source of early blight, fruit borer and white fly.
L. hirsutum f. glabratum	Leaf curl gemini virus, fruit borer and white fly.
L. peruvianum	Root knot nematode.

Biotechnology Researches for Conservation

Tomato is one of the important agricultural crops. Biotechnologically, it has been successfully used because of its highly amenable manipulation quality. Much of research has been done to improve tomato for resistance to disease and pets besides quality. Gene transfer (using *Agrobacteriu*), protoplast fusion (to produce somatic hybrids) and embryo rescue to transfer desirable traits from alien species are the major achievements in this field. Among vegetables crops tomato is the most vulnerable crop for biotechnology since it has extensive genetic and cytogenetic mapping, several genetic markers, easy transformation and regeneration of plants from explants from a number of tissues besides its short generation time and easy inter-and-intra species hybridization (Stevens, 1986). The main empasis in tomato was to develop transgenics with delayed fruit ripening and virus resistance. During 1986-1995, 353 field trials of transgenic tomato were conducted worldwide, including USA (268), developing countries (33) and other countries (52), including two trials in Australia. Among the developing countries China had 9 trials and Thailand 2 trials and a few trials in Mexico (James, 1996). Till February, 1998, in USA alone there were 377 field testing of tomato transgenics (Medley, 1998). A few field trials of tomato transgenic have also been conducted in Europe (James, 1996). A large number of improved cultivars have been developed using biotechnological know how.

Table 8.4: Tissue culture techniques for conservation of tomato

References	Work conducted
Micropropagation	
Polevaya et al. (1988)	Obtained shoot from apical and axillary meristems cultured on MS medium with added gibberellins, Kin and glucose.
Izadpanah and Khosh-Khui (1989)	Used shoopt-tip explants taken from 8-week-old greenhouse-grown tomato cultivars Cal-j, Petomech and Red Cloud and cultured on MS medium supplemented with varying levels of Kin.
Dwivedi et al. (1990)	Studied direct shoot bud differentiation, without intervening callusing. It was induced in tomato leaf segments.
Ye et al. (1994)	Cultured cotyledons from 10 to 21-day-old seedlings with MS medium supplemented with 3% sucrose, 0.8% agar and 0.5-1 mg/1 IAA and 0.5-2 mg/1 zeatin. They obtained shoots (90%) of explants in 2 weeks, with the highest shot induction rate on medium with 1 mg IAA + 1 mg zeatin/1. Plantlets were obtained and transferred to soil in pots.
Ichmiura and Oda (1995)	Studied the induction of adventitious buds on cotyledon segments of tomato cv. Zuiken cultured on MS medium containing 0.1 mg/1 zeatin and 3% sucrose, with agar, agarose or gellan gum as a gelling agent of alternatively a polyester supporting material. They observed that many adventitious buds were induced and the induced shoots grew normally on all the agars and gellan gum.
Embryo culture	
Neal and Topoleski (1983)	They excised embryos 15 days after pollination and tested on 20 media. Growth and morphological development on six of the media were better than on MS medium.
Neal and Topoleski (1985)	They reported that Kin (10^{-8} or 10^{-7} M) promoted embryo development and expansion of cotyledons but inhibited subsequent growth. Kin (10^{-8} or 10^{-7} M) in combination with either GA_3 (10^{-8} or 10^{-7} M) or IAA (10^{-9} M) showed greatest potential for inducing development and growth of embryos excised before differentiation had occurred.
Gradziel (1988)	Conducted experiments by using various techniques

References	Work conducted
	like, bud pollinations where pollen germination on immature stigmas was enhanced by applying a germination medium, multiple pollinations, auxin application, embryo culture and environmental conditioning.
Leshem et al. (1989)	Observed that the growth of embryos excised from fruits 2 weeks after anthesis was slow.
Tipirdamaz and Karakullukcu (1993)	Studied *in vitro* culture of tomato cv. Falcon embryos in saline (NaCl) condition with proline and glycinebetaine supplemented media.
Chen and Adachi (1996)	Cultured globular stage embryos 13 and 15 days after pollination (DAP) *in vitro.*
Patil et al. (1993)	Produced backcross progenies following embryo rescue of two hexaploid somatic hybrid (SH) (*L. esculentum* + *L. peruvianum* accessions 6 and 18) with diploid tomato cultivars Moneymaker (MM) and Pusa Ruby (PR).
Demirel and Seniz (1997)	Reported that the 24-day-old embryos isolated and cultured on modified MS medium showed good germination and growth.
Anther culture	
Gresshoff and Doy (1972)	Haploids could be maintained in culture on defined minimal media for at least one year
Krzysko and Rogozinska (1975)	Indicated that callus cells could be induced to develop into seedless pseudofruits
Shchereva et al. (1990)	Isolated anthers at various stages from 2 forms (homozygous and heterozygous for the male sterility gene ms 1035) from each of 9 varieties and cultured on MS medium modified in various ways (MS2 and MS2 with different growth regulators for callus induction and organogenesis and MS3 without growth regulators for rooting and regeneration).
Summers et al. (1992)	Determined the microspore developmental stage and anther length on callus production.
Lee et al.(1999)	They observed changes in the gametes during the culture of anthers with pollens at different stages of development (meiosis, tetrads, uninucleate pollen and mature pollen) and best stages for inducing embryoid formation were those involving uninucleate or binucleate pollen and mature pollen.

References	Work conducted
Protoplast culture	
Leaf discs and mesophyll protoplasts	
Shahin (1985)	Regenerated shoots were easily rooted on the basal medium used plus the appropriate auxin (dependent on cultivar) and plantlets were planted in soil within 2 months of isolation of protoplasts
Wang et al. (1989)	They isolated leaf protoplasts from 2- to 3-week-old seedlings, and cultured on MS liquid medium supplemented with 1 mg/1 2,4-D and 0.1 mg/1 BA, followed by subculture on fresh medium.
Bellini et al. (1990)	Observed that low light intensity, a long photoperiod (16 h) and 25/17°C temperature (day/night) were best for obtaining viable protoplasts by using myoinositol as an osmoticum in the digestion medium.
Pindel et al. (1998)	Isolated protoplasts from mesophyll leaf tissue of *in vitro*-grown 3- to 4-week-old seedlings of two wild tomato species *L. glanculosum* PI 731163 and *L. peruvianum* PI 128650.
Cotyledonary protoplasts	
Chen et al. (1988)	Observed that growth of seedlings at 15°C, short treatment with an enzyme mixture to isolate protoplasts.
Stommel and Sinden (1991)	Evaluated the response to three cytokinins (zeatin, BA and Kin) on combination with IAA.
Chen and Adachi (1994)	Regenerated plantles through somatic embryogenesis from cotyledonary protoplasts of tomato cv. Kyoryokutoko.
Callus from protoplasts	
Latif et al. (1993)	Established a protocol for rapid, high frequency plant regeneration from *L. chilense* protoplasts.
Patil et al. (1994)	Used three Indian cultivars (PED, S7, and S10) for shoot regeneration from protoplast-derived calli.
Other tissues	
Sink et al. (1986)	Described a protocol for the protoplast culture of *L. esculentum* cultivars and *S. lycopersicoides* and their inter-generic protoplast fusion and subsequent shoot regeneration.
Burns and Pressey (1988)	Examined cell alteration after the addition of pectin methlyesterase (PME, EC 3.1.1.11) to locular gel and pericarp tissues of ripening fruits.
Gleddie et al. (1989)	They used longitudinal sections containing cortical cells taken from stem internodes of a hybrid between *L. esculentum* and *S. lycopersicoides* as tissue sources

References	Work conducted
	for enzymatic protoplast isolation.
Protoplasts for screening for resistance	
Motoyoshi and Oshima (1975)	Developed a procedure to isolate tomato mesophyll protoplasts. The procedure is sufficient for studying virus infection and multiplication.
Shahin and Spivey (1986)	Isolated protoplasts from cotyledons of UC82, which is susceptible to *Fusarium oxysporum* f. *lycopersici* race 2 and cultured on medium containing the nonspecific toxin fusaric acid.
Protoplast fusion	
(Zelcer, 1991).	Indicated the importance of somatic hybridization. The possibility of introgressing genes from *Solanum* species into *L. esculentum* and of obtaining novel cytoplasm-nucleus combinations (cybrids) is an important stratagem to extend the genetic variation available for tomato breeding.
Lefrancois et al. (1991)	Developed a procedure to regenerate plants from protoplasts of wild species with promising traits thought to be controlled by organelles
Identification of fusion products	
Jansen et al. (1991)	Induced cell organelle markers, particularly streptomycin and lincomycin resistance, in *L. peruvianum, L. pennellii, L. esculentum* and *S. lycopersicoides* by treating protoplasts or leaf tissue with N-methyl-N-nitrosourea.
Derks et al. (1990)	Produced somatic hybrids fusing protoplasts of a cytoplasmic albino mutant of *L. esculentum* as recipient and *L. hirsutum, S. nigrum, S. commersonii* and *S. etuberosum* as donors of cytoplasm.
Intergeneric and interspecific fusions	
Wijbrandi et al. (1988)	Introduced kanamycin resistance into tomato cv. Bellina by genetic transformation and its protoplasts were fused with (1) non-irradiated or (2) gamma-irradiated protoplasts of *L. peruvianum.*
Jain et al. (1988)	Indicated that Atrzine resistance can be successfully transferred from *S. nigrum* ($2n = 4x = 48$) to tomato cv. VF36 by protoplast fusion in the following combinations: (i) *S. nigrum* with tomato, (ii) UV-irradiated *S. nigrum* with tomato, (iii) UV-irradiated *S. nigrum* with iodoacetate treated tomato, and (iv) backfusion of tomato with a somatic hybrid.
Melzer and	Analyzed using isozyme and RFLP over 30 loci in 2

References	Work conducted
O'Connell (1990)	somatic hybrids generated from a single fusion event between *L. esculentum* protoplasts and gamma-irradiated *L. pennellii* protoplasts.
Wijbrandi et al. (1990)	Produced somatic hybrids between the cultivated tomato and wild *L. peruvianum* by fusion of leaf protoplasts from both species in the presence of PEG or in an electric field. Somatic hybrids were selected on the basis of kanamycin resistance of *L. esculentum* and the plant regeneratin capacity of *L. peruvianum.*
Bonnema et al. (1991)	Regenerated cybrids following protoplast fusion of iodoacetamide-treated leaf mesophyll cells of *L. esculentum* cv. UC82 and gamma-irradiated cell suspensions of *L. pennelli,* genotype LA716.
Wachocki et al. (1991)	Compared the organization of the mitochondrial genome in somatic hybrids and cybrids regenerated following fusion of protoplasts from *L. esculentum* and *L. pennellii* to assess the role of the nuclear genotype on the inheritance of organellar genomes.
Bonnema and O'Connell (1991)	Recovered somatic hybrid plants following fusion of leaf mesophyll protoplasts isolated from tomato cv. UC82 with protoplast isolated from suspension cultured cells of *L. chilense* LA 1959.
Sakata et al. (1991)	Fused cotyledon mesophyll protoplasts of *L. esculentum* cv. Ponderosa with mesophyll protoplasts from true leaves of *L. peruvianum* using polyethylene glycol (PEG) solution and cultured the fusion products *in vitro.* Somatic hybrids were selected by morphological characteristics of shoots which differentiated on the calli.
Ratushnyak et al. (1991)	Fused protoplasts of a chlorophyll-deficient muant with protoplasts of *L. peruvianum* var. *dentatum* inactivated by gamma-irradiation, and somatic hybrids were selected on the basis of capacity for photosynthesis and high regenerative ability.
Melzer and O'Connell (1992)	Recovered asymmetric somatic hybrids following fusion of tomato leaf mesophyll protoplasts with gamma-irradiated protoplasts isolated from *L. pennellii* suspension cells.
Bonnema et al. (1992)	Used RFLP to characterize the organization of the mitochondrial and chloroplast genomes in a population (82 individuals) of symmetric and asymmetric somatic hybrids of tomato. The protoplast fusion products were regenerated following the fusion of leaf mesophyll protoplasts of *L. esculentum* (cv. UC82) with

References	Work conducted
	suspension cell protoplasts of *L. pennellii* that had been irradiated with 5, 10, 15, 25, 50 or 100 kR from a gamma source.
Li and Sink (1992)	Fused mesophyll (M) and suspension culture (S) derivedrotoplasts of both tomato and its wild relative *S. lycoperisicoides* as S + M, M + M and S + S combinations respectively to resolve the role of parental cell types in determining ctDNA transmission to intergeneric somatic hybrid plants.
Gavrilenko et al. (1992)	Obtained intergeneric somatic hybrid plans by protoplast fusion between leaf mesophyll protoplasts form two cultivars of tomato (cv. Tamina and plastome mutant P1-alb I) and mesophyll protoplasts from two representatives of the Etuberosa series of the genus *Solanum* (wild species *S. etuberosum* and sexual interspecific hybrid *S. brevidens* x *S. etuberosum*).
Sakata and Monma (1993)	suggested that the presence of aneuploidy and chromosome pairing including autosyndesis in the H_1 were the main causes of morphological variations in individuals in the H_2.
Jourdan et al. (1993).	Produced somatic hybrids between three processing tomato lines (OH7870, OH832 and OH8245) and an accession of *L. hirsutum* (PI126445). All hybrids had inherited the ctDNA from the wild species but contained mtDNA from both parents. Some of the hybrids were fertile and produced abundant fruit and seed upon selfing.
Hossain et al. (1994)	Electrofused mesophyll protoplass of tomato cv. Ohgata Zuiko and *S. lycoperisicoides* and produced fertile tetraploid somatic hybrids.
Retushnyak et al. (1993)	Regenerated thirteen nuclear asymmetric hybrids under selective conditions following fusion of chlorophyll-deficient protoplasts from cultivated tomato (*L. esculentum*) and gamma-irradiated protoplasts from the wild species *L. peruvianum* var *dentatum.*
Bonnema et al. (1995)	Identified mutants in a population of cybrids formed following protoplast fusion between tomato (cf UC82) and *L. pennellii* LA716 based on two features, a variegated sectoring of light and dark green regions on their leaves, stems and fruit and reduced growth in the field.
Bruggemann et al., (1995).	They used *L. peruvianum* for transfer of desirable traits like cold tolerance and resistance to virus diseases. The cold tolerance was introduced from *L. peruvianum*

References	Work conducted
	to the somatic hybrids.
Kobayashi et al. (1996)	Developed somatic hybridization methods to facilitate the use of *S. ochranthum* for tomato germplasm improvement.
Samoylov et al. (1996)	Studied morphology, the extent of elimination of donor chromosomes and the organelle composition of highly asymmetric somatic hybrid plants between a interspecific tomato hybrid *L. esculentum* x *L. pennellii* (EP) as donor and a *S. melongena* (E) recipient.
Brown et al. (1996)	Reported the mapping of resistance to *Meloidogyne chitwoodi* derived from *S. bulbocastanum.*
Parokonny et al. (1997)	Used genomic *in situ* hybridization (GISH) to examine genome interactions in two allohexaploid ($2n = 6x = 72$) *L. esculentum* (+) *L. peruvianum* somatic hybrids and their seed progenies originated from subsequent backcrosses to *L. esculentum.*
Matsumoto et al. (1997)	Obtained somatic hybrid plants between the F_1 of the cross *L. esculentum* x *L. preuvianum* and *S. lycopersicoides* by symmetric electrofusion.
Kochevenko et al. (2000)	Regenerated fertile cybrid plants of three subclones, B1A, B3A, B4A from the single colony obtained after fusion of mesophyll protoplasts of plastome chlorophyll deficient mutant *L. peruvianum* var. *dentatum* (line 3767 and •-irridiatee mesophyll protoplasts of *L. esculentum* (cv. Quedlinburger Fruhe Liebe).
Microplast fusion	
Ramulu et al. (1996a)	The microprotoplasts contained one or a few chromosomes. Reported microprotoplast-mediated chromosome transfer (MMCT) through fusion of small (subdiploid) microprotoplasts of a transgenic triploid potato *(Solanum tuberosum)* cell lien with leaf protoplasts of tobacco *(Nicotiana tabacum)* and the wild tomato species *Lycopersicon peruvianum.*
Ramulu et al. (1996b)	Reported the results on the transfer of single, specific chromosomes carrying kanamycin resistance (KanR) and bea-glucuronidase (GUS) traits form a transformed donor line of potato *(S. tuberosum)* to a recipient line of the tomato species *L. peruvianum* through microprotoplast fusion.
Identification of somatic hybrids	
Poulsen et al. (1980)	Examined five somatic hybrids of potato and tomato obtained by protoplast fusion, using the isoelectric focusing patterns of ribulose 1,5-diphosphate

References	Work conducted
	carboxylase as phenotypic markers of the nuclear and chloroplast genomes.
Ninnemann and Juttner (1981)	Found hybrid cell calli contained more volatile components than were present in both parents. Therefore, the technique is useful for the identification of hybrid tissue derived by cell or protoplast fusion.
Schoenmakers et al. (1991)	Tried the selection of fusion products in 2 ways, namely fusion of nitrate reductase deficient tomato with potato gave rise only to hybrid callus, parental calli being rarely obtained and not regenerating into shoots; fusion of a cytoplsmic albino mutant of tomato with potato gave rise of albino and green plants.
Jacobsen et al. (1992)	Reported one endogenous genetic marker, the amylose-free (amf) mutant of potato, to be helpful not only for the confirmation of fusion products but also for the study of genetic complementation and the segregation of amylose-free starch in microspores.
Gavrilenko et al. (1994)	Used RAPD to study the genetic heterogeneity of the somatic hybrids. The wide genotypic and phenotypic variation in the hybrids is attributed not only to interaction of the parental genomes but also to somaclonal variation.
Wolters et al.(1995)	Analyzed the ctDNA type and mtDNA composition of 17 somatic hybrids between a cytoplasmic albino tomato (ALRC × M8-7) and monoploid potato 7322 (designated as A7 hybrids) and 18 somatic hybrids between a nitrate reductase-deficient tomato (C31-244) and monoploid potato 7322 (designated as C7 hybrids).
Rutgers et al. (1997)	Used microplast fusion and reported that the integration sites and copy number of alien marker genes[neomycin phosphotransferase II *(nptII)* (kanamycin kinase) and beta-glucuronidase (*uidA*)], can be introduced into diploid potato *Solanum tuberosum* through transformation by *A. tumefaciens.*
Shikanai et al. (1998)	Constructed a physical map of the mitochondrial genome for amala-sterile tomato MSA1 which had been generated by an asymmetric cell fusion between tomato *(L. esculentum)* and wild potato *(Solanum acaule).*
Garriga-Caldere et al. (1997)	Used GISH and RFLP analyses for establishing potato lines with monosomic additions of tomato chromosomes. RFLP analysis using chromosome-specific DNA probes indicated that BC_1 progenies had retained all 12 tomato chromosomes albeit in different individual plants.

References	Work conducted
Garriga-Caldere et al. (1998)	Produced by backcrossing three BC_1 genotypes of potato (+) tomato fusion hybrids to tetraploid potato pollinators, BC_2 populations.
Cytological studies	
Jong et al. (1993)	Used three somatic hybrids resulting from protopalst fusions ofa diploid kanamycin-resistant line of tomato *(Lycopersicon esculentum)* and a dihaploid hygromycin-resistant transformant of a monophaploid potato *(S. tuberosum)* line for a cytogenetic study on chromosome pairing and meiotic recombination.
Schoenmakers et al. (1993)	Selected some 25 allotriploids from 450 somatic hybrids obtained from fusion of protoplasts of diploid tomato (genotypes C31-244 and ALRC x M8-7) and monohaploid potato (line 7322).
Wolters et al. (1994)	Found that all somatic hybrids studied at root tip mitotic anaphase showed laggards and bridges at a higher frequency in the 3 anaeuploids than in the single euploid.
Schoenmakers et al. (1994a)	Examined some 110 asymmetric fusion products obtained by fusion of hygromycin-resistant tomato protoplasts and gamma-irradiated kanamycin-resistant potato protoplasts that expressed beta-glucuronidase (GUS).
Schoenmakers et al. (1994b)	Described the aggregation of nuclei in heterokaryons of tomato and unirradiated or irradiated potato protoplasts and the effect of gamma-irradiaiton of potato and tomato protoplasts on single (SS) and double-stranded (ds) DNA fragmentation, DNA repair and DNA synthesis as revelaed by alkaline and pulsed field gel electrophoresis and an immunocytochemical technique.

Genetic Engineering

Tomato is the most vulnerable crop for biotechnology since it has extensive genetic and cytogenetic mapping, several genetic markers, easy transformation and regeneration of plants from explants from a number of tissues besides its short generation time and easy inter-and-intra species hybridization (Stevens, 1986). The main empasis in tomato was to develop transgenics with delayed fruit ripening and virus resistance. During 1986-1995, 353 field trials of transgenic tomato were conducted worldwide, including USA (268), developing countries (33) and other countries (52), including two trials in Australia. Among the developing countries China had 9 trials and Thailand 2 trials and a few trials in Mexico (James, 1996). Till

February, 1998, in USA alone there were 377 field testing of tomato transgenics (Medley, 1998). A few field trials of tomato transgenic have also been conducted in Europe (James, 1996).

The development of **Breeder friendly diagnostic kits** for RFLP analysis seems to have commercial feasibility (De Verna and Alpert, 1990). Mapping QTL by RFLP analysis in tomato has been undertaken by Jones and St. Clair (1997) for resistance to late blight. Resistance to yellow leaf curl virus in tomato has received attention o biotechnologists in recent years in USA (Lindhout, 1997). Similarly, in tomato work is in progress on spotted wilt virus (Van Staden, 1997) and tomato mottle virus (Stevens, 1997).

Table 8.5: Transgenic Tomato released for commercialization

Country	Crop	Trait	Company
USA	Tomato	Delayed fruit softening (PG antisense)	Calgene (1994) Flavr Savr
	Tomato	Delayed ripening (ACC synthase antisense)	DNAP (1995) Endless Summer
	Tomato	Delayed fruit ripening (ACC synthase antisense)	Monsanto (1995)
	Tomato	Altered ripening	Agritope (1996)
	Tomato	Delayed ripening (PGL antisense)	Zeneca/Peto (1995)
	Tomato	Virus resistance (Bt) (Lepidoptera)	Monsanto (1997)
	Squash	Virus resistance (WMV2/ZYMY)	Asgrow (1995) Freedom II
China	Tomato	Virus resistance	Public Sector (1994)
	Tomato	Delayed ripening	Huazhong Agricultural University (1997)
Mexico	Tomato	Delayed ripening (`Flavr Savr')	Calgene (1995)
Australia	Tomato	Insect resistance	Public Sector (1997)
European Union	Tomato	Delayed ripening tomato	Zeneca (1995) (Product Only)

Source: James (1996), Medley (1998), Millis (1998), and Cheng (1998).

Table 8.6: Genetic Engineering Researches on tomato

References	Work conducted
Map-based gene cloning	
Martin et al. (1992)	Identified five yeast artificial chromosomes (YACs) containg regions for *Tm2a* and *Pto* genes, from a YAC library of tomato

References	Work conducted
Martin et al. (1993a)	Presented first report of successful map-based cloning reported tomato.
Martin et al. (1993b)	Described the multi step positional cloning strategy for map based cloning of *Pto* locus.
Martin et al. (1993c)	Demonstrated it by using a cDNA clone consegregating with *Pto* locus to transform a susceptible tomato plant.
Zabel et al. (1993)	Achieved the following steps towards the map-based cloning of *Mi.* : (i) identification of the acid phosphatase gene *Aps1,* a marker tightly linked to *Mi* and amenable to cloning through a product-based cloning approach, (ii) identification of various additionally linked RFLP and RAPD markers by screening genetic stocks of tomato differing only for a small chromosomal region carrying the resistance gene, (iii) construction of a high resolution molecular linkage map to allow chromosome walking in the *Mi* region, and (iv) construction of 2 long-range physical maps of the closest flanking markers (GP79 and Aps1) which together span over 1200 kb.
Daelen et al. (1993)	Isolated *Mi* through a map based cloning approach. They used pulsed field gel electrophoresis and various rarely cutting restriction enzymes in single, double and partial digestions to establish long-range physical maps of the two closest flanking markers, acid phosphatase 1 *(Aps1)* and GP79, which span over 400 and 800 kb respectively.
Zhang et al. (1994)	isolated a set of six tomato YAC cloens encompassing jointless physical mapping of one of the clones, TY142 clone allowed locaization of the target locus to aregino of less than 50 kb within the TY142 clone.
Dixon et al. (1995)	They made efforts to isolate the tomato *Cf-2* resistance gene to *Cladosporium fulvum (Fulvia fulva)* by map based cloning selected plants recombinant for RFLP markers close to *Cf-2* by exploiting the flanking morphological markers *yv* (ywllow virescent) and *tl* (thiaminless).

References	Work conducted
Giovannoni et al. (1995)	Used a combination of pooled sample mapping and classical RFLP analysis to construct the high density genetic maps for the regions of chromosomes 5 and 10 spanning the *rin* and *nor* loci which result in significant inhibition of the ripening process in tomato.
Pillen et al. (1996)	They constructed a high-resolution map of a 4.3 cM region surrounding the *Tm2a* gene, to clone *Tm2a* by positional cloning. In total, 13 RFLP and RAPD markers were mapped in close proximity to *Tm2a* using 2112 individuals from an intraspecific *Lycopersicon peruvianum* backcross.
Alpert and Tanksley (1996)	For the first time reported precise mapping of a QTL.
Brommonschenkel and Tanksley (1997)	Isolated two YACs containing genomic DNA for tomato spotted wilt tospovirus resistance gene *Sw5* using CT220, an RFLP marker which is tightly linked to the *Sw5*
Kaloshian et al. (1998)	Used fine-structure mapping of recombinants with AFLP markers and RFLP markers derived from physically mapped cosmid subclones localized *Mi* to a genomic region of about 550 kb.
Genetic transformation	
Agrobacterium tumefaciens mediated transformation	
McCormick et al. (1986)	Observed transformation to be less efficient with binary T-DNA vectors than with cointegrate T-DNA vectors.
Fillattii et al. (1987b)	Described an efficient transformation system in which tomato cotyledons were cocultured with *A. tumefaciens* carrying a binary vector with 2 *nptII* genes and a mutant *aroA* gene (conferring tolerance to glyphosate).
Shahin et al. (1986); Morgan et al. (1987) Sukhpinda et al. (1987) Lipp-Joao and Brown (1994)	Reported root transformation by using *A. rhizogenes*
Davis et al. (1991b)	Observed that tomato cotyledons could be transformed over a longtime interval after wounding although the highest transformation competency occurred during the first 24 h.

References	Work conducted
(Davis et al., 1991a).	Studied differences in transformation of th etomato cultivars Ohio 7870, Roma and UCD82b by the wild-type *Agrobacterium* strains A6, A66 and A281 were identified in a leaf disc assay system Ohio 7870 and Roma were more readily transformed than UCD82b by all 3 strains of *A. tumefaciens.*
Davis et al. (1992)	Investigated the possibility of involvement of wound ethylene in transformation by *A. tumefaciens* by testing the effect of ethylene inhibitiors and enhancers on crown-gall tumoru formation in tomato cotyledons cultured *in vitro.*
Rőekel et al. (1993)	Studied the effect of two parameters on the transformation frequency (1) The use of feeder layers during cocultivation, (2) that *Agrobacterium* strains harbouring a 1,1-succinamopine-type helper plasmid yielded significantly higher transformation frequencies than those with octopine-or nopaline-type helper plasmids.
Hamza and Chupeau (1993)	Studied effect of age of explant, genotype, duration of preculture with feeder cells on transformation.
Lipp-Joao and Brown (1993).	Observed enhanced transformation of tomato root clones cocultivated *A. tumefaciens* in the presence of 20 μM acetosyringone.
Pollen-mediated transformation	
Chesnokov et al. (1989)	They added exogenous DNA to the pollen of the recipient before selfing to transformation the tomato Fakel with plasmids contaning a gene for neomycin phosphotransferase II. Southern blot hybridization confirmed the integration of the gene into the genome of the material obtained.
Protoplast transformation	
Jongsma et al. (1987)	Transformed protoplasts from the new tomato genotype MsK93, which carried the regeneration potential of *Lycopersicon peruvianum* with plasmid DNAs containing a chimaeric kanamycin-resistance gene and putative tomato origins of replication, using a calcium phosphate DNA-mediated transformation procedure.
Microinjection	
Toyoda et al. (1988)	Described the apparatus and method to enable callus cells to be selected, injected with foreign DNA and cultured in the same plate.

References	Work conducted
Ilyubaev et al. (1988)	Injected DNA into cells regenerated from leaf mesophyll protoplasts in TM2 medium.
Direct transformation through seeds	
Glavinich et al. (1988)	Soaked seeds of Zheltyi Grushevidnyi (Yellow Pyriform) for 24 h in DNA extracted from seedlings of Moskvich. Four plants obtained showed the determinate habit and red fruit of Moskvich rather than the normal habit and yellow fruit of Zheltyi Grushevidnyi, but they had pear-shaped or plum-shaped fruit rather than the spherical fruit of Moskvich.
Twell et al. (1989)	Transiently expressed chimaeric genes containing a pollen-specific promoter from tomato or the CaMV (cauliflower mosaic caulimovirus) 35S promoter following their introduction into tobacco pollen using high velocity microprojectiles
Microprojectile	
Eck et al. (1995)	Used bombarded cell suspensions of *Lycopersicon esculentum* cv. VFNT Cherry and *L. pennellii* with tungsten particles coated with either plasmid or yeast artificial chromosome (YAC) DNA containing the *gus* and *nptII* genes
Baum et al. (1997)	Reported an improved protocol for the ballistic transient transformation of developing tomato *(Lycopersicon esculentum)* fruits which allowed high resolution cis-analysis of fruit-specific transcriptional activation.
Electroporation	
Tsukada et al. (1989)	Achieved optimum transfer of a plasmid harbouring the gene for chloramphenicol acetyltransferase (CAT) into protoplasts from cultivated tomato leaves by electroporation at 666 or 7000 V/cm (single dischare) using a 47 μF capacitor in the presence of 10-100 μg DNA. The *cat* gene was strongly expressed when under the control of the CaMV 35S RNA promoter. CAT activity was detected in cells 10 days after electroporation.
Nakata et al. (1992)	Established a simple and effective gene transfer system for tomato cells. They studied a stable transformation of *L. peruvianum* by electroporation using the kanamycin resistance gene *(nptII)* as a marker.

References	Work conducted
Liposome-mediated transformation	
Rosenberg et al. (1990)	Investigated conditions for efficient fusion of liposomes and plant cell membranes, gene delivery and gene expression using plasmids containing either the *cat* gene or a cloned dimmer of th etomato yellow leaf curl virus (TYLCV) genome.
Gene transfer	
Disease resistance	
Tumer et al. (1987)	Constructed and introduced a chimaeric gene encoding the alfalfa mosaic virus (AMV) coat protein into tomato plants.
Nelson et al. (1987)	Produced transgenic tomato plant expressing the coat protein (CP) of the common (U1) strain of tobacco mosaic virus (TMV).
McGarvey et al. (1990)	showed that plants transformed with a single copy of a tomato necrosis-causing satellite RNA of CMV expressed the satellite sequence but showed no disease symptoms and had a normal appearance.
Saito et al. (1992)	Introduced cDNA for T73-satRNA into a binary vector (pTOK162) through a homologous recombination in an *A. tumefaciens* cell and then transferred to leaf discs of tomato.
Motoyoshi and Ugaki (1993)	Produced transgenic F_1 hybrid (*L. esculentum* x *L. peruvianum*) which carried an introduced chimaeric tobacco mosaic tobamovirus (TMV) coat protein gene cDNA expressing the coat protein under the control of the 35S RNA gene promoter from cauliflower mosaic caulimovirus.
Kim et al. (1994)	Cloned the nucleocapsid protein *(N)* gene from the tomato spotted wilt tospovirus (TSWV) Hawaiian L. isolate.
Kunik et al. (1994)	Cloned the tomato yellow leaf curl (biemini) virus *(tylcv)* gene that encodes the capsid protein (*vi*) into an *Agrobacterium Ti*-derived plasmid.
Xue et al. (1994)	Employed *Agrobacterium*-mediated transormation to develop transgenic tomato of line G-80 that contained the coat protein gene of cucumber mosaic cucumovirus white leaf (CMV-WL) strain, a member of cucumovirus subgroup II. A total of 29 independently transformed G-80 plants were obtained.

References	Work conducted
McGarvery et al. (1994)	Produced transgenic tomato plants (genotypes UC82b and CL43) expressing cucumber mosaic cuvumovirus (CMV) satellite RNA fused to a *gus* gene for using *Agrobacterium*-mediated transformation.
Ultzen et al. (1995)	Reported the transformatin of an inbred tomato line with the TSWV nucleoprotein gene cassette resulted in high levels of resistance to TSWV that were maintained in hybrids derived from the parental tomato line.
Dong et al. (1995)	Transformed cotyledon segments of tomato with *A. tumefaciens* containing CMV satellite (sat) RNA cDNA. PAGE and Northern blot analysi sdemonstrated the expression of CMV sat-RNA gene in the transgenic plants.
Ynag et al. (1995a)	Regenerated 42 transgenic tomato from leaf discs following *A. tumefaciens*-mediated transformation with a chimaeric gene containing a cloned cDNA encoding the cucumber mosaic cucumovirus (CMV) coat protein *(cmv-cp).*
Yang et al. (1995b)	Inoculated transgenic plants of tomato 8805 R_1-R_4 progeny with *cmv-cp* gene at the seedling stage.
Gielen et al. (1996)	Used the *cmv-cp* gene for generating protection to CMV infections in cultivated tomato.
Gonsalves et al. (1996a)	Developed transgenic tomatoes that express translatable or non-translatable forms of the nucleocapsid (*N*) gene of tomato spotted wilt tospovirus (TSWV).
Gonsalves et al.(1996b)	Transferred the nucleocapsid protein gene of the lettuce isolate of tomato spotted wilt tospovirus (TSWV-BL) into a tobacco msaic tobamovirus (TMV) resistant tomato line (Geneva 80) via *A. tumefaciens.*
Cheng et al. (1997)	Introduced cDNA of cucmber mosaic cucumovirus protein (CMV-CP) into tomato (cultivars Zhongshu 5, Su 8807, Lichun) plants by *A. tumefaciens*-mediated transfer.
Wang et al. (1997)	Transformed tomatoes by *A. tumefaciens* containing GUS and CMV-CP (cucumber mosaic cucumovirus coat protein) genes on the Ti-plasmid constructs.

References	Work conducted
Sree Vidya et al. (2000)	Introduced the coat protein gene of *Physalis* mottle tymovirus (PhMV) through *Agrobacterium* transformation of cotyledonary leaves. The T_1 progenies segregated in monogenic ration showing Mendelian inheritance.
Fungal diseases	
Jongedijk et al. (1995)	Reported that simultaneous expression ofa tobacco class I chitinase and a class I beta-1, 3-glucanase gene in tomato resulted in increased fungal resistance, whereas transgenic tomato plants expressinog either one of these genes were not protected against fungal infection.
Insect resistance	
Fischhoff et al. (1987)	Determined the structure of an insect control protein gene from *Bacillus thuringiensis* subsp. *kurstaki* strain HD1 and generated truncated forms of the gene.
Salm et al. (1994)	Produced transgenic tomato plants exhibiting insect resistance due to the introduction of modified *cryIA(b)* and *cryIC* genes of *B. thuringiensis.*
Rhim et al. (1995)	Used a crysxtal delta-endotoxin gene of *B. thuringiensis* subsp. *tenebrionis (Btt)* encoding a coleopteran insect-specific toxin to construct a chimaeric gene which expressed the toxin in plant cells.
Herbicide resistance	
Fillatti et al. (1987b)	Regenerated over 100 transgenic plants containing 2 *nptII* genes and mutant *aroA* gene which conferred tolerance to glyphosate and 80% of these both exhibited NPTII enzyme activity and produced the mutant AROA protein.
Field evaluation	
Nelson et al. (1988)	Produced and field evaluated transgenic tomato plans that expressed the coat protein *(cp)* of the common (U1) TMV strain.
Delannay et al. (1989)	Field-tested transgenic plants expressing a lepidopteran-specific insect control protein from *B. thuringiensis* subsp. *kurstaki* for two years.

References	Work conducted
Sanders et al. (1992)	Reported that under field conditions, transgenic tomato plants that expressed the coat protein *(cp)* gene of common (U1) strain of tobacco mosaic tobamovirus (TMV) showed a high degree of resistance to the U1 strain and to a more severe strain of TMV, PV230. Tomato fruit yields of inoculated control plants decreased by 20% with U1 and 69% with PV230, whereas the CP+ linedid not show any yield reductio after inoculation with U1 or PV230.
Asakawa et al. (1993)	Investigated the impact on the environment of the release of tomato plans with an introduced gene for TMV resistance.
Noteborn et al. (1994)	Assessed the transgenic *Bt* toxin tomatoes for food safety. *In vivo* and *in vitro* studies in mammals indicated the absence of specific binding sites for *cryla(b)* and of acute pathologic effects.
Fuchs et al. (1996)	Evaluated transgenic tomato plants expressiong the coat protein gene of cucumber mosaic cucumovirus (CMV) strain WL for resistance to CMV infections under field conditions for 2 years. Three transgenic inbred lines, 2 hemizygous and 1 homozygous and 1 transgenic hybrid were field-tested.
Transposon mutagenesis system	
Yoder et al. (1987)	Mobilized maize transposable element Ac in tomato. The Ac element was isolated from the maize wx-m7 allele and introduced into tomato via *Agrobacterium.*
Belzile et al. (1989)	Observed sexual transmission of transposed activator elements in transgenic tomatoes.
Yoder (1990)	Observed rapid proliferation of the maize transposable element activator in transgenic tomato. Based on DNA gel blot hybridization he concluded that the mechanism of Ac amplfication was associated with transposition.
Belzile and Yoder (1992)	Observed dispersed small clusters of insertion sites. Majority of the insertion sites were linked to other insertion site.
Osborne et al. (1991)	Investigated the distribution of transposed Ac (activator) elements of maize in the tomato genome.

References	Work conducted
Khush and Yoder (1992)	Used Southern hybridization to assess the activity of activator (Ac) elements in progeny plants derived from a tomato transformant carrying five Acs at two loci.
Rommens et al. (1993)	Studied transposition pattern of a modified Ds element in tomato.
Healy et al. (1993)	Used a novel, sable allele of Ac (Ac3) having a deletion of the outermost 5 bp of the 5'-terminal inverted repeat.
Knapp et al. (1994)	Obtained transgenic tomato lines containing Ds elements at defined genomic positions.
Belzile and Yoder (1994)	Examined the behaviour of the maize transposable element Ac in transgenic tomato with the goalof developing an efficient insertional mutagenesis system.
Carroll et al. (1995)	Analysed the pattern of germinal transpositions of artificial dissociation (Ds) transposons in tomato. T-DNA constructs carrying Ds were transformed into tomato, and the elements were transactivated by crossing to liens transformed with a stabilized activator (sAc) that expressed the transposase gene.
Peterson and Yoder (1995)	Observed that amplification of Ac in tomato is correlated with high Ac transposition activity.
Fruit ripening	
Grierson and Fray (1994).	Indicated that at least 25 genes showing elevated expression during ripening have been cloned and several including polygalacturonase which modifies fruit texture have been found to be ripening-specific In addition, genes have been cloned for ACC synthase and ACC oxidase which control the synthesis of ethylene, a substance critical in ripening.
Polygalacturonase	
Antisense	
Smith et al. (1988)	Introduced a plasmid containing the cauliflower mosaic virus (CaMV) 35S promotor.

References	Work conducted
Kramer et al. (1992)	Developed processing and fresh market tomato genotypes with an antisense PG construct and lines which produced fruits with PG levels reduced by more than 99%. Analysis of field-grown material demonstrated a significant increase in the serum viscosity of processed juice and paste and a significant decrease in softening during storage of fresh market fruit relative to non-transgenic controls.
Sense	
Smith et al. (1990)	Observed that in tomato plans transformed with a chimaeric polygalacturonase (*pg*) gene, designed toproduce constitutively a truncated PG transcript, expression of the endogenous *pg* gene was inhibited during ripening, resulting in a substantial reduction in PG mRNA and enzyme accumulation.
Poole (1993)	Transformed tomato lines with a chimaeric gene in which a 730 bp fragment from the 5' end of the PG cDNA clone pTOM6 including the start of the protein coding region was fused to a 528 bp fragment of the CaMV 35S promoter and a 261 bp nopaline synthase terminator sequence.
ACC synthase	
Antisense	
Oeller et al. (1991)	Reported that expression of antisense RNA to the rate-limiting enzyme in the biosynthetic pathway of ethylene, ACC synthase, inhibited fruit ripening in tomato plants.
Reed et al. (1995)	Developed and characterised tomato plants that were delayed in fruit ripening by *A. tumefacines*-mediated transfer of a gene encoding 1-aminocyclopropane-1-carboxylate deaminase (ACCd) from the soil bacterium *Pseudomonas chloroaphis* into the tomato genome.
Expression of ACC deaminase	
Klee et al. (1991)	Cloned the bacterial gene encoding ACC deaminase and introduced into tomato plants.
Quality assessment of transgenic tomato	
Schuch et al. (1991)	Analyzed fruit quality and composition in transgenic plans of tomato cv. Ailsa Craig, modified by the expression of antisense RAN to polygalacturonase (PG).

References	Work conducted
Carrington et al. (1993)	Observed that cell walls of tomato fruit exhibited increasing solubilization of pectins as ripening proceeded; this process was not evident in fruit from transgenic plans with the antisense gene for polygalacturonase (PG).
Sozzi-Quiroga and Fraschina (1997)	Evaluated sensory attributes and biochemical parameters in transgenic tomato fruits of line CR3 with reduced polygalacturonase activity.
Porretta et al. (1998)	Evaluated the physiochemical and sensory properties of tomato pulps (i.e. diced tomatoes with 30% tomato juice as packing medium), prepared from transgenic tomato fruits with reduced levels of polygalacturonase (PG) activity due to the expression of a *pg* antisesnse gene.
***In vitro* conservation**	
Zhao et al. (1993)	Studied the factors affecting the viability of tomato pollen stored in liquid nitrogen (LN). Pre-freezing at –25°C increased the viability of pollen stored in LN.

CAPSICUM

Capsicum is an important vegetable cum spice crop. The genus *Capsicum* belongs to the family Solanaceae. Paprika is the second most important spice crop with world production estimated at 60,000 t/ha (Buckenhuskes, 1999). It consists about 100 economically important species and many of the species have various botanical varieties. Two major important species are *C. frutescens* and *C. annuum.* The important species are:

C. annuum	sweet pepper or chilli
C. baccatum, C. frutescens	tobasco pepper
C. chinense and *C. pubescens*	

Capsicums (sweet or hot peppers) is sometimes confused with *Piper nigrum* (spice crop). Capsicums are grown all over the world and are predominantly tropical while paprikas are grown in cooler climates.

Much of the work has been done on anther culture. The varietiers developed through anther culture are grown in an area of 38,400 ha in China (Li et al., 1995). One of the promising varieties developed through anther culture was Haihua 3. Many diseases caused by fungus and viruses are likely to be controlled by the development of transgenics carrying genes for resistance.

Biotechnological researches should be directed to develop varieties resistant to cucumber mosaic, phytophthora diseases, fruit rot (*Colletotrichum capsici*) and thrips and white fly, Germplasm preservation, etc.

Tissue culture

Micropropagation

Micropropagation in *Capsicum* has been reported with shoot buds, nodes, axillary buds and shoot-tips. Rogozinska and Drozdowska (1996) compared two methods of pepper regeneration for microcutting production. Swamy (1983) reported that regeneration of shoot buds and multiple shoots occurred from cotyledon and shoot-tip explants respectively of *C. annuum* cultured on MS medium supplemented with various growth regulators of which BA was the most effective for regeneration. IAA, NAA and IBA induced roots from explants. Sun and Wang (1989) developed a technique for rapid mass propagation by *in vitro* culture of stem developed a technique for rapid mass propagation by *in vitro* culture of stem meristems using *C. annuum* cv. Qiemen, *C. frutescens* form 310, *C. praetermissum* form 234 and *C. baccatum* form 85008 on supplemented MS medium.

Anther Culture

Studies have been conducted on anther culture of capsicum in order to produce dihaploids with many desirable traits.

Genetics of anther-cultured plants: Genetic expression of major characters in sweet pepper lines derived by anther culture. Chen (1984) found in the study of 74 H_2 and 5 H_3 lines of *Ca. annuum* var. *grossum* derived by anther culture that there was considerable variation in the botanical and agronomic characters between lines derived from the same parent.

(Jiang and Li, (1984) distinguished the main fruit characters of lines derived by anther culture from a sweet × hot *C. annuum* F_1 hybrid and grown on up to 5 generations (H1-H5) were inherited from the parent and plants in H2-H5 generations. They found them uniform. Li and Jiang (1990) reported that Haihua 3 has a wide adaptability in China and a high tolerance to diseases. Li et al. (1995) developed 6 cultivars directly from anther culture and a further 6 hybrids with parents developed from anther culture have been cultivated on over 38,400 ha throughout China.

Venczel and Mityko (1995) used anther culture technique to produce maintenance lines of the Hungarian variety Feherozon Synthetic (FS). It was composed of 8 different genotypes but identical phenotypes which were stable for the 5 main traits (growth habit and fruit colour, flavour, shape and position) were stable and homozygous

Morrison (1987) detected gametoclonal variation in doubled haploid lines (DH) of pepper. These lines were evaluated in the field for a number of traits, like plant height, flower morphology, yield per plant and fruit quality.

Anther culture has been used to develop double haploid lines showing resistance to different diseases and pests. Abak et al. (1982) evaluated a total of 27 doubled haploid C. annuum plants from the cross PM217 (resistant) × Yolo Wonder (susceptible) for both root and stem resistance to *Phytophthora capsici.* Daubeze et al. (1989) reported that the resistance of *C. annuum* line CM334 to strain S197 of *P. capsici* and to pathotype 0 of tobacco mosaic tobamovirus. It remains stable even at temperatures >30°C. It was genetically analyzed using 2 sets of doubled haploid lines obtained by anther culture. Daubeze et al. (1989) made crosses between the line H3 (bred in Ethiopia by crossing a local population with a Kenyan cultivar) with resistance to *Leveillula taurica* and the susceptible Vania. They obtained doubled haploids from the F_1s by *in vitro* culture and colchicine treatment.

Hwang et al. (1998) screened 33 breeding lines of peppers for resistance to bacterial spot (*Xanthomonas campestris* pv. *vesicatoria*). Plants derived from anther culture of resistant lines were screened again and analyzed for inheritance of resistance.

Evaluation of Anther-cultured Plants: Anther culture technique is a useful tool for producing viable haploids and generating completely homozygous plants in *C. annuum.* However, spontaneous doubling of chromosomes is also common. Wang et al. (1981) cultured anthers of 19 cultivars subjected to anther culture and only Squarehead and Lu-tai produced embryoids from which haploid plantlets developed. Diploidization, induced by treatment with 0.2% colchicine for 24 h, resulted in flowering and seed set.

Androgenesis occurs in *C. annuum* anthers when they are incubated in a continuous warm environment (29°C) and continuous light. Munyon et al. (1989) reterived 40 plantlets and embryoids and analyzed for isoenzyme markers; 35 exhibited a single allele for markers suggesting microspore origin, while 5 were heterozygous indicating somatic tissue origin. Sixty-nine androgenic plants from 4 accessions by culturing anthers *in vitro* under carefully controlled conditions (on an average 4.7 plantlets per 100 anthers cultured) were obtained by Barcaccia et al. (1999). Thirty-three of these plants spontaneously underwent chromosome doubling and were cytologically proven to be doubled haploids. Following are the important researches made in this field:

Table 8.7: Tissue culture efforts for conservation of capsicum

Reference	Work conducted
Harn et al. (1975)	2,4-D was effective in inducing haploid callus when used in combination with NAA and BA. Haploid callus and embryoids were induced more frequently when the anthers were at the late uninucleate stage Activated charcoal has been found to be useful supplementation. 2,4-D alone or in combination with cytokinin is commonly used for induction of embryoids. The response varies between genotypes.
Vaulx et al. (1981)	Studied the effects of initial treatment of cultured anthers at 35°C in the dark for 2 days or 8 days in eight F_1 hybrids.
Morrison et al. (1986)	Prechilling of flower buds has been recommended for success in anther culture. Excised anthers from prechilled flower buds of *C. annuum* cv. Emerald Giant × *C. chinense,* cultured in the dark at 35°C for 8 days and 25°C for 4 days and then floated on liquid medium overlaying a solid medium supplemented with 2.0% charcoal.
Wu and Zhang (1986)	Studied the effect of acridine yellow on development of anthers of *C. frutescens* var. *longum* cultured *in vitro. In vitro* acridine yellow treatment of anthers of Haihua 28 at 100-1000 ppm inhibited synthesis of DNA, ATPase, proteins and polysaccharides in tapetum, tetraspore and microspore.
Yoon et al. (1991)	Cultured anthers in the mid-uninucleate to early bicellular stage, from 5 F_1 hybrids of hot pepper on Dumas de Vaulx basic medium supplemented with various growth regulators. Efficiency of anther culture was dependent on donor genotype and induction of embryoids from microspores and was best on medium containing 0.1 mg Kin and 0.1 mg 2,4-D/1 or medium containing 0.1 mg 2,4-D and 0.1 mg NAA/1.
Mak and Maheswari (1994)	Used cold and heat shock pretreatment to the anthers of 10 varieties and 1 F1 hybrid before culture and incubated at 25°C with 12 h light or 29°C with continuous light. Only 6 varieties showed a response in terms of callus, embryoid and plantlet formation (ranging from 0.7% to 9.5%).
Gonzalez-Melendi et al. (1995)	Induced pollen embryogenesis in *C. annuum* by selecting the late vacuolated microspore. It is the most suitable developmental stage for the induction.
(Qin and Rotino, 1995)	Number of guard cells in chloroplast has been used as ploidy indicator of *in vitro*-grown androgenic pepper plantlets.
Gyulai et al. (1999)	Used the spontaneous doubled haploid (DH-R1) plants developed from F_1 hybrids of *blocky type* pepper (2n=24) in anther culture and advanced generation DH-R_2 in a greenhouse for flow cytometric analysis, which revealed plans with haploid (n) and diploid (2n) genomes.

Thidiazuron (TDZ) has been reported to be very effective in inducing androgenesis. The best results on callus induction were seen in the hybrid with *C. chinense* on MS medium with TDZ on morphogenesis from callus in the hybrid with *C. eximium,* and in direct embryoid induction in Kalinkov and Byala Kapiya on MS medium with carrot extract (Pandeva et al., 1990). Cytokinins and phenyl urea derivatives, macroelements and carbon source have also been effective in induction of androgenesis. Carbon source has also been reported to influence andrognesis. Dolcet-Sanjuan et al. (1997) described a new and simple protocol for androgenesis in *C. annuum.* The initial medium, a modification of Nitsch and Nitsch H medium, consisted of a 2-phase system of semi-solid and liquid medium and contained maltose as a carbon source. The total number of embryos formed was highest with maltose at 40 g/1, but embryos developed better at 10-20 g/1. **Carbenicillin,** an antibiotic (50 mg/1) inhibited bacterial growth and positively influenced embryoid production.

Protoplast Culture

Table 8.8: Selected work on protoplast culture of capsicum

Reference	Work conducted
Saxena et al. (1981)	Isolated protoplasts from 15-day-old leaves in an MS solution containing cellulase, macerozyme and mannitol.
Donato et al. (1989)	Isolated protoplasts from young cotyledons of cv. Yolo Wonder and cultured on NT, KM or B5 liquid medium without hormones which survived for many days but did not proliferate. Auxin added to the culture medium reduced protoplast viability. Protoplasts proliferated when purified rather than crude preparations of enzymes were used for cell wall degradation.
(Murphy and Kyle, 1994)	Protoplasts can be isolated from cotyledons, hypocotyls, leaf mesophylls and callus tissues. Viable protoplast were isolated from leaf and cotyledon explants of *Capsicum*
Szasz et al. (1995)	Cultured cotyledon, hypocotyl and leaf mesophyll protoplasts from 31 genotypes of *Capsicum* (including *C. annuum, C. baccatum, C. frutescens* and *C. chacoense*) using an alginate disc embedding method.

Somatic Embryogenesis

Somatic embryogenesis in *Capsicum* has been induced directly and indirectly via callus. Various explants such as cotyledons, leaves, embryos, etc. have been used for indirect embryogenesis. But for direct embryogenesis only embryos have been used. Following aspects of somatic embryogenesis have been studied:

Table 8.9: Selected work on somatic embryogenesis of capsicum

Reference	Work conducted
Binzel et al. (1996)	Direct somatic embryogenesis and plant regeneration in pepper.
Buyukalaca and Mavituna (1995b)	Synthetic seeds have been developed using the somatic embryos.
Kintzios et al. (1998)	Investigated the effect of light on the induction, development and maturation of somatic embryos, after prolonged incubation of cultures. Cotyledon and leaf explants were cultured on solid MS medium supplemented with various plant growth regulators for the induction of callus and somatic embryos.

Production of secondary metabolites

Capsaicin is the secondary metabolite of *Capsicum* (*C. frutescens*) . It has been obtained from plant cell culture (Yoeman et al., 1982). The production of capsaicin is increase in tissue culture by adding precursor-like vanillylamine and isocapric acid, eliminating sucrose from the cultural media and decreasing the level of N to 5% in the MS solution.

In vitro root cultures of chilli pepper was employed for biosynthesis of sesquiterpenic phytoalexin capsidiol. The production of capsidiol was elicited with cellulase. The optimum concentration of cellulase alogn with sucrose for production of capsidiol was established. Maximal amount of capsidiol was secreated in the medium at 24 h after elicitation (Chavez-Moctezuma and Lozoya-Gloria, 1996).

Genetic Engineering

Biochemical and molecular markers

Multiple resistances *Capsicum* are available to several pests and diseases, but have to be transferred from neo agronomic or market type of pepper to another. Problems in selecting simultaneously for multigenic resistances and polygenic quality characters maybe eased by the development of molecular markers and a molecular linkage map for *Capsicum.*

Table 8.10: Selected work on *in vitro* conservation in *Capsicum* have (seeds and pollen grains)

References	Work conducted
Seeds	
Lanteri (1990)	Cryopreserved pepper seeds
Belletti et al. (1990)	Observed the possible damage caused by cooling and rewarming in association with liquid nitrogen storage. Therefore, he subjected 3 seeds samples of *C. annuum* at 3 water content levels (about 4, 6 and 10%) to slow and fast cooling and rewarming rates. Seed viability was checked after 1, 3, 5, 10 and 25 complete cycles. The results showed almost the same pattern both in pepper and eggplant: seed viability was not affected by slow or fast cooling and rewarming rates and the germination percentage was slightly lowered by the increase of cycle number.
Pollen	
Kristof and Branabas (1983)	Stored pollen grains of the *C. annuum* varieties Soroksari Hajtato and Hatvani Hajtato at – 196°C in liquid N2 for 3 and 10 months and their capacity for germination and fertilization was retained.
Bezdickova (1989)	Stored at 4°C for >16 days resulted in a loss of pollen viability in line CW, the male parent of Dora F1. At –20 and –50°C, pollen germination and functional capacity (tested by controlled pollination under normal growing conditions) was maintained for about 66 days, decreasing by about 50% after 120 days.
Bezdickova (1988)	Observed germination of pollen of *C. annuum* line CW in liquid medium (supplemented with sucrose and H_3BO_3) following storage at 4, -20 and –50°C. Germination capacity was unimpaired following storage at 4°C for up to 16 days, but declined rapidly thereafter.

ONION

The common bulb onion, *Allium cepa* L. (2n = 2 × = 16). It belongs to family family Alliaceae. Its history is very old, its use going back over 4000 years. It is probably a native of South Asia or Mediterranean region. In Egypt is was worshipped before the Christian era.

They thrive best in cool-moist regions with a sandy soil. It is the most economically important cultivated *Allium* species with a total world production of 32,402,000 tonnes (1995, FAO). It is used as food in most countries of the world and grown in nearly all cool regions. The leading onion producing nations; China, India, Turkey, USA, Japan, Iran, Pakistan, Egypt, Brazil, Poland, Russian Federation etc.

Onion is most widely cherished vegetable due to its flavour. It is used for flavouring or seasoning the food, both at mature and immature bulb stages, besides being used as salad and pickle. It is also used in processing industry for dehydration in the form of onion flakes and powder. It is a major commodity of international trade.

Being onion a biennial and allogamous species, homozygous line development represents a necessary step either before hybrid production or for gathering useful genes.

With the spread of onion culture the world over, cultivars evolved more and more diversity in shape, colour, flavour, keeping quality and critical adaptations to cultivation in new climates and environments. Bulbing in onion is responsive to day length and high temperatures and bolting to low temperature. These are the characteristics which onion breeders carry on to deal with. The present day onion is self compatible but is an outbreeder and is subject to inbreeding depression. It has retained the breeding system of its early ancestors and has evolved independently the contacts with other species for several thousand years. Its hybrids with related wild species are highly sterile. The important cultivar varieties of onion are as follows:

Table 8.11: Some important cultivars evolved through different methods are mentioned here:

Variety	Characteristics
Agrifound Dark Red	It was developed through selection by the National Agricultural Development Foundation from a local stock of *Kharif* onion grown at Nasik. Bulbs deep red, globular with tight skin and moderate pungency. Suitable for *Kharif* season and has best storability.
Arka Bindu	It was developed through selection at IIHR, Banglore. A rose coloured variety with high yielding potential (35 tonnes per hectare).
Arka Kalyan	Developed by mass selection from local collection (IIHR 145) at IIHR, Banglore. Bulb globular, deep red; moderately resistant to purple blotch.
Arka Niketan	Evolved through mass selection from a local collection (IIHR 153) at IIHR, Banglore, Bulbs globose with thin neck red colour and high pungency. Suitable for growing in *rabi* and *Kharif* reasons.
Arka	Evolved through selection at IIHR, Banglore. A yellow

Variety	Characteristics
Pitambar	coloured variety with high yielding potential (35 tonnes per hectare) and suitable for export.
Arka Pragati	It was evolved through seven cycles of mass selection from a local collection (IIHR 149) at IIHR, Banglore. Bulbs globe shaped, pink, thin neck, highly pungent and early maturing. Suitable for *Rabi* as well as *kharif* season.
CO 4	This is a derivative of the cross AC 863 x CO_3.
Early Grano	Bulbs large, yellow with mild pungency. Storage life is rather short. (N.S.C. India)
Makoi CR	Evolved through selection from Makoi in Hungary. Resistant to bolting.
NuMex Sundia and NuMex Suntop	These were developed through recurrent selection from Ben Shemen. These are non bolting types.
NuMex Sunlite	It was developed by selection from Texas Grano 502 PRR. It is resistant to bolting.
Punjab selection	It was developed by mass selection from indigenous material. The bulbs are red, firm with good keeping quality. Suitable for dehydration. Tolerant to purple blotch and thrips.
Pusa Madhvi	It is a selection, developed by selfing and massing method from material collected from Kairana (UP). Bulbs medium to large, light red and flatish round.
Pusa Ratnar	It was developed through selection from an exotic hybrid cultivar Red Granex from USA. Bulbs are bronze deep red, obovate to flat globular, less pungent and drooping neck with good storage qualities.
Pusa white Flat	Developed through selection at IARI, New Delhi. Bulbs white, medium to large, flat. It has good storability and suitable for dehydration.
Pusa red	Medium sized, round red bulbs, less pungent than moist red varieties. Keeps very well in storage. Characteristically free from bolting tendency. Short to intermediate day length type.
Texas Grano 1015Y	It was derived through 5 generations of selection from an original single selection from Texas Early Grano 951.
Texas Grano 1025 Y and 1105 Y	These were developed through hybridization from a cross of Texas Early Grano 502 x Ben Shemen.
Texas Grano 1030 Y	It was developed from an F_2 late maturing selection from a cross Texas Early Grano 502 x Ben Shemen.
Yibilei 50	It was selected from a cross Mako x Bernsteinfarbige in Bulgaria.

Biotechnological Approaches for Conservation

The in vitro culture of unpollinated ovaries and flowers, is becoming in this crop a very reliable technique that allows to obtain homozygous lines in only two years whereas the conventional method (selfing) requires ten years. The production of onion haploid plants through in vitro gynogenesis has been reported. Dore and Marie (1993) reported the production in planta of haploid plants after fertilization with irradiated pollen.

Recently, new results concerning chromosome doubling as well as genetic uniformity and stability have also been obtained (Campion et al., 1994). Notwithstanding all the efforts made in many directions to achieve improvements, the technique now available still shows a few limits, first of all the reduced yield of haploids obtained from many genotypes.

Gynogenesis in Onion

Most of the work done on these aspects is from the laboratory of B. Champion and M. Schiavi, Instituto Sperimentale per l' Orticoltura, Montanaso (Italy). They have presented a review article on the whole process of gynogenesis in onion, starting from the cultivation of donor plants in greenhouse till to the production of the first seed-bulk of *Doubled Haploid* (DH) lines? The best version of the whole technique, here reviewed and synthesized in this chapter, has been assembled by gathering the best world results so far achieved for each step of the process. The composition of the most important media is shown in following table (8.12).

Table 8.12: Composition of the media used for the in vitro production of onion DH lines

Media					
Basal medium*	+	+	+	+	+
Sucrose (g.1^{-1})	100	100	100	40	40
2, 4-D (mg.$^{-1}$)	2				
NAA (mg.1^{-1})		2		0.25	
6BA (mg.1^{-1})	2	2		2	

(i) BDS salts (Dunstan and Short, 1977), thiamine 2 mg.1^{-1}, pyridoxine 1 mg.1^{-1}, nicotinic acid 1 mg.1^{-1}, Ca pantothenate 1 mg.1^{-1}, myoinositol 500 mg.1^{-1}, agar-agar (Merck) 6 g.1^{-1}.

(ii) **pH of all media was adjusted to 5.9 before autoclaving at 119° C for 20 minutes.

Factors Affecting Embryo Yield

Genotype and Cultivation of Donor Plants: It is known that genotype is one of the most important factors affecting the in vitro responses. Muren (1989) classified its donor-cultivars into high, medium and low responsive when the embryo yield per 100 ovaries were 2 or more, between 1 and 2, less than 1 respectively; the frequency of embryos reported by Keller (1990) in ovary culture of 9 cultivars, ranged from 0.75 to 2.86%, whereas, that obtained in our laboratory from both ovary and flower culture of 6 cultivars, ranged from 0.1 to more than 5%.

Relation between Flower Age and Embryo Yield: A relation between flower age and embryo yield was studied by Muren (1989). He found that the most responsive ovaries were excised from flowers collected 3 to 5 days before anthesis. At this age, the embryo sac is at the megaspore mother cell stage (Guha and Johri, 1966) and, probably, all the phases of meiosis occur in vitro during the first days of culture. No studies were made to fine eventual relations between flower age and haploid/diploid (n/2n) plants obtained.

Temperature Pre treatment: In onion and Gerbera the effect of cold and heat pre treatment on gynogenesis response has been seen, however, no useful results were obtained. Only in sugar beet cold pre treatment seemed to give an increase of gynogenic embryo yield.

In vitro Culture Conditions

Most of researchers involved in haploid production, often undervalue light importance under in vitro condition. There are no data concerning the use of gro-lux light for ovary and flower culture. Gro-lux light inhibits flowering and stimulates vegetative growth in onion adult plants.

Growth Regulators, Nutrient Factors and Type of Organ Cultured

The first haploid plants of onion from unpollinated ovules were obtained by Campion and Alloni (1988, 1990), from ovaries by Muren (1989), from ovules, ovaries and flowers by Keller (1990). Substantial improvements in ovule, ovary and flower culture were later achieved in our laboratory (Campion et al., 1992) by applying changes to the composition of the media previously described. In ovule culture, BDS salts (Dunstan and Short, 1977) gave better results than MS salts (Murashige and Skoog, 1962) whereas, in flower and ovary culture, BDS salts did not statistically differ at $P = 0.05$ from B5 (Gamborg et al., 1968) salts (Campion et al., 1992). The type of auxin used in the medium was found to be very important and in relation to the type of organ cultured. NAA gave a statistically higher embryo yield than 2,4-D in ovary culture, on the contrary, it appeared statistically less efficient than 2,4-D in flower culture (Campion et al., 1992). Among the nutrients, sucrose is the factor that mostly influences embryo yield. At 100 $g.1^{-1}$ it provided a statistically highest embryo yield in ovary culture (Muren, 1989) although, in our laboratory, we could obtain in one case 5 gynogenic embryos from 800 unfertilized ovules cultured with 15 $g.1^{-1}$ of sucrose (unpublished data). The influence of this and other sugars (i.e. maltose, glucose, fructose) should be deepen understanding.

Ovules proportionally yield as ovaries or flowers but require much more hand labour to be excised. Ovary culture is simpler than ovule culture but onion although embryo yield is, in the last case, sometimes lower.

In onion, gynogenic embryos often took more than 60 days before sprouting from flowers and ovaries. An earlier embryo sprouting could be obtained by reducing to 15 days or less the time of exposure of organs (ovaries or flowers) to growth regulators in the medium without modifying the embryo yield (Campion et al., 1994a). Although at a very low percentage, we could also induce gynogenesis in ovaries and flowers on a growth regulator-free medium. this surprising finding, obtained in the cultivar Borettana (the most responding) (Campion et al., 1994), indicates that onion already has a propensity to gynogenesis which can be enhanced by growth regulators.

When haploids were induced in planta through the fertilization with irradiated pollen, apparently normal seeds were formed during 35 days but, subsequent germination had to be carried out in vitro to avoid eventual loose of gynogenic material (Dore and Marie, 1993).

Establishment of DH Line

Plant Development

Their are three distinguished steps of haploid plant production:

1. Culture of unpollinated ovules, ovaries or flowers the product of which we called embryos;
2. Development of gynogenic embryos into rooted plants 8 to 10 cm high;
3. Micropropagation and transplant to soil of gynogenic clones.

The distinction between embryos and plants, used respectively to indicate the first product of ovary or flower culture and the well developed plants, is only convenient but not always correct since, sometimes, the first visible product of ovary or flower culture is a small plantlet. For this reason, other authors preferred to use the term regenerants (Keller, 1990a,b) without distinguishing the two steps.

Ploidy Level of Gynogenic Plants

The examination of ploidy level carried out in the gynogenic plants of onion, led to discordant results. Keller (1990b) obtained 62% of haploids, Muren (1989) 70%. whereas Campion et al. (1994) contained 90%. This could be due to the different age of flowers collected since no difference was so far found between ovary and flower culture on n/2n plant production.

Cytological observation of root tips of gynogenic plants often revealed the simultaneous presence of haploid and diploid cells in the same plant or in the same meristem. Moreover, three months after transplanting to soil, many plants of a few examined lines produced only diploid roots indicating a cytological instability of the haploid condition and an evolution towards diploid condition in onion roots. Such spontaneous doubling was also expressed in the shoot apices but at much lower frequency such as 99% of gynogenic lines were sterile at first flowering. In this case, the fertility of the remaining 1% was expressed as capacity of each line to produce at least one germinable seed.

Colchicine Treatment

The time required by the gynogenic lines of onion for a complete spontaneous diploidization is excessive and does not allow an early availability of DH lines, even when they appear to be diploid based on root tip chromosome counting.

Recently, the study of fertility restoration with the in vitro induction of chromosome doubling and a new technique has been set up. The diploidization of the shoot apices was achieved in onion by culturing haploid bulb scales on M3 solid medium containing 10 $mg.1^{-1}$ of colchicine for 72 to 80 hours followed by the standard procedure of micropropagation. This technique is not to exceed 80 hours of treatment in order to avoid the formation of polyploid cells. Field data concerning the fertility of gynogenic plants treated with colchicine are not yet available.

Micropropagation of Gynogenic Material

The best method for onion micropropagation·was described by Kahane et al. (1992). They well defined three important phases and relative times, determinant to establish a cycle of shoot regeneration indefinitely repeatable. After a small change in the medium composition (BDS salts were used instead of MS slats + KH_2PO_4 300 mg. l^{-1} and myoinositol was increased to 500 mgl^{-1}) this technique could be easily applied to our gynogenic material even after colchine treatment. In this last case, we suggest to maintain treated explants on M3 medium until shoots 1 cm long are formed before to be transferred to M4 medium for plant development and rooting (unpublished data). The micropropagated plants of gynogenic lines could easily be transplanted to soil in greenhouse without problems of acclimatization. One year is needed for bulb growth and another for flowering and seed production.

Homozygosity Assessment of Gynogenic Lines

Since gametoclonal variation can be induced in vitro during any step of androgenic or gynogenic plant formation, the homozygosity degree of gynogenic onions should be assessed before plant breeding use. The first investigation, carried out in the R2 progeny of the first DH line obtained after a spontaneous chromosome doubling, gave evidence of full homozygosity.

Problems and Perspectives

In onion, androgenesis has so far not succeeded and gynogenesis is now the only way for the production of haploid plants. The major limit of this method is the low yield of gynogenic plants obtained from ovary and flower culture. The significant improvements recently achieved (Campion et al., 1992; 1994a) do not appear sufficient to consider this method highly efficient. Genotype is the factor mostly affecting embryo yield but some others as the nutritional status of donor plants or the period of donor plant cultivation in greenhouse, are more worth consideration since they can be more easily manipulated. Also the method described by Dore and Marie (1993) (induction of haploids in plants) shows some limits: it is difficult to be applied to male-fertile genotypes since they need to be emasculated before the fertilization with irradiated pollen; a second limit is the production of a low percentage of haploids versus diploids (ploidy was here analyzed at root tip level). In this case the haploid condition would be the most reliable evidence of the gynogenic origin of plants allowing to eliminate those of doubt origin.

Notwithstanding the low yield of gynogenic embryos, the in vitro technique could be successfully applied to this species to early explore the range of genetic recombination obtained from interesting genetic materials (hybrids, populations). To this purpose the induction of 50 to 60 lines would be sufficient to exhibit a good range of genetic variability present in a cultivar.

This number of lines can be obtained by culturing 1,500 to 15,000 flowers or ovaries respectively from responding to recalcitrant cultivars.

YAMS

Yams (*Dioscorea* spp.) form a staple diet in West Africa and is often considered as richman's food where an average consumption is 0.5 kg/day; the highest consumption being 1 Kg/day is the Ivory Coast. The conventional processing techniques are boiling, roasting, frying or conversion to *fufu*. Yams are primarily used for human consumption in the tropical and sub-tropical regions. Yam-bean being a tropical tuber crop, prefer a long moist climate and high intensity of sunlight.

Most of the yams produced are consumed within the country as carbohydrate food. In each country, there is a rural market which act as a primary stock collecting centre from where the collected produce is sent to the urban market and then the yams are distributed throughout the urban area within the country. They also contribute to world economy. They are vegetatively propagated from whole tubers or tuber pieces. The bulk of the world's yam production (about 90%) occurs in West and Central Africa, but the crop is also locally important in other regions (Coursey, 1967; Degras, 1993).

In recent years, annual increases in consumption of yams were notably large (second only to rice) and were matched by similar increases in the production growth rate (FAO, 1997). The basis of the popularity of yams is their fine eating quality. The edible tubers have organoleptic properties that make them a preferred carbohydrate food, thus creating consumer-led demand for the crop to which farmers respond. As a result, yams are important not only for farm household food security, but also as a source of cash income.

Biotechnological Approaches for Conservation

Germplasm Conservation

The Consultative Group on International Agricultural Research (CGIAR), IITA holds the world mandate for improvement of yams. The group is collecting and conserving yam germplasm. In this regard, IITA maintains a collection of about 2680 accessions covering six cultivated and several wild species. A total of 1,504 accessions (56% of the whole collection) is maintained in vitro. There are several other much smaller in vitro collections (maximum of 100 accessions each) held in other research institutes around the world (Ng and Ng, 1997).

As yams are a bulky crop to handle, in the field and in conventional storage, requiring annual rejuvenation by field propagation. Therefore, in vitro maintenance of germplasm is essential to duplicate the germplasm collection and efficient conservation for gene bank.

The yam collection has faced a major problem e.g., transfer of germplasm from the field to the vitro gene bank. It was seriously hampered by acute contamination problems with fungi and bacteria. This was overcome by the development and application of a triple-disinfection method coupled with meristem culture, thus allowing the direct utilization of explants from field-grown materials (Ng and Ng, 1991).

Meristem culture method was adopted for this purpose. The regenerated callus culture are then propagated in vitro and maintained at a lower incubation temperature (18-22°C) for storage (Ng and Hahn, 1985). Currently over 50% of the germplasm of *D. rotundata, D. alata,* and *D. bulbifera* maintained at IITA is stored for 11-15 months under reduced-growth incubation conditions (Ng et al., 1998).

Cryopreservation of food yams is underway at various research institutions around the world (e.g., in France, Japan, and at IITA, Nigeria). Successful cryopreservation of *Dioscorea* species in liquid nitrogen was achieved by encapsulation-dehydration of shoot tips (Mandal and Chandel, 1995).

Germplasm Movement

Tissue culture techniques in combination with virus testings is useful in the production and safe international movement of virus-indexed yam germplasm. Axillary buds and nodal culture are the commonly used culture systems for rapid clonal propagation. Ng and Mantell (1996) used optimal conditions for in vitro micropropagation, including use of aerated liquid media, required amounts and types of mineral salts, carbon sources, and cytokinins, and the critical daylength for growth of nodal cuttings. Since 1995, IITA has distributed about 4,000 plantlets annually in various world laboratories. IITA has links with other micropropagation laboratories to receive global coverage in germplasm distribution. IITA has maintain five clones of *D. rotundata.* To increase survival at the destination, a simple and low input management system for postflask rearing of yam plantlets was developed (Ng et al., 1994; Ng and Asiedu, 1998). For germplasm exchanges and conservation, in vitro yam plantlets (small tubers or microtubers) are used. Ng and Mantell established (1996) culture media composition and incubation condtions for production of microtubers in culture. They established the frequency of their formation, their size and number, as well as differential responses between some species and cultivars.

Interestingly although microtubers formed at the node where roots were initiated, aerial microtuber formation was also observed. Such tubers can be harvested easily, and the shoots can concomitantly be used for further in vitro propagation (Ng and Ng, 1997). It is also observed that microtubers often have a longer dormancy period than tubers harvested from the field, and their germination is erratic. Therefore, it is essential to capitalize on the potential usefulness of these propagules.

Embryo Culture

Embryo culture technique is used for yam improvement. It helps for rescuing immature embryos and increasing the germination of seeds. The studies have been conducted on This is particularly the case *D. alata* where fungal infections on the aerial vines jeopardize the recovery of viable seed. To confer resistance to anthracnose disease in susceptible *D. alata* cultivars rescue of hybrid seed is important. Although media for culturing immature seeds at eight weeks after anthesis have been identified, increasing the rate of embryo recovery and culturing even younger seeds are both receiving attention.

Plant Regeneration

Plant regeneration systems for yams is used for large-scale micropropagation, as well as for gene transfer through recombinant DNA technology. Yams callus cultures are obtained from young leaf petioles of *D. alata and* can established from different explants. Plant regeneration via either embryogenesis or organogenesis has been achieved only in certain cultivars of a few food yam species. Direct multiple-bud formation was achieved in *D. opposite* with immature leaves (Kohmura et al., 1995). In *D. rotundata,* Ng and Mantell (1996) obtained somatic embryos from leaves. Twyford and Mantell (1996) achieved somatic embryogenesis and plant regeneration from suspension cultures derived from roots of in vitro plants. However, without more broadly applicable systems, realization of the potential of this technology in large-scale micropropagation is still not well established.

CASSAVA

Cassava production in India is mainly done in Andhra Pradesh, Assam, Karnataka, Kerala, Meghalaya, Rajasthan, Tamil Nadu, Tripura, Andman and Nicobar Islands, Mizoram, and Pondicherry. In Kerala, Quilon, Trivandrum, Kottyam, Malappuram and Alleppy Districts; and in Tamil Nadu, Kanya Kumari, Salem and South Arcot Districts are covered with more area under cassava and found to be major producing areas.

Cassava is attacked by more than 36 diseases caused by different organisms viz, viral, fungal, bacterial and mycoplasmal origin. In India, only 14 diseases have been reported among which cassava mosaic, brown leaf spot, anthracnose and sett rot are important (Malathi and Shanta, 1983). The main diseases are Cassava mosaic disease, brown Leaf Spot (*Cercospora henningsii* Allesch), bacterial blight (*Xanthomonas manihotis*), concentric leaf spot (*Phyllosticta manihoticola*), anthracnose (*Glomerella cingulata*); *Colletotrichum* spp. The pests diseases are stem mussel scale (*Aonidomytilus albus*) Hemiptera; Diaspididae, white grab (*Leucopholis concophora*), red Spider mite (*Tetranycus cinnabarinus*), Cassava thrips (*Retithrips syriacus* Mayet), white fly (*Bemisia tabaci* Gennadius), mealy bugs (*Phenacoccus manihoti*), Cassava hornworm (*Erinnyis ello* Linn.), varigated grasshopper (*Zonocerus variegatus*). The pests damaging roots and tubers (nematods, rodents) are Root-knot nematode (*Meloidogyne incognita*), Bandicoot (*Bandicota indica*), Field rat (*Bandicota bangalensis*), Indian gerbil (*Tatera indica*), field mouse (*Mus booduga*), etc. The important storage pest is Beetle (*Aracecerus fasciculatus*).

In cassava, good progress has been made on characterization of genes for cyanogenesis and resistance to mosaic diseases, but appropriate transformation system is yet to be developed. The major problem for genetic engineering for transformation is tackled by four different research groups – CIAT, Colombia, Swiss Federal Institute of Technology, Switzerland, Agricultural University Wageningen (WAU), the Netherlands and the International Laboratory for Tropical Agricultural Biotechnology (ILTAB), USA. The ILTAB and WAU have used an embryogenic suspension culture method for genetic transformation. Although this method is quite efficient and less genotype specific, the system is very labour intensive and time consuming. It also requires expensive equipment and advanced laboratories.

This makes the technique less suitable to laboratories in developing countries. To exploit the system commercially, much work still has to be done for transformation and regeneration of cassva. Once the improved transgenics are available for example acyanogenic cassava or mosaic disease resistant cassava plants with delayed post-harvest deterioration, it will have better market potential.

Biotechnological Research for Conservation

In the light of biotechnological research and existing problems of conventional propagation of root and tuber crops, micropropagation protocols can be taken up for a large scale production. However, unlike other starchy crops commercialization of micropropagation protocols of existing varieties may not receive attention unless transgenic with value-added traits are produced. Production of transgenics in experimental level is successful in cassava. Once the improved transgenics are available for example acyanogenic cassava or mosaic disease resistant cassava plants with delayed post-harvest deterioration, it will have better market potential.

Tissue culture method for cassava was first developed in Canada and at CIAT (Kartha and Gamborg, 1975). Later researchers in cassava growing countries also developed tissue culture techniques to suit to their cultivars. The steps involved for regeneation, propagation and establishment of plants from shoot tip culture (Nair et al., 1994).

In cassava, protocols have been developed to regenerate plantlets through somatic embryogenesis in a wide range of genotypes (Szabados et al., 1987; Stamp and Henshaw, 1987; Sudarmonowati and Henshaw, 1993 and Taylor et al., 1994). Most of the researchers used 2,4-D supplemented MS medium. Of the different methodologies, the regeneration protocol developed by Taylor et al. (1994) seemed to be ideal for genetic transformation. They used picloram (5×10^{-5} M) to induce embryogenesis in friable embryogenic cell suspension which developed into embryos on transfer to auxin free medium. This protocol may facilitate artificial seed production technology in cassava.

In India, somatic embryogenesis and plantlet regeneration have been achieved by culturing immature leaf lobes (1-3 mm) of cassava M 4 H 226 cultivars successively in MS with 0.4 to 0.8 mg/litre 2,4-D and MS medium alone under continuous light (Mukherjee 1994). Somatic embryos have also developed from shoot tip cultures in Co2 and Co3 cultivars in MS medium supplemented with 0.01 mg/litre^{-1} 2,4-D, 0.1 mg 1^{-1} BAP and 1 mg/litre^{-1} GA_3. (Narayanaswamy, et al., 1994).

A molecular genetic map of cassava is under construction through collaborative programme among CIAT, the University of Georgia, and Washington University in St. Louis (Thro et al., 1994).

Cassava is a subsistence crop in many developing countries. It makes the issue of free access to all cassava science and technology. Patents have been requested for the techniques that are developed by the WAU and ILTAB. The request for patenting specific cassava transformation and regeneration protocols are pending. Different aspects of technology used are already protected as the technique of particle bombardment owned by *Dupont* and *Kannamycin* resistant by *Monsanto.* Hence marketing of transgenic cassava plans developed using these different techniques will necessarily involve negotiations with patent holders (Dijk, 1997). If intellectual property right is implemented farmer cannot export the products as the products are protected by patent in the importing country.

SWEET POTATO

Sweet Potato, botanically known as *Ipomoea batatas* (L.), belongs to the family Convolvulaceae. It is a dicotyledonous herb with trailing vines. Leaves are heart-shaped or palmately lobed; some of the roots get modified into a few medium-sized tubers by the disposition of starch. In India, it is locally called *gurmi-alu, mitha-alu, cheenikizhangu* etc.

It was originated in Central America and North-Western parts of South America. In ancient times, tropical America and Pacific Islands grew this crop extensively. According to O' Brien (1972), sweet potato has been cultivated probably from 3000 B.C. It is grown in nearly all parts of tropical and sub-tropical countries and also in the warmer areas of the temperate regions (Onwueme, 1978). It has been used as an important staple food crop in tropical and sub-tropical countries like India, China, Japan, Taiwan, South America, Indonesia, South Pacific Islands, Nigeria, Southern U.S.A. and the Philippines; between latitudes 40° N and 40° S of the equator.

Sweet Potato is a crop of tropical and sub-tropical regions and requires a warm humid climate. In warmer temperate regions, its cultivation is restricted because of frost susceptibility. Warm sunny days and cool nights are favourable for better tuber development.

In India, sweet potato is grown mostly in Bihar and Uttar Pradesh. The next in importance are Orissa, Assam, Madhya Pradesh, West Bengal, Tamil Nadu and Kerala. Other states also grew this crop to a limited extent. In Bihar, it is grown mostly in North Bihar as a rabi crop (Sept. –Oct.); While in lower lands of South Bihar it is grown after paddy harvest in January-February; but in hilly region of santhal Parganas, it is grown as a rainy season crop in July. In Uttar Pradesh, it is grown extensively in Bundelkhand region i.e., Jhansi, Bonda, Hamirpur and Jalaun Districts; and also sparsely grown in Kasganj, Etah, Farrakhabad and Mainpuri Districts. In Assam, it is mostly grown in Cachar, Goalpara, Kamrup, Darrang and Nowgang Districts. In Tamil Nadu, large areas are covered in Tiruchirapalli, South Arcot, Tirunelveli and Madurai Districts. In Kerala, it is mostly grown in

Malabar coast. In Orissa and West Beng3al, sweet potato is grown throughout the states.

These three phases may vary with cultivar and also with the environment. In some cultivars, tuber initiation begins as early as 4 weeks after planting, tuber formation occurs within 7 weeks and the tuber development takes place during the rest of the crop period. House (1908) grouped *Ipomoea batatas* with 25 wild species, included diploid, tetraploid and hexaploid species. The diploid (2n = 30) wild species include:

I. trichocarpa, I. lacunosa, I. ramoni, I. triloba, I. setifera, I. tricolor, I. palmata, I. obscura, I. biloba, I. purpurea, I. crassicanlis, I. paniculata etc.

1. The triploid (2n = 45) species are : *I. gracilis*, etc.
2. The tetraploid (2n = 60) species are: *I. tiliacea, I. littoralis, I. autorescens, I. Biloba,* etc.
3. The hexaploid (2n = 90) species include: *I. trifida*

Sweet Potato : Improvement Strategy

The crop improvement strategy is mostly confined to:

1. Establishment of indigenous breeding population through germplasm collection, conservation and evaluation.
2. Introduction of superior exotic genotypes
3. Selection and inter-varietal hybridization.

The aims of sweet Potato improvement are :

1. Desirable plant type;
2. Good cooking and keeping quality of tubers;
3. High protein and carotene content.
4. Higher tuber yield;
5. Photo-insensitivity;
6. Resistance to diseases & pests;
7. Resistance to drought;
8. Shorter growing period
9. Wider adaptability and

Hence improvement programme will be confined to the establishment of a breeding population having broad genetic base, introduction of exotic genotypes, selection, hybridization, polyploidy and mutation.

Biotechnology for Conservation

Analysis of potential opportunities and threats of biotechnological tools with regard to developing countries reveals that opportunities are more. Biotechnology offers ecofriendly agriculture which are cost effective. Integration of biotechnological tools in conventional breeding in root and tuber crops offers good scope with viable commercial prospect. Production of transgenics in experimental level is successful in sweet potato. It is necessary that intensive programme is to be conducted for sweet potato on regeneration and molecular markers.

Success in plant regeneration through organogenesis and embryogenesis have made sweet potato more amenable for genetic transformation. General cell culture techniques may facilitate to isolate useful somaclones in all these crops especially to recovery stress tolerant as has been envisaged in sweet potato and taro to isolate NaCl tolerant cell lines.

Meristem culture

The best known tissue culture work on sweet potato is meristem culture by Mori (1971). Addition of auxins and cytokinins help in speedy regeneration of meristem and thus is found most suitable. Mori (1971) has been able to establish virus-free meristem culture of sweet Potato, eliminating most prevelent virus from the crop. M S medium supplemented with adenine and kinetin are used for this purpose.

Shoot Tip Culture

Protocol to produce shoot tip culture to eliminate virusesin sweet potato have been developed (Mori, 1971; Alconero et al., 1975; Love and Rhodes, 1987; Henderson 1984; Lizzaraga et al., 1992). Procedure for rapid multiplication of disease-free plans developed for more than 200 genetic resources through shoot tip and auxillary bud cultures (Mukherjee et al., 1994).

In-vitro tubers have been developed from internodes, fibrous roots and leaf explants. Tuber formation from internode segments in presence of kinetin and auxin is reported by Houndonoughbo (1989), from fibrous root at 8% sucrose with 1 or 5 ppm BAP by Aboul and Bouwkamp (1990) and from leaf explant in MS medium supplemented with or without NAA (0.2 mg/litre) at 12 h photoperiod by Mukherjee et al. (1996).

Protoplast culture

Table 8.12: Selected work conducted on Protoplast culture sweet potato

References	Work conducted
Wu and Ma (1979); Eilers et al. (1985); Sihachakr and Ducreaux (1987); Ozias and Perera (1990); Kobyashi et al, (1990); Belarmino et al. (1994, 1996)	Reported isolation of viable protoplast from sweet potato and some of its wild progenitors.
Belarmino et al., (1996)	Regenerated plants form stem and petiole protoplasts of sweet potato and *Ipomoea lacunose.* They also reported asymmetric protoplast fusion between sweet potato and *Ipomoea trifida* by electro fusion and polyethelene glycol (PED) treatment.
Mukherjee (1998).	Reported that protoplast of young leaves have registered higher density and viability. Enzymes cellulase (4%) macerozyme (3%) and pectolyase (0.08%) were used for cell wall digestion

Embryo Culture

To recover rare hybrids, to establish and propagate exotic species, *in-vitro* embryo culture and seed germination techniques have been developed for sweet potato and its wild relatives viz. *I. trifida, I. cairica,* and *I. murocoides* (Mukherjee et al., 1991b, Mukherjee and Vimala, 1994; Mukherjee 1994a). The embryo culture in sweet potato can help to recover rare hybrids. MS medium supplemented with GA-3 and coconut milk is found essential for initial morphogenesis of the immature embryos and to enhance callus development. A combintion of NAA, BAP and GA-3 can produce multiple seedlings. Sequence transfer of embryos to the above media is ideal for the recovery of seedling, rather than hormones in combination. For the development of mature embryos, auxins and cytokinins are not essential.

Embryo Rescue Technique: Embryo culture is an aid to conventional breeding to rescue rare hybrids not obtainable by conventional methods due to post fertilization barriers like embryo-endosperm incompatibility and seedling lehality. Ng (1992) could grow mature embryos of cassava on half strength MS medium with 3% sucrose and immature embryo son further supplementation with 15% coconut water (CW), 5 mg/litre IAA and 4% sucrose.

Organogenesis or Somatic Embryogenesis

Plans can be regenerated through callusing and somatic embryogenesis from stem, leaf, petiole and root explants on subsequent cultures to 2,4-D or 2,4-D + BAP supplemented MS to plain MS medium. Embryogenic response is observed to be influenced by genotype, growth regulators as well as by explants. Variation in embryogenic resonse among the genotypes with different growth regulators is now established.

Regeneration through organogenesis or somatic embryogenesis has been developed in different cultivars of sweet potato by culturing leaf, shoot, root explants on MS medium supplemented with 2,4-D, kinetin, picoloni acid, thidiazuron and 2,4-5-T. Though high frequency regeneration protocols have been developed, most of the results are specific to certain genotypes. Optimization of the concentration of 2,4-D or 2,4-D and kinetin or addition of thidiazuron, or picolonic acid could enhance the regeneration response (Chee and Cantliffe, 1992; Desamero et al., 1994; Gosukonda et al., 1995, Mukherjee et al., 1991a, 1993, 1997). Use of 2,4-5-T reported to be more effective than 2,4-D (Almazrooei et al., 1997). Exploitation of somatic embryogenesis protocol for genetic transformation and artificial seeding depends on the homogeneity and genetic stability of the regeneration system. Studies on these aspects have been carried out in India by Mukherjee (1998).

Anther / pollen culture

Haploid breeding is recommended in sweet potato and cassava to overcome the problems associated with heterozygosity and higher ploidy. Haploid breeding work employing anther culture has been attempted by several researchers. Though information are available on callusing and regeneration (Nair et al., 1994), true haploid is yet to be developed for both the crops. To ovoid the interface of diploid tissue, work in pollen culture is going on at the CIAT (Cali, Columbia) and at CTCRI, Thiruvananthapuram (India). Pollen embryogenesis is induced in cassava in MS medium supplemented with 0.4 to 0.8 mg/litre 2,4-D (Mukherjee, 1994 b).

Somaclonal Variation

Both genetic and cytogenetic variations are common features of plant cell and tissue cultures. Genetic changes occurring during *in-vitro* regeneration are transmitted to regenerants and progenies are defined as somaclonal variations (Larkin and Scowcroft, 1981).

Somaclonal variationis likely to occur during *in-vitro* regeneration which is undesirable for propagation and storage. To asses the stability, embryogenic callus and regenerated plants are subjected to histo-morpho-cyto-biochemical and biomolecular marker evaluations.

Useful traits can be obtained in variants derived from the tissue grown by adding metabolites and other substances to the media. Salgado et al. (1985) could isolate NaCl resistant variant cells from sweet potato cell suspension. Cells have been subcultured and tested for salt resistance. Resistance to 1% NaCl has been stale for 3 passages in NaCl free medium. Regeneration of plantlets through callusing and embryogenesis was achieved from callus cells grown in 1% NaCl (Mukherjee, 1998) and the same methods can be exploited for other tuber crops. *In-vitro* selection technique for salinity tolerance in taro has been developed using callus tissue produced from shoot tip explants. Salt tolerant tissue was selected by stepwise transfer of tissue masses to successively higher concentrations of artificial sea water. Plantlets were produced up to 70% concentration of artificial sea water (Nyman and Arditti, 1984 b).

Artificial Seeds

Artificial seeds also known as synthetic seeds are good quality somatic embryos or propagules enclosed in protective nutrient coating. Artificial seed technology provides a rapid, low volume, cost competitive propagation method. *In-vitro* somatic embryogenesis facilitate the production of synseeds provided synchronization in development and germination of embryos is achieved.

Somatic embryos produced with NaCl or L-proline could be germinated 7 days after storage at 8 °C on transfer to MS medium and also on direct sowing in sterilized soil at 25 ± 2 °C under 12hr photoperiod (3000 lux). Embryos encapsulated with Na-alginate could be stored for 30 days and grow in to plants in culture medium and also on direct sowing in sterilized soil. However percentage of germination of encapsulated embryos was much higher than non-capsulated embryos *in vivo* and *in vitro.*

Production of artificial seeds in sweet potato has been attempted by Chee and Cantliffe (1992). Enhanced and synchronous embryogenic response registered in four different cultivars with addition of NaCl or L-proline in 2,4-D supplemented MS medium may favour artificial seeding in sweet potato (Mukherjee, 1998).

Germplasm Collection

Germplasm collection is an assembly of diverse genetic stocks which serves as a raw material from which the desirable ones are identified. Improvement of sweet potato through efficient exploitation of available germplasm reserves is possible. The superior indigenous and exotic genotypes are required to be collected and conserve methodically to exploit their genetic variability. Those are to be catelogued systematically, adopting prescribed parameters for classification of different genotypes assembled in the germplasm bank. While evaluating the germplasm, following morphological and quality parameters are considered:

TARO

Colocasia esculenta (L.) is an important food crop. The biotechnology researches have been made for its improvement. Much of the work has been done with regard to tissue culture, however, rsearches are on the way to improve the crop genetically.

Tissue Culture

Nair et al. (1994) successfully developed cormel tip culture protocol for regeneration and multiplication of taro is presented. *In vitro* tubes are les vulnerable to adverse environments compared to *in-vitro* plans. Transit and handling are also easy for tubers. They can serve as excellent means for propagation. Shoot regeneration protocol from meristematic cell layers with auxin and cytokinins has been developed and positive effect of NAA on embryogenesis have been explained (Gomez et al., 1989; Mix 994; Malamug et al., 1992). Development of embryogenic clusters in Indian cultivars by culturing cormel tips in MS medium supplemented with NAA (0.25 to 0.5 mg/l), BAP (0.25 to 0.5 mg/litre) and GA_3 (0.1-0.2 mg/litre) has been developed by Mukherjee et al., (1997). Subculturing of embryogenic cluster in NAA, BAP supplemented medium yielded 20-28 plantlets/cormel tip.

9

Conservation of Spices and Condiments Through Biotechnology

Spices are the varieties of dried vegetable products containing essential oil, used for flavouring food. Webster defined spices specifically as "*Any of the various vegetable production, as pepper, cinnamon, nutmeg mace, allspice, ginger, cloves etc. used in cookery to season food and to flavour sauces, pickles etc. a vegetable a condiment or relish, usually in the form of a powder, also such condiments collectively*".

India has a well-known reputation as a land of spices from time immemorial. Indian spices are offered higher prices in the world trade. Indian export accounts for 30 to 40 per cent of the world trade and nearby 20-37 per cent of the foreign exchange is obtained through pepper, the *Black Gold* among spices. Pepper 'The Kind of spices' is the largest foreign exchange earner and Indian pepper is also of the finest quality. Next to pepper, cardamom is the most important spice crop grown in India. There are other spices like chili, turmeric, ginger, aniseed, celery, coriander, caraway, cumin, dill, fenugreek garlic, onion, saffron and vanilla.

India exports oil of pepper, cardamom, ginger, celery, cumin, cinnamon leaf and oleoresins of pepper, ginger, chillies, turmeric and fenugreek to West Germany, Japan, France, Australia, U.S.A, etc. The latest trend noticed in the diversification of spices exports is in powder form, both in bulk and retail sale. Powders of chilli, turmeric, pepper etc. have started their way to foreign countries in appreciable quantities.

Table 9.1: Origin distribution and growing states in India

Spices	Origin	World	India
Allspice	West Indies, Central and South America	Jamaica, Mexico, Gautemala, Honduras	Kerala, Tamil Nadu, Karnataka, Mharashtra
Aniseed	Mediterranean region	Mexico, Spain, Germany, Turkey, Italy, Bulgaria, Cyprus	Rajasthan, Punjab, Uttar Pradesh, Orissa
Cardamom	Western ghats of India	Sri Lanka, Thailand, Indonesia, Guatemala, El-salvadar, Papua New Guinea	Kerala, Karnataka, Tamil Nadu
Celery	Southern Europe	United Kingdom	Utter Pradesh, Punjab
Cinnamon	Ceylon and Malabar coast	Seychelles, Burma, Malayan Peninsula, South America, Sri Lanka, Madagascar	Naga hills, Coastal districts of South Karnataka, Western ghats, Tamil Nadu
Clove	Moluccas	Zanzibar, Pemba, Madagascar, Indonesia, Malaysia, Sri Lanka, Brazil, Papua New Guinea	Tamil Nadu, Kerala, Karnataka, Maharashtra
Cumin	Upper Egypt, Turkey and East Mediteranean region	Turkey, Iran, Japan, Morocco, Indonesia, Southern Russia, China	Rajasthan, Gujarat
Fennel	Southern Europe and Mediteranean area	Rumania, Russia, Germany, Italy, Argentina, U.S.A	Gujarat, Rajasthan
Fenugreek	Egypt	Egypt, France, Lebanon, Argentina	Rajasthan, Gujarat
Garlic	Central Asia and Mediterranean region	U.S.A., Egypt, Bugaria, Hungary, Taiwan	Gujarat, Rajasthan, Maharashtra, Tamil Nadu
Ginger	Tropical, South East Asia	Australia, Jamaica, Japan, Nigeria, Thailand, China, Fiji, Siagra Leone	Kerala, Karnataka, Andhra Pradesh, Assam, Orissa, Himachel Pradesh,

Spices	Origin	World	India
			Tamil Nadu, Sikkim
Nutmeg	Moluccas, East Indies	Malaysia	Tamil Nadu, Kerala, Karnataka, Maharashtra
Pepper	Western Ghats, Upper Assam Forests	Sri Lanka, Malaysia, Indonesia, Brazil, Thailand, Vietnam, China, Madagascar, Mexico, Micronesia	Kerala, Karnataka, Tamil Nadu, Assam, West Bengal, Andhra Pradesh, Goa.
Saffron	Greece and Asia Minor	Italy, France, Germany, Greece, Iran, China, Spain	Jammu and Kashmir
Turmeric	South East Asia	Thailand, Peru, China, Sri Lanka, Pakistan, Bangladesh, Spain, Jamaica	Kerala, Tamil Nadu, Maharashtra
Vanilla	Mexico to Brazil	Java, Mauritius, Madagascar, Tahiti, Seycheles, Zanzibar, Brazil, Jamaica, Island of West Indies	Kerala, Tamil Nadu, Karnataka

Spices are classified in different ways. They may be grouped according to the part of the plants use such as dried flower buds of clove; fruits such as black pepper, nutmeg and vanilla; underground stems like ginger and turmeric and seeds and seed like structures which refer to cardamom, coriander, dill etc. The international standard organization lists 107 spices (ISO 676).

Biotechnology has great potential in spice cultivation and industry. Many private organizations have already taken up commercial multiplication of cardamom through tissue culture. The possibility of production of flavour and volatile constituents in culture makes nutmeg and vanilla tissue culture a profitable proposition. At the National Research Centre for Spices the priorities are refining techniques for micropropagation of free spices, *in-vitro* selection for disease resistance in black pepper and ginger, production of virus free planting material in cardamom and conversion of the valuable spice germplasm *in vitro* repositories.

Table 9.2: Botanical name, family, common English name and parts used are as follows.

Common name	Botanical Name	Family	Parts used
Asafoelida	*Ferula assa foetida*	Apiaceae	Rhizome exudates
Bengal cardamom	*Amomum kepulaga*	Apiaceae	Fruit, seed
Bilimbi, cucumber trees	*Averrhoa bilimli*	Averrhoa-cear	Fruit
Bird eyes chilli	*Capsiceno annum*	Solanaceae	Fruit
Bird eyes chilli	*Capsieum frutes cens*	Solanaceae	Fruit
Bittan famal	*Foeniculum vulgane var. vulganes*	Apiacese	Leaf, twing, freint
Black caraway	*Bunium persicum*	Apiaceae	Seed, tuber
Black caraway	*Carum bulbo eastanum*	Apiaceaae	Fruit
Black mustard	*Brassica nigra*	Brassicaceae	Seed
Cambodian cardamom	*Amomum subulatum*	Zingiberaceae	Fruit, seed
Camboge	*Gareinia canbogia*	Clusiaceae	Pericarp of fruit
Cameron cardamom	*Aframomum hamburyi*	Clusiaceae	"
Caper	*Capparis spinosa*	Capparidaceae	Floral bud
Caraway	*Carum carvi*	Apiaceae	Fruit
Celeriae	*Apium graveolens var. rapaceum*	Apioceae	Fruit, root, leaf
Celery	*Apium graveolens var. dules*	Apioceae	Fruit, root, leaf
Chineses cassia	*Cinnamomum aromaticum*	Lauraceae	Bark, leaves
Chive	*Allium galanga*	Lauraceae	Leaf
Clervil	*Anthriocus cereifolium*	Apiaceae	Leaf
Corambola	*Averrhoa carambola*	Averrhoa-coaes	Fruit
Coriander	*Coriandrum sativum*	Aiaceae	Leaf fruit
Cumin	*Cumimum cymimum*	Apioceae	Fruit

Common name	Botanical Name	Family	Parts used
Dill	*Anothum graveolens*	Spiaceae	Fruit leaf
Garden angeliea	*Angelica arechangelica*	Apiaceae	Fruit petioles, root
Gareimia, kokum	*Garcinia indica*	Clusiaceae	Pariearp of fruit
Garlic	*Allium sativum*	Clusiaceae	Bulb
Grain of Paradise	*Aframomum melegueta*	Clusiaceae	Bulb
Greater galanga	*Alpinia officinanum*	Zingiberaceae	Rhizome
Horsoradish	*Armoracia rusticana*	Apiaceae	Root
Indian dill	*Anethum sowa*	Apiaceae	Fruit
Indian leek	*Allium tuberosum*	Apiaceae	Bulb, leaf
Indian mustard	*Brassica juncea*	Brassieaceae	Seed
Indonesian cassia	*Cinnamomum bermanii*	Bauraceae	Bark, leaves
Koranima cardamom	*Aframomum koranima*	Bauraceae	Bark, leaves
Leek	*Allium porrum*	Bauraceae	Bark, leaves
Lesser galanga	*Amomum aromaticum*	Bauraceae	Bark, leaves
Madagascar cardamom	*Aframomum angustifolium*	Zingiberaceae	Fruit seed
Nepalese Cardomom	*Amomum Dubulatum*	Zingiberaceae	Fruit, seed
Onion	*Allium cepa*	Alliaceae	Fruit, seed
Potato onion	*Allium cepaaggregatum*	Alliaceae	Fruit, seed
Round cardamom	*Amomum krevanh*	Alliaceae	Fruit, seed
Saffron	*Crocud sativus*	Iridacee	Stigma
Shallot	*Allium ascalonicum*	Zingiberaceae	Bulb
Small cardamom	*Eletlaria cardamomum*	Zingiberaceaes	Fruit, seed
Srilankan Cardamon	*Elattaria cardamomum var. major*	Zingiberaceae	Fruit, seed

Common name	Botanical Name	Family	Parts used
Srilankan cinnamon	*Cinnamomum zeylanicum*	Lauraceaes	Bark, leaf
Srilankan citronella	*Cymbopojon mardus*	Roaceae	Leaf
Sweet fennal	*Foeniculum vulgares var. dulea*	Apiaceae	Leaf, twig, fruit
Sweetflag calamus	*Acorus calamus*	Araceae	Rhizome
Tajpat	*Cinnamomum tamala*	Lauraceae	Leaf bark
Tarragon	*Artemisia dracumculus*	Asteraceae	Leaf
Tsao-ko Cardamom	*Ammomum tsao-ko*	Asteraceae	Fruit seed
Turmerie	*Curcuma longa*	Zingiberaceae	Rhizome
Vietnameso cassia	*Cinna momum crureirii*	Lauraceae	Bark
Welsh onion	*Allium fistulosum*	Lauraceae	Bark
West Indian lemongrass	*Cymbopogon eitratus*	Poaceae	Leaf

Over 80 species of spices are cultivated in India. They are either annuals or perennials growing from tropical to temperate regions of the country. The most important of them are black pepper, small cardamom, ginger, turmeric, tree and seed spices. They earn foreign exchange equivalent of Rs. 300 crores annually. The annual production of these spices is almost static. The major bottle-necks in increasing the production and productivity of spices are lack of adequate disease-free planting material of the high-yielding varieties and prevalence of major diseases and pests with no effective control measures. Besides, the existing genetic variability in nutmeg and clove is insufficient for any meaningful crop improvement programme.

In spices, most of which are perennials, biotechnology comes handy to augment the conventional crop improvement programmes. Particularly through.

1. Micropropagation and rapid clonal multiplication of high-yielding genotypes to generate adequate quality and disease-free planting material
2. Somaclonal variations and utilization of somatic cell hybridization and anther culture for crop improvement

3. *In vitro* selection for resistance to biotic and abiotic stresses
4. Production of flavour and volatile constituents in culture
5. *In vitro* conservation of germplasm and its safe exchange.

CARDAMOM

Elettaria cardamomum Naton (Zingiberaceae) is popularly known as the *Queen of Spices*. It is a herbaceous perennial. A fully grown plant is about 2-4 metres high. The real stem of the plant is the underground rhizome. The aerial pseudostem is made up of leaf sheaths. Chromosome neuters of cultivated caselaman in 2n-48.

The important varieties of are Malabar, Mysore and Vazhuka (Sahadevan, 1965). These are grown in different tracts and are mainly identifiable by nature of their panicle, shape and size of fruit and other growth features.

Table 9.3: Important varieties of cardamom

Malabar Cardamom	Plants of Malabar cardamom are medium sized rarely exceeding 2.7 n in height. The leaves are 30 to 45 cm long and may be pubescent or glabrous. The panicles are prostrate and the fruits roundish or egg shaped. This type is better suited at 600-900 metres elevation, is susceptible to thrips and is commonly cultivated in Kerala, Tamil Nadu and Karnataka at lower elevations.
Mysore Cardamom	Plants of Mysore Cardamom are quite robust and grow three to four metres in height. The leaves are lanceolate and glabrous on both surfaces. The panicles are erect and the capsules bold and long. They are better adapted to an altitude ranging from 900 to 1200 metres above MSL, and thrive well under assured and well distributed rainfall conditions. This type is cultivated mostly in Kerala. (Idukki Distriet)s.
Vazhuka	It is a natural hybrid of Malabar and Mysore types and consequently, the plants exhibit various characteristics which are intermediary between the two types. They also show adaptability to a fairly wide range of environmental conditions. The plants are robust like the Mysore type and have deep green leaves; semi erect panicles and roundish to oblong capsules. This type is cultivated extensively in Kerala on elevations raging from 900 to 1200 metres above MSL. The existing plantation being mainly of seedling population, very high variability in yield and associated characters is observed. Frequency distribution of yield of green capsules in a clump in seedling population and clonal population.

The other varieties are PV-1, CCS-1, ICRI-1, ICRI-2, ICRI-3, ICRI-4.

Cardamom and cinnamon were the first to enter the spice market of the West. Cardamom found a place in Queen Sheba's gift box to king Solomon. The sweet aroma of cardamom has been praised in Indian literature. Cardamom is used for flavouring various preparations of food, confectionery, beverages and liquors, Cardamom is also used for medicinal purposes both in allopathy and ayruveda. In the Middle East countries, cardamom is used mainly for preparation of 'Gahwa' or Arab Coffee (Cardamom flavoured coffee).

Cultivation of cardamom is mostly concentrated in the evergreen forests of Western Ghats in South India. Besides India, cardamom is cultivated in Guatemala, Tanzania, Sri Lanka, Elsalvador, Vietnam, Loss, Cambodia and Papua New Guinea. Earlier India accounted for 70 per cent of the world production and now it is 41 per cent only due keen competition from Guatemala which accounts for per cent of the present world production.

Biotechnological Approaches for Conservation

The literature on biotechnological improvement of this very meager. However, some of the work has been done on tissue culture for its improvement.

Being a cross pollinated crop, clones are ideal for generating true to type planting material from high yielding clumps. Clonal multiplication of cardamom by tissue culture is reported by Nadgauda et al. (1983). They reported that young sprouted buds Cardamom (10-20 mm long) excised and cultured on MS medium, supplemented with BAP (0.5 mg/1), Kn (0.5 mg/1), IAA (2 mg/l) CW (5%), Calcium pantothenate (0.1 mg/l) and biotin (0.1 mg/l) grew to length of over 40 mm in eight weeks. Shoot tips from elongated buds or rooted plantlets when excised and transferred to the same medium gave rise to multiple shoots, which on transfer to Whites' liquid medium could be rooted. A multiplication ratio of 1:3 was observed in each subculture. Tissue culture raised plants have successfully grown in the field where growth data indicated their superiority over plants raised from seeds.

High multiplication of disease free planting material makes micropropagation more profitable than conventional methods.

Many commercial laboratories are engaged in a large-scale production of cardamom clonal planting material using micropropagation. Field evaluation of tissue-cultured plants of cardamom has been carried out at Indian Institute of Spices Research (IISR), Calicut. The results showed that the microproapgated plants performed on par with suckers of the original mother plant (Lukose, 1993).

SMALL CARDAMOM

Queen of spices, the small cardamom is the dried fruit of the perennial herb *Elettaria cardamomum* Maton.

Tissue Culture

Apical shoot tip culture and panicle culture to produce multiple shoots and plantlets and have been reported. Shoot buds and immature panicles were selected as starting material from elite clones identified in planters fields, that had traits such as compound panicles, bold capsules and high yield potential. The rate of multiplication achieved is abouf 5,000 plantlets per shoot explant in a year. The plantlets have been transferred to the field.

Clonal Multiplication

Protocol for clonal multiplication from vegetative buds in small cardamom has been standardized. An average of 6 axillary shoots could be produced within 30 days of culture. This method is extensively used for raising clones of the high-yielding Coorg cardamom `Selection 1' and for the production of `katte'- disease free nucleus planting material.

The field experiments conducted at the NRCS indicate that the tissue cultured plantlets derived from vegetative buds and the suckers produce equal yield.

Immature inflorescence could also be used to multiply cardamom clones, by converting the floral buds into vegetative buds.

Regeneration of Plantlets from Callus

The protocol for organogenesis and plant regenerations from vegetative-bud-derived callus cultures is endowed with an excellent regeneration system, that at present is used for large scale production of somaclones and isolation of useful types from them.

Inflorescence Culture

It was also possible to clonally multiply cardomon using immature inflorescence by the conservation of floral bud into vegetative buds.

Field Evaluation of Tissue Culture Plants

Field experiments conducted have revealed that the tissue-cultured plantlets derived from vegetative buds are on at par with suckers in yield performance.

BLACK PEPPER

Piper nigrum L. (Pepper) belongs to family Piperaceae). It is the most important of all the spices and popularly known as the *King of spices*. Ever since Vasco Da Gama found it in Kerala in the 15th century it had spread to the western World and it was this spice that was the cause of many expeditions and wars. It is one of the earliest spices known to mankind and for many ages a staple article of commerce between India and Europe.

Pepper or black pepper is the product of the pepper vine. The plant grows to a height of 10 m. or more. The vines branch horizontally from the nodes and do not attain length but the fully grown vines completely cover the standard presenting the appearance of a bush. The plant is perennial climbing. Leaves are dark green and shiny above and pale green underneath smooth, entire and broad. The inflorescence is a catkin, a form of spike varying according to the varieties. The fruit is a seeded berry, sessile, small, usually globose and sometimes elongated or oval. It has a thin, soft pericarp surrounding the seed.

It is cultivated as a pure crop on a plantation scale. As a mixed crop it is grown in Kerala and Karnataka States along with coffee arecanut/coconut on a large scale. In arecanut gardens the areca trees themselves serve as standards. Almost every homestead in Kerala grows pepper along with mango, jack or coconut. The plant is available throughout Malabar coast and Western Ghats. Assam, West Bengal, Karnataka and Tamil Nadu also grow pepper to a limited extent, but Kerala is the earliest to grow this spice. Outside India it is grown in Sri Lanka, Malaysia, Indonesia, Brazil, Mexico, China (PR), Vietnam, Thailand and Madagascar.

India held the monopoly in the world pepper market from the earliest times till the 19th century but lost her place to Indonesia and Malaysia. However, in India, pepper is cultivated in an area of more than 195050 hectares the annual average production being 40 to 54 thousand tonnes, craning foreign exchange of Rs. 1811 million. Kerala (Wynad, Iaukki, Kannlia, Kozikode, Kottayam, Trivandrum, Ernakulam and Quilon) accounts for more than 96% of it where it is grown in 1.80 lakh hectares. The other major pepper producing States are Karnataka and Tamil Nadu.

Table 9.4: The important varieties of black pepper are as follows:

Malabar Varieties	
Kalluvalli	Kottavally
Kalluvalli Ii	Kottavally
Uthirankotta	Karimkotta
Travancore Varieties	
Cheriyakaniyakadan	Arikottanadan
Chumala	Kaniandan
Kottanadan	Kuthiravally
Munda	Kumbhakodi
Thulakodi	Karivilanchi
Mundi	
Cultivars of North Kanara and Mysore Regions	
Malligesera	Motakara
Karemalligesera	Morate
Doddagya	Tattisara and Vakkalgooja

Related species of Piper

The *Piper beetle* is an economically important species cultivated extensively in India. The leaves of which are used as a masticatory. The *Piper longum* L. (Indian long pepper) and *Piper chaba* Hunt. (Java long pepper) are another groups of medicinally important spices. Protocols for rapid clonal multiplication of *Piper longum* from shoot tip explants are available (Sarasan *et al.,* 1993; Rema *et al.,* 1995). Micropropagation of *P. chaba* (Nirmal Babu *et al.,* 1994; Rema *et al.,* 1995), *P. betle* (Nirmal Babu *et al.,* 1992c) has been standardized. Micropropagated plantlets of these species are being evaluated for their filed performance at the IISR, Calicut.

Biotechnological Approaches for Conservation

Foot rot is the major disease affecting pepper plantations. *In vitro* cloning of black peppers has been reported using seedlings and mature shoot tip explants (Nazeem *et al.,* 1993; Philip *et al.,* 1992; Nirmal Babu *et al.,* 1993a, b) and tissue cultured plantlets were successfully established in field. The multiplication rate is around 6 shoots/culture in about 90 days. Phenolic exudates from the cut surface and bacterial contamination (Raj Mohan, 1985, Fitchet, 1988a; b) severely affect black pepper tissue culture.

Micropropagation

Using shoot explants from both mature and juvenile plants micropropagation of black pepper can now be done. The technique was perfected by the NRC. Though up to 6 suckers have been obtained in some cultures, the multiplication rate is still low (average 3) commercially.

Establishment of Callus Cultures and Regeneration of Plantlets

Callus cultures, using leaf or stem tissues, could easily be established in black pepper and its related species *Piper colubrinum* and *P. longum*. But regeneration of plantlets from callus appears to be hampered by phenolic exudations. Sporadically though, plantlets have been observed in a few cultures where the leaf tissue was used as explant. A high-frequency regeneration system once standardized will help in developing techniques for transfer of resistance to *Phytophthora* (the casual organism of *Phytophthora* foot-rot) from *Piper colubrinum* into the cultivated black pepper.

Micropropagation of other Species of Piper and in-vitro conservation of Germplasm

Related species *P. longum*, can also be micropropagated using internal explants that results in direct organogenesis. The growth in *Piper* species is slow with no necrosis symptoms. These cultures could be used for *in-vitro* conservation of germplasm. The subculture intervals should be extended substantially for any meaningful *in vitro* conservation programme.

GINGER

Ginger, the underground rhizomes of the perennial herb *Zingiber officinale* Rosc. Belongs to family Zingiberaceae. It is an important tropical spice. It is produced and exported in largest quantity from India. There is no seed propagation, therefore variability is limited. Somaclonal variation can be exploited for evolving high yielding, high-quality lines and for developing resistant types to major diseases, rhizome-rot and bacterial wilt.

Biotechnological Approaches for Conservation

Plant breeding crop improvement programmes in ginger are hampered by the absence of seed set leading to limited variability. Somaclonal variation could be an important source of variability that could be exploited to evolve high-yielding and high-quality lines and to develop lines resistant to rhizome rot and bacterial wilt. Tissue culture techniques could also be used for *in vitro* pollination, embryo rescue and possible production of seed in ginger.

Rapid Clonal Multiplication

Ginger being a vegetatively propagated crop responds readily to the tissue culture propagation method. An average of ten adventitious shoots could be obtained using the protocol standardized at the NRCS. Since major diseases spread through the contaminated seed rhizomes in ginger, using tissue culture pathogen-free planting material can be produced. The tissue cultured plants cannot be directly planted in the field. It takes around two seasons for the tissue cultured plants to develop rhizomes of sufficient size. In vitro rhizome formation has been observed in some cultures.

Clonal multiplication of ginger from vegetative buds has been reported (Hosoki and Sagawa, 1977; Nadgauda *et al.,* 1980; Pillai and Kumar, 1982; Balachandran *et al.,* 1990; Choi, 1991). Ginger cultivation is threatened by rhizome-rot disease caused by *Pseudomonas solanacearum* and *Phythium* spp. Diseases of ginger are often spread through infected seed rhizomes. Tissue culture techniques would help in the production of pathogen-free planting material of high yielding varieties.

Field evaluation of tissue cultured plants at the IISR, Calicut indicated that they cannot be used directly for commercial planting, since it takes 2 crop seasons for the micropropagated plants to develop normal-sized rhizomes that can be used as seed rhizomes.

Regeneration of Plantlets

Plants have been regenerated from leaf, ovary and vegetative bud tissues via the callus phase. These techniques are used in the production of somaclones screened for disease resistance and other useful characters.

In vitro Selection

The high-frequency regeneration system developed at the NRCS is being used for *in vitro* selection of resistant types to *Pythium aphanidermatum* and *Pseudomonas solanacearum*, the causal agents of dreaded rhizome-rot and bacterial wilt. The crude culture filtrates extracted from these pathogens are being used as selecting agents.

Inflorescence cultures and in vitro development of fruit

In ginger there is no report of fruit set. However, by supplying the required nutrients to the immature inflorescence in culture, it was possible to make the ovary develop into fruit. By culturing immature inflorescence in a suitable medium, direct development of clonal plantlets could be achieved by the conversion of floral buds into vegetative buds and their subsequent development into complete plants. In single flower cultures also, complete plantlets could be developed directly from the ovary tissues.

TURMERIC

Turmeric of commerce is the dried rhizome of *Curcuma longa* Linn. India is one of the major producers and exporters of this spice. Young vegetative buds from the germinating rhizomes as explants were used for micropropagating turmeric. These buds readily respond to the culture condition putting out 8-10 adventitious shoots in 40 days of culture. This technique may be used for the production of disease-free planting material. Callus differentiation and plantlet formation has been achieved in turmeric.

NUTMEG

Nutmeg (*Myristica fragrans* Houtt.), cinnamon (*Cinnamomum verum* resl.), clove (*Syzygium aromaticum* Linn.) Merr & Perry and allspice (*Pimenta dioica* Linn.) Merr. are the tree spices of relevance in the Indian context. It was introduced to India towards the end of 18th century and in India, it is grown in certain pockets of Kerala, Tamil Nadu and Karnataka. More specifically, it is grown near Courtallam of Tirunelveli district and round about the Burliear zone on the eastern slopes of the Nilgiris, and Kaladi in Kerala. It is also reported to grown in Cochin, Trivandrum, Kottayam and Calicut.

Nutmeg is a handsome, evergreen tree with dense foliage. Trees are 10-12 metres high, branches are spreading with dark grey bark; leaves are shiny and oblong to ovate in shape; flowers appear in cymes, each cyme has several branches on which are born a number of flowers which hang down, flowers are small, pale yellow and bell shaped, slightly aromatic; fruit is fleshy, globose in shape and lemon yellow to light brown in colour; inside the fruit is the single glossy brown seed with a brittle shell over which is the beautiful, brilliant, scarlet net like membrane or aril known as mace which is strongly fragile and aromatic. Nutmeg trees are dioccious in nature and is diploid (24-42). Nutmeg is generally propagated though seeds.

The important diseases are fruit rot, leaf and shot hole, Dieback and Thread Blight. A scale insect *Saissetia nigra* has been observed occasionally to infest the shoots.

The nutmeg of commerce is the ovoid kernel, which is hard and brown enclosing which is a thin shell. Surrounding this shell is the aril, scarlet in colour, which furnishes the mace of commerce. Medicinally nutmeg acts as stimulant and carminative. Its use as abortificient also reported. An essential oil is extracted for use in medicine, toilet soaps, dental pastes, for flovouring chewing gums and chewing tobacco. It contains 25 to 40 per cent of fixed oil known as Oleum myristica expressum. The aromatic oil is of butter like consistency and is orange in colour. The oil is used for flavouring backed goods, cakes, cookies, puddings and pickles.

Biotechnological Approaches for Conservation

Micropropagation

Shoot-tip cultures of nutmeg, cinnamon and allspice have been established. Growth of these cultures was satisfactory in allspice, but slow in both nutmeg and cinnamon. In cinnamon, rooting of shoot-tip explants was observed in a few cultures.

Callus cultures from leaf and shoot explants were established in cinnamon, nutmeg and allspice.

Nutmeg and mace are the two important spices obtained from the nutmeg tree.

TAMARIND

Tamarindus indica (tamarind) is believed to be native of tropical Africa and probably also to some parts of South India. It is a slow growing tree and attains a very large size and lives for more than 100 years. The tree produces fruits with acid pulp which is largely used in curries and chutney. The pulp is therefore, a commercial stem in many tropical countries. The tree has also many medicinal uses.

Biotechnological Approaches for Conservation

Micropropagation

Kopp and Natarajan (1990) described a method for the *in vitro* propagation from shoot-tips. Jaiswal and Gulati (1991) using entire cotyledon excised from 12-day-old seedling, raised *in vitro* from seeds, reported that in a period of four months 34 to 55 shoots/cotyledon were regenerated.

Ganga and Balakrishnamurthy (1997a) used axillary buds (collected in May) as the best explants for clonal propagation. They produced 16.40 plantlets per culture. However, when the axillary buds were collected and culture in the month other than May (December, January, February, March, September and October) did not show any response. They also reported that the stem collected from *in vitro* germinated seedling was capable to differentiate into shoots. They further reported that MS medium fortified with 6% sucrose, 0.5 mg/l BA and 0.5 mg/l GA_3 was very effective for multiple shoot induction, including number of shoots/culture and length of microshoot.

The microshoots produced from axillary buds in the primary culture when transferred to MS medium supplemented with 5.0 mg/l BA and 0.5 mg/l GA_3 gave higher bud break (82.50%); earlier bud break (9.33 days), and 6.40 shoots/cultivar (Ganga and Balakrishnamurthy, 1979b). They further indicated that stem bits cultured in MS medium supplemented with 2.5 mg/l BA and 0.5 mg/l BA recorded 80 per cent shoot-bud differentiation and 4.40 shoots/culture in period of 15-26 days. Root initiation appeared to be related to the media composition and auxin concentration in the medium (Ganga and Balakrishnamurthy, 1997b).

Research on biotechnology of tamarind is limited. There is a need to explore the literature on characterization of germplasm. Considering the wide diversity present in developing world, such a study would go a long way in protecting these materials with authentication for conservation. It is a crop that holds promise for the commerce and biotechnological research.

10

Conservation of Flower Crops Through Biotechnology

Floriculture is one of the branches of agriculture. The commercial flower forcing is an important segment of the floriculture industry. It is a production or manufacturing business. The products are flowers and ornamental plants, and they are produced in closely controlled conditions-to a great extent in greenhouses. The individuals who work in the floriculture industry could be called floriculturists, horticulturists, or agriculturists, but frequently they are known as flower and plant producers or florists.

Floriculture has by far, a greater annual growth potential of 25 to 30 per cent which is 25 to 30 times more than that of cereals or any other agricultural produce. India being specially, well placed to meet the international demand of cut-flowers which peaks during winter months. Apart from this, there are certain compelling reasons for India to venture into international trade. Being a tropical moderate country, India is a treasure house of ornamental plants which will buy us a special niche in the market. The government has approved 24 projects with a total foreign investment of Rs. 8.47 crores cleared to boost flower export. Most of the projects involve either technical collaboration or financial collaboration with well-established foreign companies. These projects will become functional in the next two years. These measures should, to a certain extent put India on the global floriculture market. The corporate who have already taken a plunge have so in collaboration with the seasoned Dutch horticulture firms.

The per capita consumption of flowers is going up around the world. The business of floriculture is surging forward, the latest estimates of the annual global flower trade is worth 2, 500 million US dollars and the Indian contribution at present is about $0.36 million. Indian participation will in due course show a marked difference for the better in the next two years.

The international floriculture market trade is estimated to be 40 billion US dollars (Rs. 125,000 crores) of which cut-flowers account for nearly 60 per cent i.e., 25 billion US dollars (Rs. 80,000 crores). Countries like Columbia, Israel, Kenya and Italy have made an entree and have created a modicum of competition in a field not even recognized as a major industry. But the world floriculture trade is still controlled by Holland which has the lions share of 67 per cent to its credit. Amongst the many changes which were ushered in after and due to the Indian economic reforms is the budding of the floriculture industry. The **bloom boom** has caught on the imagination of the entrepreneurs.

Most of the floriculture units are situated in south part of India with a few scattered in the north India., Bangalore and Hyderabad have become the hub of activity and will be cultivating roses of every conceivable variety on a massive scale. There are also some projects in the country which are venturing into cultivation of more exotic and expensive flowers, companies like Natural Synergies grow Orchids, while Oriental Floritech is exporting Carnations.

In India, about 26 projects with foreign collaborations have been approved between August 1991-July 1994. The capital cost of every hectare works out to between Rs. 1.5 to 2 crores, with the minimum project requirement of around three hectares, each project cost works out to a prohibitive Rs. 5 crores.

Following are the reasons for the luke warm response to the floriculture industry:

1. Cost of establishment is quite high.
2. The industry requires a technologically advanced infrastructure to ensure quality.
3. The one competitive factor which also determines the prices clocked at the auctions is the quality.
4. The absolute necessities to ensure quality are specialized transport vehicles, refrigerated trucks, precooling units and cold storage at air terminals along with regular international flights. None of these are readily available at present in India.
5. Flowers end up with a rather low priority in air cargo as they are very bulky, voluminous and perishable, the freight component is very high. The minimum freight per stem of rose works out to Rs. 2.30 to Rs. 3.50.

6. The Indian exports are levied an import duty of 15 per cent which can go up to 25 per cent during peak season.

Central Government has sponsored certain schemes for setting up model floriculture centre in the Eighth Plan. The Rs. 10 crore scheme involves setting up of ten model centres in the public sector at Mohali in Punjab, Calcutta, Lucknow, Bangalore, Srinagar, Trivandrum, Gangtok, Pune and Himachal Pradesh. Also eight centres will be set up in the private sector in all these places except Srinagar and Himachal Pradesh.

Nearly 30 per cent of the outlay has been earmarked for the private sector under the scheme. These model centres are expected to be the focal units for introducing new varieties and their multiplication. These centres will also supply basic materials to registered private nurseries along with imparting training in flower production and post-harvest handling of cut-flowers. Each of these units will be attached to a tissue culture unit to enable research and development for the rapid production of the desired quality of flowers.

Considering the phenomenal rate at which the industry is growing, the flower council of Holland projects a 15 per cent growth by 1997, floriculture can prove to be the one possibility of obtaining large exports in a short time. This fact clubbed with the scheduled opening of the world's second flower exchange in Singapore in 1996 is going to herald the boom in demand for fresh and cut flowers from that region.

The lucre of the industry is such that other administrative bodies are also promoting the same. The National Bank for Agricultural and Rural Development (NABARD), for instance, has already sanctioned an export-oriented floriculture scheme in Maharashtra. There are other projects involving the cultivation of Australian wild flower - Beraldton Wax, which have been sanctioned by NABARD to be set up in Haryana. The deal includes the company Plantex Australia Pvt. Ltd. collaborating for technical assistance and planting material.

A Dutch development bank has agreed to give non-refundable aid to two Hyderabad based companies and a Hague based Netherlands Financierings-Maatschappij Voor Ontiwikkelingsladen, specializes in extending long term credit to viable private sector companies is developing countries.

Floriculture projects have been sprouting up on a large scale and each project has an impressive outlay. Amongst those which have registered offices in Hyderabad are Nagarjuna Agritech, Bommidala Florex, Classic Biotech & Exports, Sri Vasavi Florex, Aditya Florex, Suvarna Florex, Neha International, Jagadambay Agri Genetics, Greenhouse Flori and TMT India are notable.

BIOTECHNOLOGICAL APPROACHES FOR CONSERVATION

Developments in plant biotechnology has revolutionizing plant breeding and propagation. Ornamentals and flower crops had been benefited by the biotechnological impact. Several techniques are still on the way of development to evolve, examine and refined to understand the genetical and physiological mechanisms of plant cells, tissues and organs at molecular levels for transforming the vast scope and opportunities. Techniques of plant biotechnology are being used by various corporate bodies and private multinational research organization is engaged in extensive research work specially *in vitro* cloning of crops for mass multiplication of true to type and vigorous plant propagules and availability of disease free planting material through micropropagation. They are developing new and attractive varieties of floricultural crops for gaining rich profits.

Somaclonal variation and *in vitro* selection, *in vitro* breeding by production of androgenic plants, gynogenesis, somatic embryogenesis, synthetic seed technology, protoplast fusion, genetic engineering and transformation are the latest developments. The available information revealed that floricultural crops have complemented conventional plant breeding techniques to a great extent.

Tissue culture

Micropropagation

About three-fourths of the researches have been done on micropropagation of ornamentals (Murashige, 1990). The earliest success with the rapid propagation of orchids (Morrel, 1960), is the inspiration behind the successful application of tissue culture for mass multiplication of many ornamental and flower crops. Tissue culture technique like he choice of the explant, sterilization techniques, media composition, cultural conditions and hardening techniques vary widely within and among the varieties of a floricultural crops. Protocols have been developed for many commercially important flower crops. Many of them are now exclusively propagated by tissue culture methods only. Genetically engineered floricultural crops have to be propagated only by *in vitro.*

About 300 commercial companies are engaged in tissue culture. They are producing approximately 400 million plants annually in USA, England. The Netherlands, France, Germany and Australia. In Singapore and Thailand tissue culture in ornamentals has been adopted as cottage industry. In India, about 40 companies are now engaged in tissue culture (Chaturvedi *et al.*, 1995). The reader may refer the reviews in Ammirato *et al.* (1990) and by Chaturvedi *et al.* (1995) on many ornamental crops, Rout and Das (1996) on Chrysanthemum, Giresback (1986) and Singh (1995) on orchids and Kim and De Hertog (1997) on bulbuous ornamentals.

Somaclonal Variation and In vitro *Selection*

Table 10.1: Somaclonal variation in ornamental plants.

Flower	Work conducted	Reference
Carnations	Developed plants with red flowers had been regenerated from callus (White Sim' cultivar).	Hacket and Anderson(1967)
	Reported a frequency of 0.083% variation in callus cultures using petals as explant compared to 0.66% and 5.34% variation from non-irradiated meristems and X-ray irradiated cuttings. They also advised a complementary approach involving regeneration procedure and induction of mutagenesis to obtain higher degree of variation for further selection in carnations.	Buiatti et al. (1986) Buiatti (1989)
	Obtained 2 dwarf regeneration from 2 genotypes of Diantini tetraploid Carnation. They further indicated that the inheritance of this trait was found to be either additive and polygenic or monogenic with an intermediate heterozygote. The degree and frequency of dwarfing depended upon explant, genotype and /or environment.	Sparnaiij *et al.* (1990)
Chrysanthemum	They noted delayed flowering in regenerated long-term cultures.	Sutter and Longhands (1981)
	Shoots derived from ray florets of chrysanthemum has been found to be more floriferous with higher frequency of abnormalities than when regenerated from vegetative parts.	Bush et al. (1976); De Jong and Custers (1986); Malaure et al; (1991)

Flower	Work conducted	Reference
Cyclaman persicum Mill	Recorded 4.5-36.3% variation in growth habit, leaf shape, flower shape, flower colour and chlorophyll pattern in 6,000 plants obtained from *in vitro* cultures of 8 diploid genotypes.	Schwenkel and Grunewaldt (1990)
Day lily (*Hemerocallis lililo-asphodetus*	A dwarf somaclonal variant was reported	Griesbach (1989)
Fuchsia	Reported variation in colour, shape and size of flowers using callus of stem, leaves and flower parts.	Bouharmant and Dabin (1986)
Geranium	The tissue-cultured cultivar, Velvet Rose, has been released. Having improved sturdiness, vigour and attractiveness, it is the first named cultivar derived from tissue culture	(Skirvin *et al.*, 1993). Geier and Reuther (1981) and Geier (1982)
Anthurium scherzerianum	Observed a very low level variation in embryogenic callus derived from spadix tissues	
Haworthia	Reported somaclonal variants	Yamabe and Yamada (1973)
Kalanchoe	Recorded 0.2-23.4% variation in 4 cultivars cultured from microcuttings or shoot tips or leaf explants.	Schwaiger and Horn (1988)
Lisianthus (*Eustoma grandiflorum*)	Reported a dwarf basal branching somaclonal variant.	Griesbach and Semenuik (1987); Griesback et al. (1988)
Poinsettia	Recorded increased instability in calli and bioreactor cell suspension cultures.	Geier *et al.* (1992)
Tuberose	Reported considerable variation in floral traits in eight of the 35 plants regenerated from embryogenic callus obtained by microspore culture	Giant and Tsay (1989)
Wiegela	Reported somaclonal variants	Duron and Decourtye (1986)
Wishbone flower (*Tolernia fournierii*	Obtained a somaclonal variant of white coloured flowers and compact growth habit. It was named as *Uconn White*.	Brand and Bridgen (1989)

The occurrence of variations in cell cultures of ornamentals and flower crops are rare, therefore, the chances of selecting a useful natural somaclonal variant is laborious and rare, remote especially if such changes are recessive and occur at a low frequency. The selection of a variant plant is observed only at a later stage when the plants are fully grown. The selection of desirable cell lines or tissues at an earlier stage is always advantageous. This can avoid the selection of non-variant or less useful plants. The genes governing a particular trait may not able to express in cell cultures. This may be due to unavailability of specialized or organized structures in the culture or may be due to under regulation by other genes. In some cases it has been observed that these variant cell lines are not stable. Hence, induction of variation in cell cultures intentionally by in vitro mutagenesis is now being resorted increasingly to create variability for further screening and selection. Somaclonal variation can be, however, useful in crops with narrow genetic base and to a certain extent for in vitro selection against biotic and abiotic factors.

Varied *in vitro* selection techniques are useful for recovery of cell lines resistant to biotic or biotic stresses in many agricultural crops to. Selection may be done by screening cell lines for resistance to salinity by using NaCl or for mineral stresses (both deficiency and toxicity) by employing a mixture of salts as selecting agents. In many cases, the resistance to mineral stresses in under monogenic control which is quite encouraging. Using screening method against pathotoxins or other substances it is also possible to select against fungal pathogens which produce a similar effect on plants Duncan (1997).

In Vitro Mutagenesis

The technique of *in vitro* mutagenesis is an important method which helps to improve vegetatively propagated plant materials, especially when one or few characters of a desirable cultivar are to be modified. (Broertjes and van Harten, 1988). Exposure of the *in vitro* cultures to either physical or chemical mutagens can lead to higher recovery of variants but, when the *in vitro* cultures are exposed to mutagenic treatments, it will be difficult to determine whether the resultant variation is induced or spontaneous.

The newer cultivars can be produced using *in vitro* mutagenesis, but in many ornamentals, this technique is yet to be employed or has failed to yield useful mutants. This technique can be a possible approach when conventional approaches of improvement is cumbersome and protocols for advanced techniques like protoplast fusion or genetic engineering has not been standardized.

Table 10.2: *In Vitro* Mutagenesis in important ornamentals.

Flower	Work conducted	Reference
Alpinia purpurata (an important commercial cut flower exported from West Indies to Europe and North America)	Induced mutations in by exposing *in vitro* plantlets to 15-40 kR gamma radiation	Fereol *et al.* (1996)
Chrysanthemum	Subjected the explants to 5-25 gy X-ray and gamma ray irradiation and subsequent *in vitro* culture obtained 12 new cultivars of, several chimeric mutants with change in inflorescence colour.	Jerzy and Zalewska (1996)
Cut rose cv. Ilesta	Subjected the basal segments of *in vitro* derived microshoots of 25 – 60 gy X-ray dose and subsequent culture *in vitro*. They reported 73% flower mutants (size, colour and number of peals), 14% with changed growth types and 13% with modified leaves.	Walther and Sauer (1989),
Iris	Observed that gamma irradiation of 1-2 kR and 4-8 kR is the best for inducing mutations in callus and adventitious buds of organ cultures.	Liu Quiglin et al., 1995
Lisianthus	Obtained 3 desirable mutants (tentatively named as Purple Robin, Purple Fantasy and Red Robin) by exposing 2 cultivars, *viz.*, Pastel Murasaki and Morgenrot to 0.25 – 15 gy gamma rays for 90 days and by subsequent *in vitro* culture.	Nagatomi *et al.* (1996)
Pink gerbera cultivar	Obtained 15% variants for flower colour from a following 2 cycles of *in vitro* culture of shoot explants subjected to 20 gy gamma rays. Both intra-plant and intra-inflorescence chimerism were reported by them.	Laneri *et al.* (1990)

Somatic Embryogenesis and Synthetic Seeds

Somatic embryogenesis is the process by which a single somatic cell or group of somatic cells undergoes a developmental pathway leading to production of regenerable non-zygotic embryos capable of germinating and developing into whole plants. The technique is useful for large scale clonal propagation in many crops and for gene transfer systems because it facilitates *in vitro* selection of transformed totipotent cells and reduces the chances of production of chimeras.

Table 10.3: Somatic embryogenesis in important ornamentals

Ornamental	Explant used	Reference
Antirrhinum	Leaf protoplast	Poirier Hamon et al., 1974
Anthurium andreanum	Lamina	Kuehnle et al. 1992
A. Scher zerianum	Spadix	Geier, 1982
Begonia gracilis	Lamina Segments	Castillo and Smith, 1997
Carnations	Internodal callus	Frey et al., 1992
	Leaf	Nakano and Mii, 1993
Chrysanthemum	Leaf disc	Pavingerova et al., 1994
Crotons	Endosperm	Bhojwani, 1966
Euphorbia pultcherrima	Seed	Nataraja, 1974
Fressia refracta	Inflorescence	Wang et al., 1990
Hedera helix	Stem section	Banks and Hackett, 1978
Iris	Shoot apex callus	Reuthner, 1978
	Leaves, flowers	Jehan et al., 1994
Kalanchoe	Leaf sections	Mohan Ram and Wadhi, 1965
Mesembryanthemum	Hypocotyl, leaf, Shoot, root, tip	Mehra and Mehra, 1972
Pelargonium hortorum	Seeds	Quershi and Sexena, 1992
	Cotyledons	Murthy et al., 1996
Paperomia longifolia	Stem sections	Scaramella, 1978
Petunia	Leaf section, stem Internode	Rao et al., 1973
Rose	Adventitious roots	van der Salm et al., 1996
	Callus	Jeyashree et al., 1997

Source: K. Soorianathasundaram and I. Irulappan, Floriculture, In: *Biotechnology and Its Application in Horticulture*, S. P. Ghosh (ed.) 1999, Narosa Publishing House, New Delhi, pp 202-243.

Delivering the somatic embryos through encapsulation is tricky. It has been perfected in only a few crops. Usually, calcium aiginate is used for encapsulation (refer to the Chapter Synseed in this book).

Haploid Cell Lines

Haploid cell lines can be also subjected to mutant selection using selective markers in the culture media or genetic transformation. Some of the problems associated with haploid production include, failure of anthers to grow *in vitro* and abortion of embryo, chimeral cell production, unwanted proliferation of diploid and polypoid tissues along with haploid tissue, difficulties in doubling the haploids to achieve homozygosity, etc. In many of the ornamental crops, haploid production has been reported through androgenesis and insome (e.g., Gerbera) haploid production was achieved primarily through gynogenesis (Soorianathasundaram and Irulappan, 1999).

Table 10.4: In vitro production of haploids in some ornamental/flower crops

Crop/Species	Explant	Reference
African violet	Anther	Weatherhead *et al.*, 1982
Begonia X Hiemalis	Anther	Khoder *et al.*, 1984
Calla lily	Anther	Ko Jeong *et al.*, 1996
Fressias	Anther	Bajaj and Pierik, 1974
Pelargonium	Anther	AboEl Nil and Hilderbrandt, 1971
Ipomea	Anther	Sangwant and Norreel, 1995
Lilies	Anther	Sharp *et al.*, 1971
		Arzate-Fernandez *et al.*, 1997
		Han Dong Sheng *et al.*, 1997
	Ovaries	Gu and Cheng, 1983
Petunia	Anther	Sangwan and Norreel, 1975
		Raquin, 1982
Gerbera	Anther	Paraeil *et al.*, 1977
	Immature	Ahmin and Vieth, 1986;
	Ovules	Meynet and Sibi, 1984; Sibon, 1981; Cappadocia *et al.*, 1988; Tosca *et al.*, 1990; Miyoshi and Asakura, 1996
Rose	Anther	Tabaeezadeh and Khosh-Khui, 981
	Microspore	Wissemann *et al.*, 1996

Source: K. Soorianathasundaram and I. Irulappan, Floriculture, In: *Biotechnology and Its Application in Horticulture*, S. P. Ghosh (ed.) 1999, Narosa Publishing House, New Delhi, pp 202-243.

Polyploidisation *In Vitro*

In vitro polyploidization has been employed frequently by adding colchicine to the *in vitro* cultures.

Polyploidisation *in vitro* has following advantages:

1. *In vitro* polyploidization of haploids helps in achieving rapid homozygosity in one-step.
2. It helps in heterosis breeding of ornamentals.
3. When two distant species are gametically or somatically hybridized very often it results in sterility, or when a foreign gene construct is introduced

it may result in sterility. Doubling of chromosomes in these cases can overcome sterility.

4. Polyploidization helps to obtain to generate tetraploids from diploids. Hybridization of diploid-tetraploid crosses can be attempted thento generate triploids. Triplod plants are seed sterile. In many plant species their performanceis were far better than diploids or tetraploids.

Table 10.5: Important reports on in vitro polyploidization:

Ornamentals	Work conducted	Reference
Alstroemeria	Used 02-06% colchincine *in vitro Alstroemeria* cultures of *A. Aurea x A. caryophyllaea* crosses to overcome sterility.	Lu Chen Shing and Bridgen (1997)
Calla lily (*Zantedescia aethiopica*)	Obtained tetraploids by treating rapidly multiplying shoot cultures with 0.05% colchicine for 1-4 days.	Cohen and Yao (1996)
R. wichuriana	Induce tetraploidy from in vitro cultures of diploid by adding colchicine or 3H thymidine or colcemid to the culture media by filter sterilization	Roberts et al. (1990)

Protoplasmic Fusion

Protoplast fusion technique is a difficult task in ornamentals. It needs the establishment of standard protocols, especially of its regeneration. Protoplast regeneration techniques helps in fusion of protoplasts from two different plants or even species to evolve a new somatic hybrid. Process generally involves isolation of protoplasts, fusion regeneration of somatic hybrid, screening regenerants for morphogenetic variation and selection of somatic hybrid (also refer to Chapter Protoplast Technology in this book). Not much work has been done on ornamentals protoplast culture. The following important reports are as follows:

Table 10.6: Protoplast culture of some ornamentals

Species	Reference
Callistephus chinensis	Pillai *et al.*, 1990
Centauria cyanus	Pillai *et al.*, 1990
Dendrathema	Otsuka *et al.*, 1985
Eustoma gradiflorum	Murayama *et al.*, 1996
Forsythia X *intermedia*	Ochatt, 1994
Haemerocallis	Ling Jing Tyan and Sauve, 1995
Iris germanica	Shimuzu *et al.*, 1996
Lilium X *formolingii*	Godo *et al.*, 1996
Lilium X *elegans*	Godo *et al.*, 1996
Lilium speciorubel	Sugura, 1993
Limonium perezii	Kunitake and Mii, 1990
Liriodendron tulipifera	Merkle and Sommers, 1981
Lonicera nitida	Ochatt, 1991
Matthiola incana	Hosoki and Ando 1989
Rosa persica X *R., xanthina*	Mathew *et al.*, 1991
Senecio X *hybrida*	Pillai *et al.*, 1990
Weigela X *florida*	Ochatt, 1993

Source: K. Soorianathasundaram and I. Irulappan, Floriculture, In: *Biotechnology and Its Application in Horticulture*, S. P. Ghosh (ed.) 1999, Narosa Publishing House, New Delhi, pp 202-243.

Table 10.7: Protoplast fusion in some ornamental and other plants

Plants	Work conducted	Reference
Nicotiana and Petunia	Successful protoplast fusion has been successfully reported between *Petunia hybrida and P. axillaries; P. hybrida; P. parodii, P. Parodii* with *Pinflata*	Power et al. (1976, 1979, 1980); Cocking (1983).
Phalenopsis orchids	protoplasmic fusion	Chen et al. (1995
Dianthus barbatus and *Gypsophila paniculata*	Used Hypocotyls derived protoplasts of *Dianthus barbatus* were made to fuse with cell suspension culture derived protoplasts of *Gypsophila paniculata* by electrofusion technique to obtain somatic hybrids.	Nakano et al. (1996).

Information on these aspects on ornamentals are still lacking though there exists a vast potential to be explored and under exploited till date.

Embryo Rescue

This technique has been employed in various crops to circumvent post fertilization barriers leading to ovule abortion especially in wide hybridization or to culture immature seeds with well-developed endosperm. Following are the important reports on embryo rescue of ornamentals plants:

Table 10.8: Embryo rescue of ornamentals plants

Ornamentals	Work conducted	References
Chrysanthemum	Reported a system for embryo rescue and seed set in which the rescued seedlings registered earliness and vigour over the non-rescued seedlings.	Anderson *et al.* (1990)
Rose	Reported a major problem in rose breeding programmes the seeds (achenes) fail to germinate or have a low percentage of germination (less than 20%). They lack of germination and is attributed to mechanical restriction of embryo expansion by presence of a thick pericarp of growth-inhibitor induced dormancy.	Jackson and Blundell (1963)
Rosa hybrida cultivars	In roses, Reoprted scuccessful germination of seed under in vitro condition and the possibility of embryo rescue technique to recover potential crosses.	Gudin (1994)
	Able to rescue and culture 4-5 weeks old embryos after exposing them to cold treatment at 4ºC for one month from two crosses in which otherwise, hips failed to develop after crossing or only a low percentage of mature achenes.	Marchant et al. (1994)
Lily	In this plant hybridization	Asano (1982); Asano

Ornamentals	Work conducted	References
	followed by embryo rescue techniques is commonly employed in breeding. Two new lily *Northern Beauty* and *Starburst Sensation* developed by such distant crosses and embryo rescue have been released for commercial cultivation	and Myodo, (1980) van Tuyl et al. (1986); Shen Ge zhi et al. (1997); Collicut and Ronald, 1995, 1996).
Alstromeria	Embryo rescue techniques had been successfully employed.	Kristiansen, 1995
Gypsophila	Embryo rescue techniques had been successfully employed.	Kishi *et al.*, 1994
Limonium	Embryo rescue techniques had been successfully employed.	Burge *et al.*, 1995
Pelargonium	Embryo rescue techniques had been successfully employed.	Cassels *et al.*, 1995
Hibiscus syriacus (2x × 4x crosses)	Embryo rescue techniques had been successfully employed.	Kim *et al.*, 1996
Tulips	Embryo rescue techniques had been successfully employed.	van Creij, 1997

Incorporation of resistant genes and creation of novel flower colours and plant types by wide hybridization is still and attractive proposition to many plant breeders and embryo rescue technology can be a handy tool to meet this objective.

Cryopreservation

The conventional conservation of germplasm has certain limitations. Seed materials cannot be stored for a long time. Field genebanks are costly to maintain the subject to damage by adverse biotic and abiotic conditions or vandalism. Cryopreservation technique, in which in vitro regenerable tissues are safely freeze preserved, in liquid nitrogen in majority of the cases can be efficient alternative cost effective means of germplasm conservation. Fukai *et al.* (1988) reported a method of cryopresevation of shoot tips of Chrysanthemums with 10% DMSO solution + 3% sucrose cooled gradually at the rate of 2°C/min. from 0 to 40°C.

This technique is helpful to conserve male sterile parental lines, maintainer lines, species or varieties with poor seed viability and germination. Kobayashi et al. (1990) indicated that with the increasing interest in genetic engineering of plants, preservation of cultured cells with unique attributes has assumed much importance. Selection of surviving cell lines when cryoprotectants are not used, can be also freeze tolerant. We need simple and reliable methodologies and a routine laboratory protocol for cryopreservation.

Table 10.9: Reports on cryopreservation of ornamental plants

Ornamentals	Work conducted	Reference
Chrysanthemum	Reported high viability of shoot tips in liquid nitrogen even after 8 months period.	Fukai and Oe (1990)
Carnation	Cryopreservation of shoot tips by encapsulation and dehydration has been described	Lynch et al. (1996); Tannoury et al. (1995)
Camellia japonica	Embryonic axes of was preserved by a simple technique dessication in laminar flow hood and direct immersion in liquid nitrogen. But the somatic embryos could not be cryopreserved even after encapsulation.	Janiero et al. (1996)

Glossary

Aberrant: A growth that deviates from the normal or usual type; as exceptional growths **(aberrations)** which occur in some tissue cultures.

Accession: Plant or seed sample, strain or population held in a genebank or breeding programme for conservation and use.

Active Genepool Collection: A collection of germplasm used for regereration, multiplication, distribution, characterization and evaluation. Ideally, germplasm in the active collection should be maintained in sufficient quantity to be available on request. Active collection germplasm is commonly duplicated in a base collection and is often stored under medium to long-term storage.

Anther culture: Refers to culture of single pollen grains or of the anther, containing the male gametophytes or micropores, with the objective of producing monoploid plants.

Artificial seed: Encapsulated or coated somatic embryos (embryoids) that are planted and treated like seed.

Auxin: One of a large class of plant hormones (phytohormones) allegedly produced in the growing tips of stems and roots. It causes cell enlargement and elongation but in excised, cultured systems they may promote cell division.

Axenic culture: A culture without foreign or undesired life forms. An axenic culture may include the purposeful co-cultivation of different types of cells, tissues or organisms.

Batch culture : A cell suspension grown in liquid medium of a set volume. Inocula of successive subcultures are of similar size and cultures contain about the same cell mass at the end of each passage. Cultures commonly exhibit five distinct phases per passage; a lag phase follow inoculation, then an exponential growth phase, a linear growth phase, next a deceleration phase and finally a stationary phase.

Biodiversity convention: Biodiversity convention took place on June 5, 1992 during **Earth summit** held at Rio de Janeiro (Brazil. The objectives of the convention are: conservation of biological diversity, sustainable use of its components and the fair and equitable sharing of the benefits arising out of the utilisation of genetic resources, including by appropriate access to genetic resources and by appropriate transfer of relevant technologies, taking into account all rights over these resources and to technologies, and by appropriate funding.

Biological diversity: Biodiversity. It refers to the variety and variability among living organisms and the ecosystem complexes in which they occur. In the simplest sense, biodiversity means the sum total of species richness, i.e. the number of species of plants, animals and microorganisms occurring in a given habitat.

Biological resources: Genetic resources, organisms or parts thereof, populations, or any other biotic components of ecosystems with actual or potential use or value for humanity.

Biosphere reserves: These are undisturbed natural areas suitable for scientific study as well as areas in which conditions of disturbance are under control. They have been set aside for ecological research and habitat preservation. There are 14 sites proposed as biosphere reserves in India- Nilgiris, Namdapha, Nanda Devi, Uttarkhand, North Islands of Andamans, Gulf of Mannar, Kaziranga, Sunderbans, Thar desert, Manas, Kanha, Nokrek, Little Rann of Kutch and Great Nicobar Island.

Biotechnology: Bioengineering. The employment of biochemical processes on an industrial scale, most notably recombinant DNA techniques to make or modify products or processes for specific use.

Biotechnology: The industrial use of biological processes; as yeast fermentation for alcohol production or plant cell culture for extraction of secondary products. The scope of biotechnology has been dramatically increased through genetic engineering.

Blue book: UNEP has compiled endangered species of the world under the title, 'Blue Book'.

Callus culture: The cultivation of callus, usually on solidified medium and initiated by inoculation of small explants or sections from established organ or other cultures (inocula). Callus may be maintained indefinitely by regular subdivision and subculture. It may be used as the basis for organogenetic (shoot, root) cultures, cell cultures or proliferation of embryoids.

Callus, pl. *calli* or calluses: 1. Wound tissue, tissue formed on or below a wounded surface. **2.** Disorganized tumor-like masses of plant cells that form in culture. These proliferate in an irregular tissue mass, and vary widely in texture, appearance and rate of growth; a function of the tissue type (species and explant) and the composition of the medium. the process of callus formation is **callogenesis.**

Cell culture: The culture of single or groups of cells on solid or dispersed in liquid nutrient media (cell suspension cultures) usually following a set of defined growing conditions (protocols).

Cell hybridization: Formation of synkaryons, viable cell **hybrids** produced through cell fusion. they are identifiable by their increased chromosome number compared to the parent cells and the possession of characters found in one or another of the parental cells.

Cell line: Developmental history or descent, through cell division from a single original cell; as callus may constitute a group of cell, all descendants from a single cell plating. Numerous **cell lines** may be present in a culture. Any deviation in culture technique may favor one cell line over another.

Cell suspension: Cells and small aggregates of cells suspended in a liquid medium; as in cell suspension cultures. Explants, or callus derived from them are transferred to liquid medium and the cultures are then agitated on a mechanical shaker. The ensuing single cells and small cell clusters are used for a number of purposes in plant tissue culture; as in single cell cloning.

Centrifuge: An apparatus used for separating particles from suspension using centrifugal force. Balanced tubes containing the suspension are rapidly rotated causing sedimentation of particles to the bottom of the tubes (pelleting) based on their weight and the density of the suspending medium.

CITES: The Washington Convention on International Trade in Endangered Species of Wild fauna and flora. The convention seeks to provide protection for certain species (e.g. the Peregrine) against over-exploitation through international trade.

Clone: 1. A basic category of cultivar. The original cultivar or variety. Not to be confused with source clone, indicating different originals within the clone. Members have a common origin, are the extension of a single cell (via mitosis) or plant and have been produced by vegetative means only. Genetic uniformity is accepted. A clone may have one or more sources. **2.** While the expectation is that members of a clone are both phenotypically and genotypically identical, this assumption is not always true and members may not be genetically or phenotypically equivalent (varients). the objective of much tissue culture propagation is to propagate very large numbers of selected plants with the same genotype. The process is **cloning.** The prefix meri reflects the explant used to initiate the clone (meristem tip source clone). In a **calliclone** cloning involves the callus stage. **3.** Specific DNA sequences are said to be **cloned** when isolated and propagated.

Co-culture: The joint culture of two or more types of cells; as a plant cell and a microorganism or two types of plant cells; as is done in various dual culture systems or the nurse culture technique.

Cryoprotectant: An agent able to prevent freezing and thawing damage to cells as they are frozen or defrosted. These substances have high water solubility and low toxicity. They are classified either as permeating (glycerol and dimethyl sulfoxide) or non-permeating (sugars, dextran, ethylene glycol, polyvinyl pyrolidone and hydroxyethyl starch) agents.

Culture medium: A prepared nutrient solution (substrate) that may be chemically defined for growing plant tissues or other organisms in vitro.

Cybrid: A cell or plant cytoplasmic hybrid (heteroplast) with the nucleus of one and chtoplasmic organelles of another, or of both cells or plants.

Cytoplasmic variant: A maternally inherited change in a cellular trait.

Direct embryogenesis: Embryoid formation directly on the surface of zygotic or somatic embryos or on seedling plant tissues in culture, without an intervening callus phase.

Direct organogenesis: Organ formation directly on the surface of relatively large intact explants, without an intervening callus phase.

Dry ice: Frozen (solid) carbon dioxide (CO_2). It is commonly used as a refrigerant.

Dual culture: A culture system that includes plant tissue and one organism (such as a nematode species) or microorrganism (such as a fungus). Dual cultures are used to study host-parasite interactions or in the production of axenic cultures for a variety of purposes. Usually the microorganism selected is an obligate parasite.

Ecosystem diversity: Existence of different landforms in a ecosystem, each of which supports different and specific vegetation.

Ecosystem: Dynamic complex of plant, animal and microorganisms and their non-living environment interacting as a functional unit.

EIA: (Environmental Impact Assessment. A report, based on detailed studies that discloses the environmental consequences of a course of action as an aid to decision making.

Embryo culture : 1. Denotes a culture in which the explant was n embryo. **Embryo cultures** have been used to obtain viable offspring from seeds with a tendency for embryo abortion or when viable.seed are limited in number. **2.** Cultures in which embryos are induced to form (embryogenesis) whether in suspension or on a variety of explants or cultures on solidified media. More correctly, these nonzygotic or somatic embryos are termed embryoids.

Endangered species: The plant and animal species which are in danger of extinction.

Endemic species: The plant and animal species confined to a given region and having originated there (or a species which occurs continuously in a given area.

Ex vitro: (Latin; *from glass*). Organisms removed from culture and transplanted, generally to soil or potting mixture.

Explant: The excised plant portion used to initiate a tissue culture. The process of dissection and removal to culture of these small organs or tissue sections is **explantation.** Explant choice, the timing of excision and pretreatment are important determinants of culture success.

***Ex-situ* conservation:** The conservation of components of biological diversity outside their natural habitats.

Extinct species: Species that are no longer known to exist in wild.

GA_3: Gibberellic acid.

Generation time: The time between successive generations of cells or organisms within a population.

Genetic diversity: It refers to the variation of genes within species.

Genetic engineering or recombinant DNA technology: Technology involving man-made changes in the genetic constitution of cells (apart from selective breeding). This technology usually employs a vector (such as the Ti plasmid of *Agrobacterium tumefaciens*) for transferring useful genetic information from a donor organism into a cell or organism that does not possess it. Genetic engineering has many potential uses.

Genetic material: Any material of plant, animal, microbial or other origin containing functional unit of heredity.

Genetic resources: Genetic material of actual or potential value.

Genetic selection: Selection of genes, cells (cell selection), clones, etc., by man within populations or between populations or species. The usual purpose is to alter a specific phenotypic character. Such selection usually results in differential success rates of the various genotypes, reflecting many variables including selection pressure and genetic variability in populations.

Genetic transformation: The transfer of extracellular DNA (genetic information) among and between species; as, for example, with the use of bacterial or viral vectors.

Genetic variation: Differences in individuals derived from the same genotype in distinction to differences caused by the environment. **Genetic variance** denotes the proportion of pohenotypic variance caused by differences in the genetic make-up of an individual.

Genome: The basic haploid chromosome set of an individual; the sum of its **genes.**

Genotoxic: Carcinogenic; toxic to the chromosomes.

Geotaxis: Plant orientation with respect to gravity.

Germplasm or germ plasm: 1. The reproductive body tissues distinct from somatic (nonreproductive tissues). The genetic material, basis of heredity of an organism, passed on through previous generations. **2.** An individual representing a type of species or culture that may be held in a repository for agronomic, historic (or other) reasons.

Green book: It lists rare plants growing in protected areas like Botanic gardens.

Habitat: The place or type of site where an organism or population naturally occurs.

In situ: In the natural, original place or position; as in the location of the explant on the mother plant prior to excision.

In vitro: (Latin = *in glass*). Experimentation on organisms or portions thereof in glassware or culture; growing under artificial conditions as in tissue culture. The antonym of in vivo.

In vivo: (Latin = *in life*) Experimentation on organisms under natural conditions within intact living organisms. the antonym of in vitro.

Indirect embryogenesis: Embryoid formation on callus tissues derived from zygotic or somatic embryos, seedling plant or other tissues in culture. The antonym is direct embryogenesis.

Indirect organogenesis: Organ formation on callus tissues derived from explants. the antonym is direct organogenesis.

***In-situ* conservation:** The conservation of ecosystems and natural habitats and the maintenance and recovery of viable populations of species in their natural surroundings and, in the case of domesticated or cultivated species, in the surroundings where they have developed their distinctive properties.

Isolation medium: A medium suitable for explant survival and development. It may be synonymous with Stage I medium, or contain antioxidant(s), reduced hormone concentration, bacterial indicators or other addenda, and precedes Stage I culture.

Liquid culture: The culture of plant cells on liquid medium, in suspension or on supports. Cultures are either held steady (stationary culture) or are shaken (agitated culture or shake culture).

Liquid medium: Medium not solidified with a gelling agent. **Liquid media** are used for suspension cultures and for a wide range of research purposes. They are also useful for Stage I culture in some micropropagation protocols. Usually a support structure or wick is used to hold the tissue above the nutrient medium.

Medium: The substrate for plant growth; as nutrient solution, soil, sand, etc. This is a general term for the liquid or solidified formulation upon which plant cells tissues or organs develop in plant tissue culture.

Meristem tip culture: Cultures derived from meristem tip explants. They are excised for virus elimination or axillary shoot proliferation purposes, less commonly for callus production.

Meristem tip: A common explant comprised of the meristem (meristematic dome) and usually one pair of leaf primordia. Also refers to apical (apical meristem tip) or lateral (lateral or axillary meristem tip) origin. The term meristem tip is often confused with the term shoot tip, which is much larger and usually has more immature leaves and some stem tissue.

Micropropagation: Refers to propagation in culture by axillary or adventitious means. It is a general term for vegetative (asexual) in vitro propagation. It sometimes refers specifically to axillary bud proliferation.

Monoculture: A one-crop agriculture system. An intensified monoculture could refer to a clonal (one genotype) agricultural system.

MS: An abbreviation for the **Murashige, T. and F. Skoog (1962)** medium formulation.

National Parks: A relatively large area of land set aside (under some Act) for its features of predominantly unspoiled natural landscape, flora and fauna, and permanently dedicated for public enjoyment, education and inspiration and protected from all interference. There are about 75 National Parks in the country.

Nucellar embryony: The process by which individual cells of the nucellus give rise to somatic embryos (embryoids). Plants on which this occurs naturally are considered good candidates for tissue culture of somatic embryoids.

Ontogeny: The developmental (**ontogenetic**) life history of an organism, its development from fertilized egg (zygote) to seed-forming adult (reproductive phase), with emphasis on embryonic development.

Organ culture: The growth in aseptic culture of plant organs such as roots or shoots, beginning with organ primordia or segments and maintaining the characteristics of the organ.

Organized growth: The in vitro development of organized explants such as meristem tips or shoot tips, floral buds or organ primordia, or their de novo formation from **unorganized** tissues.

Organogenesis: The initiation (de novo) and growth of organs (roots and shoots, usually) from cells or tissues; as in organ culture. Organs may form on the surface of explants (direct organogenesis) or upon an intervening callus phase (indirect organogenesis).

Ovule culture: A culture derived from an explanted ovule; which may be fertilized in culture (in vitro fertilization). This technique is used to study development of zygotes and young embryos and is sometimes used to rescue embryos susceptible to abortion when embryo culture is not possible.

Papain: A water soluble proteolytic enzyme (protease) extracted from papaya fruit and used especially as a meat tenderizer.

Plant tissue culture: A general term encompassing the in vitro culture of plant cells, tissues, organs and whole plantlets.

Pollen culture: The culture of pollen grains, which germinate in vitro. Such cultures may eventually form monoploid callus, from which shoots or embryoids develop into monoploid plants.

Population density: Cell number per unit medium area or medium volume.

Project Tiger: It is the long-term strategy proposed to save declining population of tigers (*Panthera tigris*) from extinction. The Project was launched by Mrs. Indira Gandhi, the then Prime Minister on April 1, 1973 following the recommendations of a special task force of the Indian Board of wildlife. Initially 9 Tiger Reserves were created in nine different states of the country with a total area of 13,017 km^2 and tiger population of 268. Now there are 23 Tiger Reserves in the country: Corbett, Sariska, Dudhwa, Ranthambore, Valmiki, Buxa, Nandapha, Manas, Bandhavgarh, Palamau, Kanha, Pench, Tadoba Andheri, Melghat, Simplipal, Sunderbans, Indravati, Nagarjuna sagar, bandipur, Kalakad, Periyar, Panna and Dhampha.

Propagule: The form or portion of an organism used for reproduction or propagation; as new shoots or callus derived from explants are subdivided into **propagules** and re-cultured for further multiplication.

Protected area: A geographically defined area which is designated or regulated and managed to achieve specific conservation objectives.

Protoplast culture: The isolation and culture of plant protoplasts by mechanical means or by enzymatic digestion of plant tissues or organs, or cultures derived from these. Protoplasts are utilized for selection or hybridization at the cellular level and for a variety of other purposes.

Protoplast fusion: The coalescence of the plasmalemma and cytoplasm of two or more **protoplasts** in contact with one another. Initial adhesion is a random process but coalescence may be promoted in various ways (induced fusion). When adhesion occurs between adjacent protoplasts during enzymatic wall degradation or between freshly isolated protoplasts in the absence of a fusion agent, it is termed spontaneous fusion.

Rare species: The species whose population numbers are decreasing so that they are likely to become more severely threatened with time in near future.

Red data book: It is the name given to the book dealing with threatened plants or animals of any region. On the global level, the IUCN published Red Data Book in two volumes. The BSI has complied three volumes of Red Data Book having information on the Indian endangered plant species.

Rejuvenation: 1. Synonymous with dedifferentiation. **2.** Treatment that leads to culture invigoration (such as subculture) or revival (dormancy breaking).

Selection culture: Utilizes difference(s) in environmental conditions or more usually in culture medium composition, such that preferred variant cells or cell lines (presumptive or putative mutants) are favored over other variants or the wild-type.

Serial float culture: Sunderland's (1977-1979) technique of floating anthers on liquid medium and subculturing them to new medium at several day intervalsan anther dehiscence, pollen release and development occur, increasing anther productivity.

Shake culture: An agitated suspension culture. Usually a flask (commonly an Erlenmeyer flask) containing the culture is attached to a horizontal or platform shaker, or agitated with a magnetic stirrer, to provide adequate aeration for cells in the liquid medium.

Shoot tip graft or micrograft: The grafting of a very small shoot tip or meristem tip onto a prepared seedling or mcropropagated rootstock in culture. Meristem tip grafting is used for in vitro virus elimination with Citrus and for other plants as an alternative to grafting in the greenhouse.

Somaclone: A plant regenerated from a tissue culture originating from somatic tissue.

Somatic cell embryogenesis: The production of embryos from somatic cells of explants (direct embryogenisis) or by induction on callus formed by explants (indirect embrygenesis). These two processes may not be materially different in results.

Somatic cell variant or embryoid: An organized embryonic structure morphologically similar to a zygotic embryo but initiated from somatic (non-zygotic) cells. These develop into plantlets in vitro through developmental processes that are similar to those of zygotic embryos.

Somatic hybrid: A cell or plant product of somatic cell fusion; as the result of cell or protoplast fusion and implying genomic integration. The process is **somatic hybridization.**

Species diversity: It refers to the variety of species within a region.

Stages of culture (I-IV): Stage I: Aseptic explantation or establishment of the explant in culture. **Stage II:** Multiplication of the propagules. **Stage III:** Rooting of the progagules and preparation for transplant to soil. **Soil IV:** Establishment of Stage II or III propagules ex vitro in soil or potting mix.

Subculture or passage: A culture derived from another culture or the aseptic division and transfer of a culture or a portion of that culture (inoculum) to fresh nutrient medium. Sub-culturing is usually done at set time intervals, the length of which is called the subculture interval or passage time.

Suspension culture: Cells and group of cells (aggregates) dispersed in an aerated, usually agitated, liquid culture medium. These are obtained by adding friable callus to the medium. The plant species, explant type and treatment; the composition of the nutrient medium; and many other features determine the size and nature of the cell aggregates, which may appear different during each culture growth phase. These cultures are used to study cell division, differentiation and metabolism, in secondary product synthesis or form the basis for single cell lines, callus cultures, somatic cell embryogenesis and many other culture purposes.

Sustainable use: The use of components of biological diversity in a way and at a rate that does not lead to the long-term decline of biological diversity, thereby maintaining its potential to meet the needs and aspirations of present and future generations.

Synchronous culture: A plant cell or microbial culture treated in such a way as to have all (or most) cells or individuals in the same stage of development or mitosis. This can be achieved in various ways including via temperature variation and nutrient limitation.

Tissue explant: An excised plant portion of tissue used to initiate a culture.

Totipotency: the potential **(totipotential)** or inherent capacity of a plant cell or tissue to develop into (recreate) an entire plant if suitably stimulated. Totipotency implies that all the information necessary for growth and reproduction of the organism is contained in the cell. Although theoretically all plant cells are **totipotent** the meristematic cells are best able to express it.

Unorganized growth: In vitro formation of tissues with few differentiated cell types and lacking recognizable structure; as with many calli.

Virus-tested or virus-free: A plant that appears healthy and repeatedly tests negatively for the presence of one or more identifiable viruses. Such a plant may then be used as a stock or donor plant (explant source) for propagation purposes, and may be certified as virus tested (certified virus tested). The term virus-free is incorrect in most cases, as such a plant may contain one or more viruses which have not been assayed.

Vulnerable: Species whose population numbers are decreasing so that they are likely to become more severely threatened with time in near future.

Woody plant medium (WPM): A modified MS (1962) medium developed, for woody plant species by G. Lloyd and B. McCown (1980). It has less nitrogen (both ammonium and nitrate), sodium, potassium and chloride than does MS (1962) medium. It has been increasingly used for the commercial propagation of ornamental trees and shrubs.

Selected Bibliography

Abelsen, P.H. 1984. Biotechnology and Biological Frontiers. Amer. Asso. Adv. Sci (AAAS), USA, publ. No. 84-8 pp. 516.

Achmus, H. and Brinks, T. 1996. Yams. (In) Application of Biotechnology in Selected Crops. pp. 110-118, Jacogsen, H.J. and Wolpers, K.H. (Eds.). GTZ. GmbH. Eschborn. Germany.

Ahmad, M., and D.L. McNeil. 1996. Comparison of crossbility, RAPD, SDS-PAGE and morphological markers of revealing genetic relationships among *Lens* species. Theor. Appl. Gen. 93: 788-793.

Ahuja, M.R. 1986: Application of biotechnology to forest tree species and problems involved. In: IUFRO Proc. Biochemical Genetic and Legislation of Forest Reproductive Material. H.-J Mujs (Ed.), Mitt. Bfh, Hamburg, No. 154, 187-199.

Ahuja, M.R. 1987: Somaclonal Variation. In: Cell and Tissue Culture in Forestry, Vol. 1, J. M. Bonga, D.J. Durzan (Eds.), Martinus Nijhoff Publishers, Dordrecht, 272-285.

Akula, A. and Dodd, W.A. 1998. Direct somatic embryogenesis in a selected tea clone, TRI-2025 (*Camellia sinsnsis* (L) O. Kuntze) from nodal explants. *Plant Cell reports* 17: 804-809.

Allen, O.N. and E.K. Allen, 1981. The Leguminosae, University of Wisconsin Press, Madison, Wis.

Altieri, M.A., D.L. Glaser and L.L. Schmidt, 1990. Diversification of Agroecosystems for Insect Pest Regulation : Experiments with Collards. In : S.R. Gliessman (Ed.. Agroecology : Researching the Ecological Basis for Sustainable Agriculture, pp. 70-82. Springer-Verlag, New York.

Ammirato, P.V. 1989. Recent progress in somatic embryogenesis. LAPTC Newslett. 57:-2-16

Ando, A., Comm. J., Polasky, S. and A., Solow. 1998. Species distribution land values and efficient conservation Science, 279: 2060-2061.

Anitha Karun and Sajini, K.K. 1994. Short term storage of coconut embryos in sterile water. *Curr. Sci.* 67(2): 118-120.

Anitha Karun, Shivashankar, S., Sajini, K.K. and Saji, K.V. 1993. Field collection and *in vitro* germination of coconut embryos. *Journal of Plantation Crops.* 21(Suppl): 291-294.

Anmirato, P.V. 1987. Organisational events during somatic embryogenesis in Plant Tissue and Cell Culture. Alan R. Liss. New York, P. `57.

Anon. 1992. Jhum: is There A Way Out? The Price of Forests (Ed. A. Agarwal), 304-310, CSE, New Delhi.

Anonymous 1980. Village Development Boards - Model Rules, 1980 (Revised. Department of Rural Development, Government of Nagaland, New Delhi.

Anonymous, 1971. Census of India. Govt. of India Press, New Delhi.

Anonymous, 1977. National Atlas of India. National Atlas Organization, Calcutta 234 pp.

Anonymous, 1979. Tropical Grazing Land Ecosystems. A state of knowledge report prepared by UNESCO/UNEP/FAO.

Argustine, P.C.; H.D. Danforsh and M.R. Bakst 1986. Biotechnology for solving agricultural problems. Martinus Nijhoff Pub., New York.

Arkin, P.J., and W.R. Scowcroft. 1981. Somaclonal variation-A novel source of variability from cell cultures for plant improvement. Theor. Appl. Genet. 60:197-214.

Arora, G.S. and Julka, J.M. 1993. Status report on biodiversity conservation : Western Himalayan ecosystem. IIPA, New Delhi (in Press.

Arumugam, N. and Bhojwani, S.S. 1990. Somatic embryogenesis in tissue cultures of *Podophyllum hexandrum*. Can.J. Bot.68: 487-491

Ashwood Smith M. and Grant, E., 1997. Genetic stability in cellular systems stored in the frozen state. In K.E. Iliot and J. Whelan (Editors), The freezing of Mammalian Embryos Elsevier, Amsterdum pp 251-268.

Askins, R.A. 1995. Hostile landscapes and the decline of migratory songbirds. Science. 267: 1956-1957.

Bademes. M.L., M.J. Asins, E.A. Carbonell, and G. Glacer. 1996. Genetic diversity in apricot, *Prunus armeniaca,* aimed at improving resistance to plum pox virus. Plant Breed. 115: 133-139.

Bajaj, Y.P.S. 1991. Automated micropropagation for *en masse* production of plants. (in) *Biotechnology in Agriculture and Forestry,* Vol. 17, pp. 4-16. Bajaj, Y.P.S. (Ed.). Springer-Verlag, Berlin.

Balakrishna, P. 1999 Molecular diversity, Molecular taxonomy and DNA finger printing. Current Science 76 : 129-131

Balakrishna, P. 1999. Molecular diversity, Molecular taxonomy and DNA finger printing. Current Science 76 : 268-269.

Barton, J.H. and W.E. Siebeck 1994 Material Transfer Agreements in Genetic Resources Exchange-the case of the International Agricultural Research Centers. Issues in Genetic Resources No.1. IPGRI, Rome.

Bateson, J.M., Grout, B.W.W. and Lane, S. 1987. The influence of container dimensions on the multiplication rate of regenerating pant cell cultures. In G., Duczate, M., Jacks and Siemean, (Editors), Plant Micropropogation in the Horticulture Industries, Belgium. pp. 275-277.

Baumann, T.W., Neuenschwander, B. 1990. Tissue culture in coffee biotechnology *Café-Cacao-the* (France) 34(2): 159-164.

Belarmino, M.M., Abe, T. and Sasahara, T. 1994. Plants regenerated from stem and petiole protoplasts of sweet potato *(Ipomoea batatas). Plant-Cell-Tissue-Organ-Culture;* 37(2): 145-50.

Belcher. A.R., Abbott, A.J., Hall, K.C. and Jackson, M.B. 1987. Gaseous constituent of culture in-vitro. Report of the long Ashton Research Station 1986. AFRC, U.K.

Benson, E.E. 1990. Free radical damage in stored plant germ plasm. IBPGR, Rome

Benson, E.E. and Withers, L.A. 1987. Gas chromatographic analysis of volatile hydrocarbon production by cryopreserved plant tissue culture, a non destruction method for assessing stability, cryoletters, 8 : 35-46.

Berwick, S.H., 1974. The Community of Wild Ruminants in the Gir Forest Ecosystem, India. A Thesis for the Degree of Doctor of Philosophy, p 226.

Bhandari. S.C. and L.L. Somani, 1990. In: Biofertilizers. L.L. Somani, S.C. Bhandari, S.N. Saxena and K.K. Vyas (eds.. Scientific Publishers, Jodhpur, India. pp. 355-361.

Bhansali, R. R. 1995. A Manual of Plant Tissue Culture (Ed.). Department of Biotechnology, Ministry of Science and Technology, Govenment of India, New Delhi.

Bhansali, R. R. 1988. Advance researches on somatic embryogenesis technique for micropropagation and genetic manipulation in Prunus sps. Fruit Biotech. 1-25.

Bhansali, R. R. 1990. Somatic embryogenesis and regeneration of plantlets in pomegranate. Annals of Botany. 66: 249-253.

Bharadwaj, K.K.R. 1975. Survial and symbiotic characteristics of Rhizobium in saline alkali soils. *Plant and Soil* 43(2) : 377-385.

Bharadwaj, K.K.R. 1975. Survial and symbiotic characteristics of Rhizobium in saline alkali soils. *Plant and Soil* 43(2) : 377-385.

Bhatti, M.H., Percival, T., Davey, C.D.M., Henshaw, G.G. and Blakesley, D. 1997. Cryopreservation of embryogenic tissueof a range of genotypes of sweet potato (*Ipomoea batatas* (L.) Lam.) using an encapsulations protocol. *Plant Cell Reports.* 16: 802-806.

Bhojwani, S.S. and M.K. Rajdan 1983. Plant Tissue Culture: Theory and Practice. Elsevier, Amsterdam.

Bhojwani, S.S. and M.K., Razdan 1996. Plant Tissue culture : Theory and Practice, an revised Edition. Elsevier Amsterdam pp. 563-588.

Bhojwani, S.S., V. Dhawan, and E.C. Cocking. 1986. *Plant Tissue Culture: A Classified Bibliography.* p. 789. Elsevier Science Publishers, Inc., New York.

Biswas, S. & Ghosh, A.K., 1976. Impact of Shifting-cultivation on Wildlife in Meghalaya. In : Shifting Cultivation in North-east India (Ed. B. Pakem *et al.*), 77-79, NEICSSR, Shillong.

Blake, J. and Eeuwens, C.J. 1982. Tissue of economically important plants. Proceedings of International symposium on *Costed ANBS, Singapore* 145-148.

Blaustein, A.R. and Wake, D.B. 1995. The puzzle of declining amphibian populations. Scientific American, April 1995.

Bloom, D.E., 1995. International Public Opinion on the Environment. Science, 269, 354-357.

Brockwell, J., P.J. Bottomley and J.E. Thies, 1995. Plant and Soil. 174 : 143.

Brookfield, H. and C. Padoch, 1994. Appreciating biodiversity: A look at the dynamism and diversity of indigenous farming practices. Environment, 36 : 6 - 11, 37 - 45.

Brown, C.M., I.Cambell and F.G. Priest 1987. Introduction to Biotechnology. Blackwell Scientific Publications, Oxford.

Brown, L.R., Moyle, P.B. and Yoshyama, R.M. 1994. Historical decline and current status of coho salmon in California. Nor. Am. J. of Fish. Managt. 14(2 : 237-161.

Bubble, S.P. & R.B. Foster, 1986. Commonness and Rarity in Neotropical Forests: Implications for Tropical Tree Conservation. pages 205-231 in M.E. Soule editors Conservation Biology: The Science of Scarcity. and Diversity. Sinauer Associates, Inc. Publ. Sunderland, Massachusetts.

Bulla, L.A. 1973. Regulation of Insect Population by Microorganisms. New York: New York Academy of Sciences.

Burley, J., 1994. World Forestry: The Professional Scientific Challenges. The Leslie L. Schaffer Lectureship in Forest Science, Vancover, B.C., Canada.

Cai, W. Q., Gonsalves, C., Tennant, P., Fermin, G., Sonza, M., Sarindu, N., Jan, F. J., Zhu, H. Y. and Gonsalves, O. D. (1999) *In vitro Cell. Develop. Biol. Plant,* 35 : 61-69.

Campion B, Bohanec B, Javornik B (1994b) Gynogenic lines of onion (*Allium cepa* L.): evidence of their homozigosity. Submitted

Cappadocia, M., Cheretien, L.and Laublin, G. 1988. Production of haploids in *Gerbera jamesonii* via ovule culture: Influence of fall versus spring sampling on callus formation and root generation. *Canadian Journal of Botany.* 66: 1107-1110.

Castri, F. Di, J.R. Vernhes, & T. Younes, 1992. (Eds.) Inventorying and Monitoring Biodiversity. Pages 1-27 in Report of an ad hoc group of IUBS-SCOPE-UNESCO, Programme Function of Biodiversity (30-31 Jan.), Paris.

Champion, H.G. & Seth, S.K., 1968. A Revised Survey of Forest Types of India, FRI, Dehradun.

Chan, J.L., Saenz, L., Talavera, C., Hornung, R., Roberg, M. and Oropeza, C. 1998. Regeneration of coconut (*Cocos nucifera* L.) from plumule explants through somatic embryogenesis. *Plant Cell Reports* 17: 515-521.

Chaterjee, S., 1995. Global 'Hot Spots' of Biodiversity. Current Science, 68, 12, 1178-1179.

Chaturvedi, C. 1995. In: Microbes and Man. S. Chandra, K.K. Khanna and H.K. Kehri (eds. BSMPL Publishers. Dehra Dun, India. pp. 151-166.

Chau, N. H. (1996) Plant Biotechnology (Editorial over review). Current opinion in Biotechnology, 7 : 127-129.

Chavan, S.A., Gogate, N.S. & Patel C.D., 1991. Endangered Ecosystem of Shoolpaneshwar Sanctuary, Rajpipla, Gujarat. Paper presented at Session of Indian Science Congress, Vadodra, India.

Chazdon, R.L. 1998. Tropical forests Log Em or Leave Em. Science. 281 : 1295-1296.

Chen. W.X., G.H. Li and Y.L. Qi, 1991. Int. J. Syst. Bacteriol. 41 : 275.

Choudhury, N. 1991. Status of wild elephants (*Elaphas maximus* Linn. in Cachar and North Cachar hills, Assam- a preliminary investigation. J. Bombay Nat. Hist. Soc. 88(2): 215-221.

Chouhan, A.S. & Singh, D.K. 1989. Changing patterns in the flora due to deforestation. Environmental Conservation and Wasteland Development in Meghalaya (Ed. A. Gupta & D.C. Dhar), 75-107, Meghalaya Science Society (MSS), Shillong.

Chowdhury, M.K.U., Ghulam Kadir, A., Parvez and Saleh, N.M. 1997. Evaluation of five promoters for use in transformation of oil palm (*Elaeis guinneensis* Jacq.) *Plant cell Reports* 16: 277-281.

Cocking, E.C. 1960. A method for isolation of plant protoplasts and vacuoles. Nature 187: 962-963.

Cody, M. L., 1986. Diversity, Rarity, and Conservation in Mediterranean Climate Region. Pages 112-152. In M.E. Soule editors Conservation Biology of Science of Scarcity and Diversity. Sinauer Associates, Inc. Publ. Sunderland, Massachusetts.

Cohen, J.E., 1995. Population Growth and Earth's Human Carrying Capacity. Science, 269, 341-346.

Cohen, J.I. et.al., 1991. Ex-situ conservation of plant genetic resources; global development and environmental concern. Science 253: 866-872.

Collicut, L. and Ronald, W.B. 1995. `Northern Beauty' Lily. *Hort. Science* 30 (3): 652-653.

Copping, L.G. and P. Rodgers 1985. Biotechnology and its application to Agriculture Monograph 32. British Crop Protection Council, England.

Courteny-Gutterson, N., C. Napoli, C., Lemieux, A. Morgan, F. Firoozabady and K.E.P. Robinson. 1994. Modification of flower colour in florists chrysanthemum: productionof a white flowering variety through molecular genetics. *Biotechnology* 12: 268-271.

Craig, J.R., Vaughan, D.J. and Sikinner, B.J., 1988. Resources of the Earth. Prentice-Hall, Eagle wood Cliffs, N.J.

Crespi, R.S. 1991. Biotechnology and Intellectual Property Part-I Patenting in Biotechnology. TiBTECH 9 : 117-121.

D.O., EN, 1994. Final Consolidated Report on all Indian Coordinated Project on conservation of endangered plant species, Seed Biology and Tissue culture Programme pp. 1-79. Published by the Development of Environment and Forests. Govt. of India, New Delhi.

Dabadghao, P.M. and K.A. Shankarnarayan, 1973: The Grass Covers of India. I.C.A.R, New Delhi.

Daily, G.C., 1995. Restoring Value to the World's Degraded Land. Science, 269, 350-354.

Daily, G.C., Ehrilich, P.R. and Haddad, N., 1993. Double Keystone Bird in A Keystone Species Complex. Proc. Natl. Acad. Sci., USA, 90, 592-594.

Dalton, H. 1980. In: Methods for evaluating biological nitrogen fixation. F.J. Bergersen (ed.. John Wiley and Sons, N.Y. pp. 13-64.

Daniel, J.C., 1983. The Book of Indian Reptiles. Bombay Natural History Society and Oxford University Press, India 141 p.

Daniels, R.J. Ranjit, M. Hegde, N.V. Joshi, M. Gadgil, 1991. Assigning Conservation Value: A Case Study From India. Conservation Biology 5(4):464 475.

Daniels, R.J.R., Hegde, M. and Gadgil, M. 1990. Birds of the man made ecosystems: The plantations. Proc. Indian Acad. Sci. (Anim. Sci. 19 1 79-89.

Darlong, V.T. & Alfred, J.R.B., 1982. Difference in Arthropod Population Structure in Soils of Forest and Jhum Sites of North-east India. Pedobiologia 23: 112-119.

Darlong, V.T. & Alfred, J.R.B., 1989. Effect of Shifting Cultivation (Jhum on Soil Fauna with particular reference to the Earthworms in North-East India. In: Advances in Management and Conservation of Soil Fauna (Ed. G.K. Veeresh, D. Rajagopal and C.A. Viraktamath), 299-308, Oxford & IBH, New Delhi.

Darlong, V.T. & Alfred, J.R.B., 1993. Micro-arthropod diversity in some soils of North-east India with special reference to effect of shifting cultivation. In : Himalayan Biodiversity: Conservation Strategies (Ed. U. Dhar), 331-323. GBPIHED, Almora.

Darlong, V.T., Choudhury, D., Sati, J.P. & Alfred, J.R.B. 1989. Wildlife and the Vanishing Forests: An Appraisal. Environmental Conservation and Wasteland Development in Meghalaya (Ed. A. Gupta & D.C. Dhar), 108-126, MSS, Shillong.

Darlong, V.T., Choudhury, D., Sati, J.P. & Alfred, J.R.B., 1989. Wildlife and the Vanishing Forests : An appraisal. In : Environment Conservation and Wasteland Development in Meghalaya (Ed. A. Gupta and D.C. Dhar), 108-127, Meghalaya Science Society, Shillong.

Das, P. and Kapoor, D. 1996. National Bureau committed to conserving India's aquatic resource base. Diversity., 12 (3) : 33-34.

Dazhong, W. and D. Pimentel, 1990. Energy flow in agroecosystems of northeast China. In : S.R. Gliessman (Ed. Agroecology : Researching the Ecological Basis for Sustainable Agriculture. pp. 322 - 336. Springer-Verlag, FAO/SIDA, 1974.

Report on Regional Seminar on Shifting Cultivation and Soil Conservation in Africa. FAO, Rome. 248 pp.

de Jong, J., Rademaker, W. and van Wordragen, M.F. 1993. Restoring adventitious shoot formation of chrysanthemum leaf explants following co-cultivation with Agrobacterium tumefaciens. *Plant Cell Tissue and Organ Culture.* 32: 263-270.

De Winnaar, W. (1987) *Information Bulletin, Citrus and Subtropical Fruit Research Institute,* South Africa, 177 : 1-2.

Deb Roy, R., 1986 : Studies on some aspects of renewable energy (fuel and pasture production in an ecosystem of *Albizzia procera* and *A. lebbeck* with grass and legumes. Ph.D. Thesis. Jiwaji University, Gwalior, M.P., India.

Deb Roy, S. & Jackson, P. 1993. Mayhem in Manas : the Threats to India's Wildlife Reserves. The Law of the Mother (Ed. E. Kemf), 156-161, Sierra Club Books, San Francisco.

Demurin, Y.,D. Skoric, and D. Karlovic. 1996. Genetic variability of tocopherol composition in sunflower seeds as a basis of breeding for improved oil quality. Plant Breed. 115: 33-36.

Dhanze, J.R. and Dhanze, R., 1998. Post impoundment impact on the biodiversity of Western Himalayan river system - A case study. Proc., Qcad. Environ. Biol. 71 : 11-16.

Dhar, U., Rawal, R.S. and S.S. Samant 1996. Endemic plant diversity in Indian Himalayas-III Brassicaceae. Biogeograghica, 72 : 19-32 1996.

Dhar, U., Rawal, R.S. Samant 1997. Structural diversity and representativeness of forest vegetation in a protected area of Kumaun Himalayas, India : Implications of conservation. Biodiversity an Conservation. 6 : 1045-1062.

Dhar, U., Rawal, R.S., Samant, S.S., Airi, S. and J. Upreti 1999. People participation in Himalayan biodiversity. conservation. A practical approach. Current Science 76 : 36-40

Dhiman, M., Moitra, S., Singh, M.N. and S.P. Bhatnagar 1998. In vitro studies on some non-coniferous gymnosperms. Current Science, 75 : 1113-1115.

Diaz, J.,P. Schmiediche, and D.F. Austin, 1996. Polygon of crossability between eleven species of *Ipomoea:* Section *Batatas* (Convolvulaceae). Euphytica 88: 189-200.

Diekmann, M. 1997. The use of biotechnology for the safe movement of coconut germplasm. Paper presented at International Symposium on coconut biotechnology held at Merida, Yuc. ,Mexico during December 1-5 1997 pp. 30.

Dijk, A.V. 1997. Developments in cassava research. Biotechnology and Development Monitor, UvA and DGIS, The Netherlands. 30 : 16-18.

Dobriyal, R.M., Singh, G.S., Rao, K.S. and K.G. Saxena 1997. Medicinal plant resources in Chhakinal watershed in the North Western Himalaya, J. Herbs, Spices and Medicinal plants 5 : 15-27.

Dolgov, S.V., Mityshikina, T.U., Rutavtsova, F.B. and Buryanov, Y. 1995. Produciton of transgenic plans of Chrysanthemum morifloium Ramat, with the gene of Bacillus turingensis and endotoxin. *Acta Hortiucltural.* 420: 46-47.

Dove, A. 1998 Botanical gardens cope with bio-prospecting loop hole. Science 281: 1273.

Dreyfus, B., J.L. Garcia and M. Gillis, 1988. Int. J. Syst. Bacteriol. 38 : 89-98.

Ehrilich, P.R. and E.O., Wilson 1991. Biodiversity studies; Science and Policy. Science 253: 758-762.

Ehrlich, P.R. 1994. Energy use and biodiversity loss. Phil. Trans. R. Soc. London B 344: 99-104.

Elmes, G.W. and Thomas, J.A. 1992. Complexity of species conservation in managed habitats : Interaction between Maculinea butterflies and their ant hosts. Biodiversity and Conservation 1: 155-169.

Favell., R.B. 1989. Plant biotechnology and its application to agriculture Phil. Trans. Royal Soc. Lond. B. 324: 525 -535.

Fay, P. 1983. The Blue-Greens. pp. 88. Edward Arnold, London.

Ferguson, M.E., and L.D. Robertson, 1996. Genetic diversity and taxonomic relationships. Euphytica. 91: 163-172.

Fitch, M.M.M. (1993) *Plant Cell Tissue Organ Cult.*, 32 : 205-212.

Flint, Michael. 1991. Biological Diversity & Developing Countries. A Synthesis Paper. ODA.

Fraley, R.T., 1994. The contributions of plant biotechnology to agriculture in the coming decades, In: "Biosafety for sustainable agriculture: sharing biotechnology regulatory experiences of the Western Hemisphere" pp. 3-28, Krattiger, A.F. and Rosemarin, A. (Eds.). ISAAA : Ithaca & S.E.I. : Stockholm.

Froese, R. and Pauiy, D. (Editors), 1997. Fish Base 97. Concepts, design and data sources. CD ROM. International Center for Living Aquatic Resources Management (ICLARM), European Commission, and Food and Agricultural Organization of the United Nations. 256 p.

Gadgil, M. and Meher-Homji, V.M. 1990. Ecological diversity. In: J.C. Daniel and J.S. Serrao ed. Conservation in developing countries : Problems and Prospects. Proceedings of the Centenary Seminar of the Bombay Natural History Society: pp. 175-198.

Gadgil, M., 1993. Forestry with A Social Purpose. In: People's Rights and Environmental Needs. W. Fernandes and S. Kulkarni (Eds. Indian Social Institution, New Delhi pp 111-130.

Gadgil, Madhav, 1994. Reckoning with Life. The Hindu Survey of the Environment.

Gautam, P.L. 1997. Agro-Biodiversity in the Indian Gene Centre and Conservation Strategies - Keynote Address. NASS-NBPGR (ICAR Workshop on National Concern for Management, Conservation and Use of Agro-Biodiversity, October, 15-16, 1997, Shimla, India. 24p. NBPGR.

Ghosh, A.K. & TiwarI, K.K., 1984. Faunal Resources of North-east India. In : Resource Potentials of North-east India Vol.II (Living Resources (Ed. R.S. Tripathi), 105-109, Meghalaya Science Society, Shillong.

GOI, 1992. Tradition, Concerns & Efforts in India. National Report to UNCED, Ministry of Environment & Forests, Govt. of India.

GOI, 1994. Conservation of Biological Diversity in India: An Appraisal. Ministry of Environment and Forests, pp. 13.

GOI, 1994. Conservation of Biological Diversity in India: An Approach. Ministry of Environment & Forests, Govt. of India, New Delhi, pp. 48.

Gosukonda, R.M., Probodessai, A., Blay, E., Prakash, C.S. and Peterson, C.M. 1995. Thidiazuron-induced adventitious shoot regeneration of sweet potato *(Ipomoea batatas). In-vitro Cell Dev. Biol.* 31: 65-71.

Green, J.B., 1993. Natural Resources of the Himalaya and the Mountains of Central Asia. Pages 137-290, IUCN Publ.

Greene, S.L., and G.A. Pederson, 1996. Eliminating duplications in germplasm collections: A white clover example. Crop Sci. 36: 1398-1400.

Gudin, S. 1994. Embryo rescue in Rosa hybrida L. *Euphytica* 72: 205-212.

Gupta, S.D. and Rath, S.C. 1933. Cryogenic Preservation of Carp Milt and its Utilization in Seed Production. In : The Third Indian Fisheries Forum Proceedings 11-14 October, 1993, Pantnagar, pp 77-79.

Harrison, J., Miller, K., Mc Neely, J., 1982. The World Coverage of Protected Areas. IUCN World National Parks Congress, Bali - Indonesia.

Henning, D.H. and Mangum, W.R., 1989. Managing the Environmental Crisis. Durham University Press, Durham & London.

Hsia ChiNi and S.S. Korban 1996. Organogenesis and somatic embryogenesis in callus cultures of Rosa hybrida and Rosa chinensis minima. *Plant Cell, Tissue and Organ Culture* 44(1): 1-6.

Ishwaran, N., 1992. Biodiversity, Protected Areas & Sustainable Development. Nature & Resources, Vol, 28, No 1, 1992.

IUCN / UNEP / WWF, 1991. Caring for the Earth. A Strategy for Sustainable Living. Gland, Switzerland.

IUCN, UNEP and WWF, 1980. : World Conservation Strategy.

Jain, S.K., 1994. Biodiversity: Some Perspective in Study and Conservation. Reg. Conv. Min. Env. For., G.I. Lucknow.

Jayashraee, N., Devi, B.P., and Reddy, P.V. 1997. Production of synthetic seeds and plant regeneration in *Rosa hybrida* cv. Kings Ranson. Indian *J. Expt. Biology* 35(3): 310-312.

Jean, M. and Cappadocia, M. 1992. Effects of some growth regulators on *in-vitro* tuberization in *Dioscorea alata. Plant Cell Reports* 11: 34-38.

Jehan, H., Courtois, D., Ehret, C., lerch, K. and Petiard, V. 1994. Plant regeneration of Lamand *Iris jermanica* L via somatic embryogenesis from leaf species and young flowers. *Plant Cell Reporter.* 13: 671-675.

Jha, P.K., K.K. Shrestha, M.P. Upadhyay, D.P. Stimart, and D.M. Spooner, 1996. Plant genetic resources of Nepal; A guide for plant breeder of agricultural, horticultural and forestry crops. Euphytica 87: 189-210.

Jones, L.H. 1984. Propagation of clonal oil palms by tissue culture. *Oil Palm News* 17: 1-8.

Jones, L.H. 1990. Endogenous cytokinins in oil palm (*Elaeis guineensis* L.) callus embryoids and regenerant plants measured by radiomimmunoassay. *Plant cell Tissue Organ Culture* 20(3): 202-209.

Kaur, G., Rathore, T.S., Rama Rao and N.S., Shekhawat 1992. In vitro micropropagation of Caralluma edulis (Edgew Benth and Hook. F. -A rare edible plant species of Indian desert Indian J. Plant Genetic Resources, 5:51-56.

Khoshoo, T.N., 1992. 2nd Pdt. Govind Ballabh Pant memorial Lecture, G.B.P.I.H.E.D., Kosi Almora, India.

Khoshoo, T.N., 1994. Census of India's Biodiversity: Tasks Ahead. Curr. Sci. 67, 577-582.

Kim, K.W. and De Hertogh, A.A. 1997. Tissue culture of Ornamental flowering bulbs (Geophytes). (in) Horticultural Reviews Vol. 18 pp. 87-169. J. Janick (Ed.) John Wiley & Sons.

Krul, W.R., and Mowbray, G.H. 1977. Formation of adventitious embryos in callus cultures of `Seyval' a French hybrid grape. *J. Am. Soc. Hort. Sci.* 102: 360-363.

Manshardt, R. M. (1992) In : *Biotechnology of Perennial Fruit Crops* (Hammerschlag, F. A. and Litz, R. E. eds.) pp. 489-511.

Marin, M.L., Mafla, G., Roca, W.M. and Withers, L.A. 1990. Cryopreservation of cassava zygotic embryos and whole seeds in liquid nitrogen. *Cryo-Letters* 1: 257-264.

Marx., J.L. ed. 1989. A Revolution in biotechnology. Cambridge Univ. Press, Cambridge.

Matthews, D., Mottley, J., Horan, I. and Roberts, A.V. 1991. A protoplast to plant system in roses. *Pl. Cell Tissue and Organ Culture.* 24: 173-180.

May, R.M., 1992. How Many Species Inhabit the Earth? Scientific American, Oct., 18-24.

McNeely, J.A. 1994. Lessons From the Past: Forests and Biodiversity. Biodiversity and Conservation 3 : 3-20.

Mehra, K.L. and M.L. Magoon, 1974: Collection, conservation and exchange of gene pools of forage grasses. Indian J. Genet. 34A:26-32.

Meyers, N. and Ayensu, E.S., 1983. Reduction of Biological Diversity and Species Loss. Ambio, 12:72 - 74.

Mukherjee, A. 1998. *In-vitro* propagation studies in sweet potato (*Ipomoea batatas* L.) Ph.D. Thesis. Utkal University.

Murashige, T., and Skoog, F. (1962). A revised medium for rapid growth bioassays with tobacco tissue cultures. *Physiol. Plant.* 15: 473-97.

Navarro L. and Juarez J., 1977. Tissue culture techniques used in Spain to recover virus-free Citrus plants. Acta Hortic. 78, 425-453.

Naveen, K.S., Krishna, G., Mamatha, H.N., Muniswamy, and Sreenath, H.L. 1998. Cultured zygotic embryos of *Coffee arabica* L. Paper presented at commercial aspects of plant tissue culture, molecular biology and medicinal plant biotechnology, held at Jamia Hamdard (Hamdard University) New Delhi from Feb. 25-27 1998. Abstract No. R-20 pp32.

Nayar, M.P. and Sastry, A.R.K. 1987, 1991, 1992. Red data book of Indian plant Vol. 1,2,3 (Botanical Survey of India.

NBFGR 1997. Annual Report 1996 - 1997, 78 p.

NBFGR 1998 a. Fish biodiversity of India. NBFGR special publication no. 2 (in press).

NBFGR 1998 b. Fish chromosome Atlas NBFGR special publication no. 1, 332p.

NBFGR 1998 c. Annual report 1997 - 1998.

Nevo, E. 1990. Molecular evolutionary genetics of isozymes: pattern, theory, and applications. In : Isozymes : Structure, Function, and Use in Biology and Medicine. Wiley-Liss Inc. NY.

NRCC, Puttur. 1998. Research Highlight 1997-98. p. 9, National Research Centre for Cashew, Puttur.

Osifo, E.O. 1988. Somatic embryogenesis in *Diosorea. Journal of Plant Physiology.* 133: 378-380.

Otsuka, H., Suematsu, N. and Toda, M. 1985. The culture and plant from mesophyll protoplast of chrysanthemum. *Bull. Shizuoka Agricultural Experiment Station, Japan* 30: 25-33.

Pant., D.D. 1999. Biodiversity conservation and evolution of plants. Current Science. 76: 21-23.

Paroda, R.S. 1997. Emerging Concern for Agro-Biodiversity in the Indian National Context: An Introspection - Keynote Address. NAAS-NBPGR (ICAR Workshop on National Concern for Management, Conservation and Use of Agro-biodiversity, October, 15-16, 1997, Shimla, India. 20p. NBPGR.

Paroda, R.S., V.L. Chopra, Mangala Rai, P.L. Gautam, Suman Sahai, Devendra Sharma and Sudhir Kochhar. 1998. Conservation, Management and Use of Agro-Biodiversity - Policy Paper. National Academy of Agricultural Sciences (NAAS), INDIA, New Delhi. 16p. NAAS.

Ponniah, A.G. and Lal, K.K. 1996. Mini Gene Bank of NBFGR. In : Symposium on fish Genetics and Biodiversity Conservation for Sustainable Production, at NBFGR Lucknow. 26-27 September, 1996. Abstract. Das, P., Ponniah, A.G., Lal, K.K. and Pandey, A.K. (eds. p 39).

Ponniah, A.G., Gopalakrishnan, A. Lal, K.K. and Srivastava, S.K. 1997. Use of fertilization protocols to enhance hatching percentage with cryopreserved milt of *Cyprinus carpio*. Nat. Acad. News Letters (press).

Purohit, S.D. and K.Tak 1992. In-vitro propagation of an adult tree *Feronia limonia* L. through axillary branching. Indian J. Expt. Biology, 30: 377-379.

Purohit, S.D., Kukda, G., Sharma, P and K. Tak. 1994. In-vitro propagation of an adult tree *Wrightia tomentosa* though enhanced axillary branching. Plant Science, 103: 67-72.

Quershi, J.A. and Saxena, P.K. 1992. Adventitiouis shot induction and somatic embryogenesis with intact seedlings of several hybrid seed geranium *(Pelargonium hortorum* Bailey) varieties. *Plant Cell Reports* 11(9): 443-448.

Rathore, T.S., Singh, R.P. and N.S. Shekhawat 1991. Clonal propagation of desert teak (*Tecomella undulata* through tissue culture. Plant Science, 79 : 217-222.

Ravel, C., and G. Charmet. 1996. A comprehensive multisite recurrent selection strategy in perennial ryegrass. Euphytica 88: 215-226.

Razdan., M.K. 1995. An introduction to plant tissue culture: Germplasm conservation : 327 – 335.

Reinert, J. and M.M. Yeoman 1982. Plant cell and Tissue culture: a laboratory mannual Springer-Verlag, Berlin.

Rout, G.R. and Das, P. 1997. Recent Trends in Biotechnology of Chrysanthemum: A. Critical Review. *Scientia Hort.* 69: 239-257.

Ruckelshaus, W.D., 1994. Towards a Sustainable World. Scientific American, 166-176, September, 1994.

Sachs, I., 1992. Transition Strategies for the 21st Century. Nature & Resources, Vol. 28, No. 1.

Sathiamoorthy, S., Jawaharlal, M. and Kumar, S. (1995) In : *Hi-tech Agriculture,* Tamil Nadu Agricultural University Press, Coimbatore, pp. 26-29.

Singh, Samar 1994. Conservation of biodiversity. Environ. Vol. -2, Science 281:1295-1296.

Singh, V., 1985. Threatened taxa and scope of conservation in Rajasthan. J. Econ Tax. Bot. 7 (3): 573-577.

Swaminathan, M.S. 1992. Biodiversity: Implications for global food security Macmillan, Madras.

Swaminathan, M.S. and S., Jana. eds. 1992. Biodiversity: Implications for Global food security. Macmillan, Madras.

Swaminathan., M.S. 1983. Genetic conservation microbes to man. Presidential address XV the International Congress of genetics. New Delhi, December 12-1.

Szabados, L., Hoyos, R., and Roca, W. 1987. *In-vitro* somatic embryogenesis and plant regeneration of cassava. *Plant Cell Report.* 6: 248-251.

Tabeezadeh and Khosh-Khui, M. 1981. Anther culture of Rosa. *Sci. Hort.* 15: 61-66.

Takeda, K., and C.L. Chang. 1996. Inheritance and geographical distribution of phenol reactionless varieties of barley. Euphytica 90: 217-221.

Taylhardat., A. and R.A., Zilinskas 1992. AGENDA 21: Biotechnology at the United Nations Conference on Environmental and Development. Biotechnology 10: 402-404.

Taylor, N.J., Edwards, M. and Henshaw, G.G. 1994. Production of friable embryogenic calli and suspension culture systems in two genotypes of cassava. Second International Scientific Meeting of Cassava Biotechnology Network (CBN), held during 22-26 August 1994, Indonesia, working doc No. 150, Vol.-I, CIAT, Colombia, 229-240.

The Consultative Group on International Agricultural Research, 1999. http://www.cgiar.org; verified July 27, 2000.

The Food and Agriculture Organization of the United Nations. 1998. The state of the world's plant genetic resources for food and agriculture, FAO, Rome.

The System-Wide Information Network for Genetic Resources, 1999. http://www.singer.cgiar.org; verified July 27, 2000.

Thimapaiah and Sherly, R.S. 1996. Micropropagation studies in Cashew (*Anacardium occidentale* L). Paper presented at National Symposium on Horticultural Biotechnology. October 28 to 30, 1996, Bangalore. Abstract no. II-10. p.5.

Thro, A.M., Bonierbale, M., Roca, W.M., Best, R. and Henry, G. 1994. Why, what, who and when : An overview of the cassava biotechnology network (CBN) and current cassava biotechnology research worldwide. Second International Scintifici Meeting of Cassava Biotechnology Network (CBN), held during 22-26, August, 1994 at Indonesia, working doc. No. 150, V-I, CIAT, Colombia.

Thursell, J. and J. Harrison, 1992. National Parks and Nature Reserve in Mountain Environment and Development. Geo Journal 271)113-126.

Tilak, K.V.B.R. 1988. In : Pulse crops (Grain legumes. B. Baldev, S. Ramanujam and H.K. Jain (eds.. Oxford and IBH Publ. Co. Pvt. Ltd., New Delhi, India. pp. 373-411.

Tillman, B.L., S.A. Harrison, C.A. Clark, E.A. Milus, and J.S. Russian. 1996. Evaluation of bread wheat germplasm for resistance to bacterial streak. Crop Sci. 36: 1063-1068.

Tohme, J., D.O. Gonzalez, S. Beebe, and M.C. Duque, 1996. AFLP analysis of gene pools of a wild bean core collection. Crop Sci 36: 1375-1384.

Twyford, C.T. 1993. Somatic embryogenesis in the foo dyam *Dioscorea alata* L., cv. Oriental Lisbon, Ph.D. thesis.

UNFTD, 1997a. United Nations Floriculture Trade Data for 1995. *Flora Culture International* 7(3): 33.

Upadhyay, V.S., P.M. Dabadghao and K.A. Shankarnarayan, 1971: Annual Report. I.G.F.R.I., Jhansi.

UPASI. 1997. Seventy first Annual Report of The United Planters Association of Southern India. p. 61-63.

Upreti., J. and U. Dhar 1996. Micropropagation of *Bauhinia vahlii* Wight and Arnott - a leguminous liana. Plant Cell Reports 16 : 250 - 254.

Usher, M.B. 1986. Wildlife Conservation Evaluation: Attributes, Criteria and Values. Pages 3-44. In M.B. Usher, Editors Wildlife conservation evaluation. Chapman and Hall, London.

Vam Jaars. Veld, A.S., Freitage, S., Chown, S.L., Muller, C., Koch, Hull, H., Bellamy, C., Kriiger, M., Endrody, M., Endrody - Younga., S., Mansell., M.W. and C.H., Scholtz 1998. Biodiversity assessment and conservation strategies. Science 279 : 2106-2106.

Van der Salm., T.P.M., van der Toorn, C.J.G., Hanisch Cate, C.H. and Dons, H.G.M. 1996. Somatic embryogenesis and shoot regeneration from excised adventitious roots of the root stock *Rosa hybrida*. L. `Money Way. *Plant Cell Reports* 15(7): 522-526.

Viana, A.M. and Mantell, S.M. 1989. Callus induction and plant regeneration of the seed-propagated yams *Dioscorea composita* Hemsl. and *D. cayenesis* Lam. *Plant Cell Tiss. Org. Cult.* 16 : 113-122.

Vij, S.P., Sharma, M.L., Kaur, P and A. Gupta 1999. Plant genetic diversity : Evaluation and conservation. Nation Seminar (Abstracts Botany Department, Punjab University, Chandigarh. Feb 22-23, 1999.

Wilson, E.O., 1988. The Current State of Biological Diversity. In: Biodiversity. National Academy Press, Washington D.C., pp. 3-5.

World Conservation Monitoring Centre, 1993. Global Biodiversity. Chapman and Hall, London.

World Resource Institute 1989. World resources 1989-90. Oxford University Press. New York.

World Resource Institute 1990. World resources 1990-91. Oxford University Press, New York.

World Resource Institute 1992. World resources 1989-93. Oxford University Press, New York.

World Resource Institute, 1992. World Resources, 1992-93. Oxford University Press, New York.

Wu, Y.W., and Ma, T.P. 1979. Isolation, culture and callus formation of *Ipomoea batatas* protoplasts. *Acta Bot. Sin.* 21: 334-338.

Yeoman, M.M. 1986. Plant cell culture Technology. Botanical Monographs vol. 23 Blackwell Scientific Publications. Oxford.

Appendix-1

Biodiversity: Institutional Mechanism and Modalities

The Institutional mechanism at the national level is provided by the Ministry of Agriculture in conjunction with the other Ministries, such as, law, Commerce, Finance, Forestry and Environment, etc. The ICAR-DARE, as the nodal organization, spearheads, the planning, coordination and implementation of all activities on agro-biodiversity in the public sector and maintains linkages with other Government Departments, Communities, NGOs and private sector. Factors such as, environment, cropping system, socio-economics have been integrated in the system through appropriately organizing a balanced system of conservation, sustainable development and progressive farming.

The ICAR has been providing a major funding support to state agricultural universities (SAUs) and also carrying out research, research management, research co-ordination, training and agricultural education activities as its own institutional set-up to bride the gaps. It is an old and experienced organizational and is a leading National Agricultural Research System (NARS) globally. The national needs of agricultural research, extension and education have been diligently and meaningfully catered by the Council all the while, despite the fact that `agriculture' is a `State Subject' as per the Constitution of India. There is a further need to self-determine and fulfill such responsibilities for the system, giving due importance, opportunities and avenues to emerging new linkages.

The ICAR has been providing a major funding support to state agricultural universities (SAUs) and also carrying out research management, research co-ordination, training and agricultural education activities as its own institutional set-up to bride the gaps. It is an old and experienced organization and is a leading National Agricultural Research System (NARS) globally. The national needs of agricultural research, extension and education have been diligently and meaningfully catered by the Council all the while, despite the fact that `agriculture' is a `State Subject' as per the Constitution of India. There is a further need to self-determine and fulfill such responsible for the system, giving due importance, opportunities and avenues to emerging new linkages.

The medium to long term perspective planning of the ICAR amplifies Council's concern for achieving a combination of sustainability and advancement in the field of agricultural research, including conservation and rational use of agro-biodiversity. The ICAR provides funds and manpower support to several university based centres to work on underutilized crops realizing scientific need to develop technology for new crops of future potential. Also, in order to strengthen research and advance technology base for rapid characterization for diversity in plant samples, a National Research Centre (NRC) on DNA finger printing has been established, besides and NRC on plant biotechnology.

THE INDIA NATIONAL PLANT GENETIC RESOURCES SYSTEM (IN-PGRS)

The India national plant genetic resources system (IN-PGRS), under the aegis of the Indian Council of Agricultural Research (ICAR) and spearheaded by the National Bureau of Plant Genetic Resources (NBPGR) is among the most dynamic systems in the world which now hold a prominent place and a leading gene-bank. The NBPGR has been entrusted with the national responsibility to plan, conduit, promote, co-ordinate and take lead in activities concerning germplasm collection, introduction, exchange, evaluation, documentation, conservation and sustain able management of diverse germplasm of crop plants and their wild relatives with a view to ensure their availability for use over time to breeders and other researchers. It includes NBPGR network of 10 regional stations/base centres/ quarantine centres over different phyto-geographic zones of the country and active collaboration and linkages with over 30 National Active germplasm Sites (NAGS.

DARE/ICAR is the umbrella organization while NBPGR is the nodal agency. It controls 11 Regional Stations located all across the country as per agro-ecological analogues. Apart from it NBPGR also have National Active Germplasm Sites (NAGS), ICAR insatiate/project, Directorates/National Research Centres/all India Coordinated Research Projects/State Agricultural Universities, Other National Stake Holder and have International Collaboration.

The IN-PGRS is responsible for conservation and management of PGR and their use through an effective collaboration between the NBPGR and the user agencies. NBPGR is the nodal organization operating the Indian PGR system and has the authority to import and export of plant germplasm for research purpose through the single window system. It is also fully equipped with very effective and efficient plant quarantine facilities. NBPGR has assisted over a dozen collaborating institutes/centres in establishing medium-term seed storage; computer and data documentation faculties besides imparting need-based, on job training to their PGR personnel.

NBPGR is helping the collaborating institutions in developing medium term seed storage faculties and also computer faculties for documentation. Bureau also provides training to their scientist and technicians.

The emerging role of non-governmental organizations (NGOs) in managing agro-biodiversity for its conservation and sustainable and equitable use, through imparting awareness to indigenous communities and farming and families and developing indigenous knowledge base banks is being widely recognized. This adds to the responsibility of the ICAR, being nodal organization in the ICAR, being nodal organization in the country's agricultural research system, to perceive, anticipate, plan, and frame appropriate policy issues and also to develop model/pilot scale projects related to agro-biodiversity management, conservation and sustainable use. Issue of compensation to farming communities may also be addressed as and where applicable. This, obviously calls for concerted efforts towards understanding the following components:

Classification, indexing and inventorization of traditional agricultural systems, still in vogue, with respect to their eco-systems, crop combinations, seasons any/or system of cultivation.

PGR ACTIVITIES IN WILD GENEPOOL

The National Bureau of Plant Genetic Resources (NBPGR) has been collecting wild relatives of crop plants. General and specific (*Abelmoschus*, *Solanum*, temperate fruits, Cucurbits) explorations have been undertaken. Inventorization of species wealth available in wild relatives of crop plants has been published which is being updated, giving a broader and wider perspective on its distribution and extent of diversity, and conservation and use of this diversity. The information points out to much less utilization of wild species while these have great direct and indirect potential in crop improvement, and even as new domesticates.

The ICAR's new initiatives in perspective planning include agricultural research information system (ARIS), agricultural human resource development (AHRD), National Agricultural technology project (NATP) for technology generation and transfer, institution -village linkage programme (IVLP), interface with Ministerial Departments/Ministries of Govt. of India, such as, DoAC DAH&D, DBT, CSIR, DRD, CAPART, MoEF, etc., and establishing partnership with NGO and private sectors. Besides involvement in several regional for a and programmes through bilateral agreements and MOUs the Council is posed well to play major role in establishing the country's lead for agricultural development and management of its agro-biodiversity.

Appendix - 2

Genetically Modified Organisms (GMOs) Promise and Danger for Biodiversity

Genetic engineering has the potential to revolutionize Indian agriculture but there is an urgent need of public debate on terminator genes, genetically modified crops and *Frankenstein foods*, etc. People do not trust on their Governments on their release, whether they are safe or not?. India, like other countries, stands to benefit enormously from harnessing the power of genetic engineering and other developments in biology for agriculture, animal husbandry and treatment of disease. However, national interest also dictates that the country has a sound and credible regulatory system to keep a close watch over the application of modern biotechnology and prevent its misuse.

Government of India already has regulations governing import, research, field trials and ultimate commercial use of genetically modified organisms or GMOs (organisms which carry an artificially introduced gene. There is a need of transparency in the functioning of the committees which perform these roles. India needs mechanisms to monitor biological patents being applied for and granted abroad as well as in India. This should not be left wholly to the Government and Indian NGOs need to keep an independent watch. The Indian Patent Office needs considerable improvement so that it can play its role of often being the first line of defense.

The scientific community and NGOs should became aware of the Terminator genes issue. North American NGO, blew the whistle on a U.S. patent innocuously titled **Control of Plant Gene Expression**. The technology provides a genetic method of making seeds sterile so that farmers can not save and replant them. Any patent granted in the U.S. and Europe provide warning signals each product, while India is dependent on foreign organizations.

Bollgard Cotton a Carrier of Bt-genes

The ruckus over the Terminator genes and the way Monsanto's field trials in India of its Bollgard genetically engineered cotton is a matter of controversy. Bollgard cotton carries the Bt gene taken from a soil bacterium. The gene makes the plant's cells produce a substance which kills certain insect pests but is harmless to the plant itself, other insects, animals and humans. The presence of the Bt gene and the substance it produces considerably reduce the use of insecticides and increase yield. Considering the well-known environmental impact of chemical insecticides, there appears to be carrying the Bt gene. Indeed, several Indian laboratories are working on introducing one of the many Bt genes into different crops and vegetables such as cotton, rice, tobacco, mustard, potato, tomato, brinjal, cauliflower and cabbage.

The Bollgard cotton field trials seem to have drawn flak because Monsanto is buying up Delta and PIne Land, the U.S. company which, along with the U.S. Department of Agriculture, owns the Terminator genes patent. It is essential to test Bollgard cotton in India which carry the Terminator genes. Therefore, genetically engineered cotton varieties should be tested in India, whether these are suitable for India.

The use of GMOs in India is governed by the *Rules for the Manufacture, Use/Import/Export and Storage of Hazardous Microorganisms/Genetically Engineered Organisms or Cells* promulgated in December 1989 by the Union Government's Ministry of Environment and Forests under the powers bestowed on it by the Indian Environment (protection) Act of 1986.

As the Terminator genes patent demonstrates vividly, the patent title can be innocuous and even the abstract gives little clue to its possible application. These rules established the following committees:

Committee	Governed by	Function
Recombinant DNA Advisory Committee (RDAC)	Department of Biotechnology, Government of India	To recommend appropriate safety regulations for genetic engineering work.
Review Committee on Genetic Manipulation (RCGM)	Department of Biotechnology, Government of India	To monitor on going research. Every organisation handling genetically engineered organisms has to set up an institutional bio-safety committee
Genetic Engineering Approval Committee (GEAC)	Department of Environment, Forest and Wildlife, Government of India	to clear any large-scale use of genetically engineered organisms. These committees have been set up.
Technology Information, Forecasting and Assessment Council	An autonomous body under the Department of Science and Technology	To notice an application of patent of terminator genes in India too.

Looking into the present situation and hazards of GMOs, there is a need to establish Biotechnology Coordination Committee in each state of India, headed by the Chief Secretary, with powers to inspect investigate and take punitive action in case of violations of statutory provisions. The rules also require district level committees, headed by the district Collector, to watch over issues of safety and meeting emergencies. In Karnataka, even the State Committee was established only after the Terminator gene controversy erupted.

According to Dr. Manju Sharma, Secretary to the Department of Biotechnology, the Review Committee on Genetic Manipulation (RCGM) only authorizes research work on transgenics and genetically modified organisms. But research on transgenic plants requires laboratory as well as field experiments in contained glass houses and in small plots in the field. In plant experiments, without field trials in small plots it would not be possible to conclude how the surrounding flora and fauna would react to the transgenics. Once the green house experiments, limited experimental field trials and other experiments were completed, the scientific information from them would be compiled. The REGM would then forward this, with its own observations, to the Genetic Engineering Approval Committee (GEAC) for a decision, said Dr. Sharma.

A petition has already been field in the supreme Court, questioning the RCGM's authority to sanction field trials.

There exits two fundamental problems with the functioning of these committees:

1. None of committees release information about the basis on which they have taken decisions about permitting or disallowing genetic engineering work. For example, when there are reports that Bollgard cotton has failed in other countries, information should be made available to scientists/general public and the reasons should be known to every body that why the tests in India were permitted. This question was raised by Dr. Pushpa Bhargava former Director of the Centre for Cellular and Molecular Biology.
2. There is no procedure for the committees to secure public comments. In contrast, the U. S. Department of Agriculture's Animal and Plant Health Inspection Service (APHIS) publishes a notice in the Federal Register (the U.S. Equivalent of the Indian Government's gazette) and seeks public review before it deregulates any genetically engineered plant or microorganism. When APHIS decides to go ahead and deregulate an genetically modified organism, its detailed *determination* of whether this organism could become a plant pest as well as an equally detailed *environmental assessment* of the potential impact of not regulating its use is made public. The APHIS Determination and Environment Assessment documents of Monsanto's Bollgard cotton are thus readily available.

In early 1999, M.S. Swaminathan Research Foundation hold a meeting on this issue and suggested that, in addition to the existing structure, there should also be a National Commission on Genetic Modifications of Crop Plants and Farm Animals, and should be headed by an independent and professionally qualified chairperson. He will coordinate and enforce a precautionary package for the safe and beneficial use of genetic engineering.

Much also needs to be done to improve the operations of the Indian Patents Office. A few years back, the patent Office actually cleared a patent which gave a U.S. company, Agracetus, rights over genetically engineered cotton when Indian law did not permit patenting of life forms. Once again, it was RAFI which drew attention to the patent! When the Indian patent law is modified to conform to the TRIPS requirements, it will need to allow genetic patenting. The role of the Patent Office will then become even more important.

According to Cornell's International service for the Acquisition of Agri-Biotech Applications (ISAAA) pointed out that in China, with 1.8 million hectares land is under various genetically engineered crops, and is second only to the U.S. in this respect. According to ISAAA, the virus resistant tobacco introduced by China had 5-7 per cent greater yield than non-genetically modified tobacco and saved 2-3 applications of insecticide. When Bt cotton was grown on 1.8 million acres in the U.S. in 1996, it saved the use of about 2.5-lakh gallons of insecticide, says the ISAAA report. Bt cotton gave average yield increases of 7 per cent compared to non-be cotton and the yield increase could be as high a 20 per cent.

Appendix-3

Convention on Biological Diversity

[Agreed Text of the Convention]

PREAMBLE

The Contracting Parties,

- Conscious of the intrinsic value of biological diversity and of the ecological, genetic, social, economic, scientific, educational, cultural, recreational and aesthetic values of biological diversity and its components,
- Conscious also of the importance of biological diversity for evolution and for maintaining life sustaining systems of the biosphere,
- Affirming that the conservation of biological diversity is a common concern of human kind,
- Reaffirming also that States are responsible for conserving their biological diversity and for using their biological resources in a sustainable manner,
- Concerned that biological diversity is being significantly reduced by certain human activities,
- Aware of the general lack of information and knowledge regarding biological diversity and of the urgent need to develop scientific, technical and institutional capacities to provide the basic understanding upon which to plan and implement appropriate measures,
- Noting that it is vital to anticipate, prevent and attack the causes of significant reduction or loss of biological diversity at source,
- Noting also that where there is a threat of significant reduction or loss of biological diversity, lack of full scientific certainty should not be used as a reason for postponing measures to avoid or minimize such a threat,
- Noting further that the fundamental requirement for the conservation of ecosystems and natural habitats and the maintenance and recovery of viable populations of species in their natural surroundings.
- Noting further that *ex-situ* measures, preferably in the country of origin, also have and important role to play,
- Recognizing the close and traditional dependence of many indigenous and local communities embodying traditional lifestyles on biological resources, and the

desirability of sharing equitably benefits arising from the use of traditional knowledge, innovations and practices relevant to the conservation of biological diversity and the sustainable use of its components,

- Recognizing also the vital role that women play in the conservation and sustainable use of biological diversity and affirming the need for the full participation of women at all levels of policy-making and implementation for biological diversity conservation,
- Stressing the importance of, and the need to promote, international, regional and global cooperation among State and intergovernmental organizations and the non-governmental sector for the sustainable use of its components,
- Acknowledging that the provision of new and additional financial resources and appropriate access to relevant technologies can be expected to make a substantial difference in the world's ability to address the loss of biological diversity,
- Acknowledging further that special provision is required to meet the need of developing countries, including the provision of new and additional financial resources and appropriate access to relevant technologies
- Noting in this regard the special conditions of the least developed countries and small island States,
- Acknowledging that substantial investments are required to conserve biological diversity and that there is the expectation of a broad range of environmental, economic and social benefits from those investments,
- Recognizing that economic and social development and poverty eradication are the first and overriding priorities of developing countries,
- Aware that conservation and sustainable use of biological diversity is of critical importance for meeting the food, health and other needs of the growing world population, for which purpose access to and sharing of both genetic resources and technologies are essential,
- Noting that, ultimately, the conservation and sustainable use of biological diversity will strengthen friendly relations among States and contribute to peace for human kind,
- Desiring to enhance and complement existing international arrangements for the conservation of biological diversity and sustainable use of its components, and
- Determined to conserve and substainable use of biological diversity for the benefit of present and future generations,

Have agreed as follows:

Article 1: Objectives

The objectives of this Convention, to be pursued in accordance with its relevant provisions, are the conservation of biological diversity, the sustainable use of its components and the fair and equitable sharing of the benefits arising out of the utilization of genetic resources, including by appropriate access to genetic resources and by appropriate transfer of relevant technologies, taking into account all rights over those resources and to technologies, and by appropriate funding.

Article 2: Use of Terms

For the purpose of this Convention:

1. *"Biological diversity"* means the variability among living organisms from all sources including, *inter alia*, terrestrial, marine and other aquatic ecosystems and the ecological complexes of which they are part; this includes diversity within species, between species and of ecosystems.

2. *"Biological resources"* includes genetic resources, organisms or parts thereof, populations, or any other biotic component of ecosystems with actual or potential use or value for humanity.

3. "Biotechnology" means any technological application that uses biological systems, living organisms, or derivatives thereof, to make or modify products or processes for specific use.

4. *"Country of origin of genetic resources"* means the country which possesses those genetic resources in *in-situ* conditions.

5. *"Country providing genetic resources"* means the country supplying genetic resources collected from *in-situ* sources, including populations of both wild and domesticated species, or taken from *ex-situ* sources, which may or may not have originated in that country.

6. *"Domesticated or cultivated species"* means species in which the evolutionary process has been influenced by humans to meet their needs.

7. *"Ecosystem"* means a dynamic complex of plant, animal and micro-organism communities and their non-living environment interacting as a functional unit.

8. *"Ex-situ conservation"* means the conservation of components of biological diversity outside their natural habitats.

9. *"Genetic material"* means any material of plant, animal, microbial or other origin containing functional units of heredity.

10. *"Genetic resources"* means genetic material of actual or potential value.

11. *"Habitat"* means the place or type of site where an organism or population naturally occurs.

12. *"In-situ* conditions" means conditions where genetic resources exist within ecosystems and natural habitats, and in the case of domesticated or cultivated species, in the surroundings where they have developed their distinctive properties.

13. *"In-situ conservation"* means the conservation of ecosystems and natural habitats and the maintenance and recovery of viable populations of species in

their natural surroundings and, in the case of domesticated or cultivated species, in the surroundings where they have developed their distinctive properties.

14. *"Protected area"* means a geographically defined area which is designated or regulated and managed to achieve specific conservation objectives.

15. *"Regional economic integration organization"* means an organization constituted by sovereign States of a given region, to which its member States have transferred competence in respect of matters governed by this Convention and which has been duly authorized, in accordance with its internal procedures, to sign, ratify, accept, approve or accede to it.

16. *"Sustainable use"* means the use of components of biological diversity in a way and at a rate that does not lead to the long-term decline of biological diversity, thereby maintaining its potential to meet the needs and aspirations of present and future generations.

17. *"Technology"* includes biotechnology.

Article 3: Principle

States have, in accordance with the *charter of the United Nations* and the principles of international law, the sovereign right to exploit their own resources pursuant to their own environmental policies, and the responsibility to ensure that activities within their jurisdiction or control do not cause damage to the environment of other States or of areas beyond the limits of national jurisdiction.

Article 4: Jurisdictional Scope

Subject to the rights of other States, and except as otherwise expressively provided in this Convention, the provisions of this Convention apply, in relation to each Contracting party:

(a) In the case of components of biological diversity, in areas within the limits of its national jurisdiction; and

(b) In the case of processes and activities, regardless of where their effects occur, carried out under its jurisdiction or control, within the area of its national jurisdiction or beyond the limits of national jurisdiction.

Article 5: Cooperation

Each Contracting Party shall, as far as possible and as appropriate, cooperate with other Contracting Parties, directly or, where appropriate, through competent international organizations, in respect of areas beyond national jurisdiction and on other matters of mutual interest for the conservation and sustainable use of biological diversity.

Article 6: General Measures for Conservation and Sustainable Use

Each Contracting Party shall, in accordance with its particular conditions and capabilities:

(a) Develop national strategies, plans or programmes for the conservation and sustainable use of biological diversity or adapt for this purpose existing strategies, plans or programmes which shall reflect, *inter alia*, the measures set out in this Convention relevant to the Contracting Party concerned; and

(b) Integrate, as far as possible and as appropriate, the conservation and sustainable use of biological diversity into relevant sectoral or cross-sectoral plans, programmes and policies.

Article 7: Identification and Monitoring

Each Contracting Party shall, as far as possible and as appropriate, in particular for the purposes of Articles 8 to 10:

(a) Identify components of biological diversity important for its conservation and sustainable use having regard to the indicative list of categories set down in Annex 1;

(b) Monitor through sampling and other techniques, the components of biological diversity identified pursuant to subparagraph (a) above, paying particular attention to those requiring urgent conservation measures and those which offer the greatest potential for sustainable use;

(c) Identify processes and categories of activities which have or are likely to have significant adverse impacts on the conservation and sustainable use of biological diversity, and monitor their effects through sampling and other techniques; and

(d) Maintain and organize by any mechanism data derived from identification and monitoring activities pursuant to subparagraphs (a) (b) and (c) above.

Article 8: In-situ Conservation

Each Contracting Party shall, as far as possible and as appropriate:

(a) Establish a system of protected areas or areas where special measures need to be taken to conserve biological diversity;

(b) Develop, where necessary, guide-lines for the selection, establishment and management of protected areas or areas where special measures need to be taken to conserve biological diversity;

(c) Regulate or manage biological resources important for the conservation of biological diversity whether within or outside protected areas with a view to ensuring their conservation and sustainable use;

(d) Promote the protection of ecosystems, natural habitats and the maintenance of viable populations of species in natural surroundings;

(e) Promote environmentally sound and sustainable development in areas adjacent to protected areas with a view to furthering protection of these areas;

(f) Rehabilitate and restore degraded ecosystems and promote the recovery of threatened species, *inter alia,* through the development and implementation of plans or other management strategies;

(g) Establish or maintain means to regulate, manage or control the risk associated with the use and release of living modified organisms resulting from biotechnology which are likely to have adverse environmental impacts that could affect the conservation and sustainable use of biological diversity, taking also into account the risks to human health;

(h) Prevent the introduction of control or eradicate those alien species which threaten ecosystems, habitats or species;

(i) Endeavour to provide the conditions needed for compatibility between present uses and the conservation of biological diversity and the sustainable use of its components;

(j) Subject to its national legislation, respect, preserve and maintain knowledge, innovations and practices of indigenous and local communities embodying traditional lifestyles relevant for the conservation and sustainable use of biological diversity and promote their wider application with the approval and involvement of the holders of such knowledge, innovations and practices and encourage the equitable sharing of the benefits arising from the utilization of such knowledge, innovations and practices,

(k) Develop or maintain necessary legislation and/or other regulatory provisions for the protection of threatened species and populations;

(l) Where a significant adverse effect on biological diversity has been determined pursuant to Article 7, regulate or manage the relevant processes and categories of activities; and

(m) Cooperate in providing financial and other support for *in-situ* conservation outlined in subparagraphs (a) to (l) above, particularly to developing countries.

Article 9: *Ex-situ* Conservation

Each Contracting Party shall, as far as possible and as appropriate, and predominantly for the purpose of complementing *in-situ* measures:

(a) Adopt measures for the *ex-situ* conservation of components of biological diversity, preferably in the country of origin of such components;

(b) Establish and maintain faculties for *ex-situ* conservation of and research on plants, animals and micro-organisms preferably in the country of origin of genetic resources;

(c) Adopt measures for the recovery and rehabilitation of threatened species and for their reintroduction into their natural habitats under appropriate conditions;

(d) Regulate and manage collection of biological resources from natural habitats for *ex-situ* conservation purposes so as not to threaten ecosystems and *in-situ* populations of species except where special temporary *ex-situ* measures are required under subparagraph (c) above; and

(e) Cooperate in providing financial and other support for *ex-situ* conservation outlined in subparagraphs (a) to (d) above and in the establishment and maintenance of *ex-situ* conservation facilities in developing countries.

Article 10: Sustainable Use of Components of Biological Diversity

Each Contracting Party shall, as far as possible and as appropriate:

(a) Integrate consideration of the conservation and sustainable use of biological resources into national decision-making;

(b) Adopt measures relating to the use of biological resources to avoid or minimize adverse impacts on biological diversity;

(c) Protect and encourage customary use of biological resources in accordance with traditional cultural practices that are compatible with conservation or sustainable use requirements;

(d) Support local populations to develop and implement remedial action in degraded areas where biological diversity has been reduced; and

(e) Encourage cooperation between its governmental authorities and its private sector in developing methods for sustainable use of biological resources.

Article 11: Incentive Measures

Each Contracting Party shall, as far as possible and as appropriate, adopt economically and socially sound measures that act as incentives for the conservation and sustainable use of components of biological diversity.

Article 12: Research and Training

The Contracting Parties, taking into account the special needs of developing countries, shall:

(a) Establish and maintain programmes for scientific and technical education and training in measures for the identification, conservation and sustainable use of biological diversity and its components and provide support for such education and training for the specific needs of developing countries;

(b) Promote and encourage research which contributes to the conservation and sustainable use of biological diversity, particularly in developing countries, *inter alla*, in accordance with decisions of the Conference of the Parties taken in consequence of recommendations of the Subsidiary Body on Scientific, Technical and Technological Advice; and

(c) In keeping with the provisions of Article 16, 18 and 20, promote and cooperate in the use of scientific advances in biological diversity research in developing methods for conservation and sustainable use of biological resources.

Article 13: Public Education and Awareness

The Contracting Parties shall:

(a) Promote and encourage understanding of the importance of, and the measures required for, the conservation of biological diversity, as well as its propagation through media, and the inclusion of these topics in educational programmes; and

(b) Cooperate, as appropriate, with other States and international organizations in developing educational and public awareness programmes, with respect to conservation and sustainable use of biological diversity.

Article 14: Impact Assessment and Minimizing Adverse Impacts

1. Each Contracting Party, as far as possible and as appropriate, shall:

(a) Introduce appropriate procedures requiring environmental impact assessment of its proposed projects that are likely to have significant adverse effects on biological diversity with a view to avoiding or minimizing such effects and, where appropriate, allow for public participating in such procedures;

(b) Introduce appropriate arrangements to ensure that the environmental consequences of its programmes and policies that are likely to have significant adverse impacts on biological diversity are duly taken into account;

(c) Promote, on the basis of reciprocity, notification, exchange of information and consultation on activities under their jurisdiction or control which are likely to significantly affect adversely the biological diversity of other States or areas beyond the limits of national jurisdiction, by encouraging the conclusion of bilateral, regional or multilateral arrangements as appropriate;

(d) In the case of imminent or grave danger or damage originating under its jurisdiction or control to biological diversity within the area under jurisdiction of other States or in areas beyond the limits of national jurisdiction, notify immediately the potentially affected States of such danger or damage, as well as initiate action to prevent or minimize such danger or damage;

(e) Promote national arrangements for emergency responses to activities or events, whether caused naturally or otherwise, which present a grave and imminent danger to biological diversity and encourage international cooperation to supplement such national efforts and, where appropriate and agreed by the State or regional economic integration organizations concerned, to establish joint contingency plans.

2. The Conference of the Parties shall examine, on the basis of studies to be carried out, the issue of liability and redress, including restoration and compensation, for damage to biological diversity, except where such liability is a purely internal matter.

Article 15: Access to Genetic Resources

1. Recognizing the sovereign rights of States over their natural resources, the authority to determine access to genetic resources rests with the national government and is subject to national legislation.

2. Each Contracting Party shall endeavour to create conditions to facilitate access to genetic resources for environmentally sound uses by other Contracting Parties and not to impose restriction that run counter to the objectives of this Convention.

3. For the purpose of this Convention, the genetic resources being provided by a Contracting Party, as referred to in this Article and Article 16 and 19, and only those that are provided by Contracting Parties that are countries of origin of such resources or by the Parties that have acquired the genetic resources in accordance with this Convention.
4. Access, where granted, shall be on mutually agreed terms and subject to the provision of this Article.
5. Access to genetic resources shall be subject to prior informed consent of the Contracting Party providing such resources, unless otherwise determined by that party.
6. Each Contracting Party shall endeavour to develop and carry out scientific research based on genetic resources provided by other Contracting Parties with the full participation of, and where possible in, such Contracting Parties.
7. Each Contracting Party shall take legislative, administrative or policy measures, as appropriate, and in accordance with Articles 16 and 19 and where necessary through the financial mechanism established by Articles 20 and 21 with the aim of sharing in a fair and equitable way the results of research and development and the benefits arising from the commercial and other utilization of genetic resources with the Contracting Party providing such resources. Such sharing shall be upon mutually agreed terms.

Article 16: Access to and Transfer of Technology

1. Each Contracting Party, recognizing that technology includes biotechnology, and that both access to and transfer of technology among Contracting Parties are essential elements for the attainment of the objectives of this Convention, undertakes subject to the provision of this Article to provide and/or facilitate access for and transfer to other Contracting Parties of technologies that are relevant to the conservation and sustainable use of biological diversity or make use of genetic resources and do not cause significant damage to the environment.
2. Access to an transfer of technology referred to in paragraph 1 above to developing countries shall be provided and/or facilitated under fair and most favourable terms, including on concessional and preferential terms where mutually agreed, and where necessary in accordance with the financial mechanism established by Articles 20 and 21. In the case of technology subject to patents and other intellectual property rights, such access and transfer shall be provided on terms which recognize and are consistent with the adequate and effective protection of intellectual property rights. The application of this paragraph shall be consistent with paragraphs 3, 4 and 5 below.
3. Each Contracting Party shall take legislative, administrative or policy measures, as appropriate, with the aim that Contracting Parties, in particular those that are developing countries, which provide genetic resources are provided access to and transfer of technology which makes use of those resources, on mutually agreed terms, including technology protected by patents and other intellectual property rights, where necessary through the provisions of Articles 20 and 21

and in accordance with international law and consistent with paragraphs 4 and 5 below.

4. Each Contracting Party shall take legislative, administrative or policy measures, as appropriate, with the aim that the private sector facilitates access to, joint development and transfer of technology referred to in paragraph 1 above for the benefit of both governmental institutions and the private sector of developing countries and in this regard shall abide by the obligations included in paragraphs 1, 2 and 3 above.

5. The Contracting Parties, recognizing that patents and other intellectual property rights may have an influence on the implementation of this Convention, shall cooperate in this regard subject to national legislation and international law in order to ensure that such rights are supportive of and do not run counter to its objectives.

Article 17: Exchange of Information

1. The Contracting Parties, shall facilitate the exchange of information, from all publicly available sources, relevant to the conservation and sustainable use of biological diversity, taking into account the special needs of developing countries.

2. Such exchange of information shall include exchange of results of technical, scientific and socioeconomic research as well as information on training and surveying programmes, specialized knowledge, indigenous and traditional knowledge as such and in combination with the technologies referred to in Article 16, paragraph 1. It shall also, where feasible, include repatriation of information.

Article 18: Technical and Scientific Cooperation

1. The contracting parties shall promote international technical and scientific cooperation in the field of conservation and sustainable use of biological diversity, where necessary through the appropriate international and national institutions.

2. The Contracting Parties shall promote international technical and scientific cooperation with other Contracting Parties, in particular developing countries, in implementing this convention, inter *alia*, through the development and implementation of national policies. In promoting such cooperation, special attention should be given to the development and strengthening of national capabilities, by means of human resources development and institution building.

3. The conference of the Parties, at its first meeting, shall determine how to establish a clearing-house mechanism to promote and facilitate technical and scientific cooperation.

4. The Contracting Parties shall, in accordance with national legislation and policies, encourage and develop methods of cooperation for the development and use of technologies, including indigenous and traditional technologies, in pursuance of the objective of this convention. For this purpose, the Contracting

Parties shall also promote cooperation in the training of personnel and exchange of experts.

5. The Contracting Parties shall, subject to mutual agreement, promote the establishment of joint research programmes and joint ventures for the development of technologies relevant to the objectives of this Convention.

Article 19: Handling of Biotechnology and Distribution of its Benefits

1. Each Contracting Party shall take legislative, administrative or policy measures, as appropriate, to provide for the effective participation in biotechnological research activities by those Contracting Parties, especially developing countries, which provide the genetic resources for such research, and where feasible in such Contracting Parties.
2. Each Contracting Party shall take all practicable measures to promote and advance priority access on a fair and equitable basis by Contracting Parties, especially developing countries, to the results and benefits arising from biotechnologies based upon genetic resources provided by those Contracting Parties. Such access shall be on mutually agreed terms.
3. The Parties shall consider the need for an modalities of a protocol setting out appropriate procedures including, in particular, advance informed agreement, in the field of the safe transfer, handling and use of any living modified organism resulting from biotechnology that may have adverse effect on the conservation and sustainable use of biological diversity.
4. Each Contracting Party shall, directly or by acquiring any natural or legal person under its jurisdiction providing the organisms referred to in paragraph 3 above, provide any available information about the use and safety regulations required by that Contracting Party in handling such organisms, as well as any available information on the potential adverse impact of the specific organisms concerned to the Contracting Party into which those organisms are to be introduced.

Article 20: Financial Resources

1. Each Contracting Party undertakes to provide, in accordance with its capabilities, financial support and incentives in respect of those national activities which are intended to achieve the objectives of this Convention, in accordance with its national plans, priorities and programmes.
2. The developed country Parties shall provide new and additional financial resources to enable developing country Parties to meet the agreed full incremental costs to them of implementing measures which fulfill the obligations of this Convention and to benefit from its provisions and which costs are agreed between a developing country Party and the institutional structure referred to in Article 21, in accordance with policy, strategy, programme priorities and eligibility criteria and an indicative list of incremental costs established by the Conference of the parties. Other Parties, including countries undergoing the process of transition to a market economy, may voluntarily assume the obligation of the developed country parties. For the purpose of this

Article, the conference of the Parties, shall at its first meeting establish a list of developed country Parties and other parties which voluntarily assume the obligations of the developed country parties. The conference of the Parties shall periodically review and if necessary amend the list. Contributions from other countries and sources on a voluntary basis would also be encouraged. The implementation of these commitments shall take into account the need for adequacy, predictability and timely flow of funds and the importance of burden-sharing among the contributing Parties included in the list.

3. The developed country Parties may also provide, and developing country Parties avail themselves of financial resources related to the implementation of this Convention through bilateral, regional and other multilateral channels.

4. The extent to which developing country Parties will effectively implement their commitments under the Convention will depend on the effective implementation by developed country Parties of their commitments under the Convention related to financial resources and transfer of technology and will take fully into account the fact that economic and social development and eradication of poverty are the first and overriding priorities of the developing country Parties.

5. The Parties shall take full account of the specific needs and special situation of least developed countries in their actions with regard to funding and transfer of technology.

6. The Contracting Parties shall also take into consideration the special conditions resulting from the dependence on distribution and location of biological diversity within developing country Parties, in particular small island States.

7. Consideration shall also be given to the special situation of developing countries, including those that are most environmentally vulnerable, such as those with, arid and semi-arid zones, coastal and mountainous areas.

Article 21: Financial Mechanism

1. There shall be a mechanism for the provision of financial resources to developing country Parties for purposes of this Convention on a grant or concessional basis the essential elements of which are described in this Article. The mechanism shall function under the authority and guidance of, and be accountable to, the Conference of the Parties for purposes of this Convention. The operations of the mechanism shall be carried out by such institutional structure as may be decided upon by the Conference of the Parties at its first meeting. For purposes of this convention, the Conference of the Parties shall determine the policy strategy, programme priorities and eligibility criteria relating to the access to and utilization of such resources. The contributions shall be such as to take into account the need for predictability, adequacy and timely flow of funds referred to in Article 20 in accordance with the amount of resources needed to be decided periodically by the conference of the Parties and the importance of burden-sharing among the contributing Parties included in the list referred to in Article 20, paragraph 2. Voluntary contributions may also be made by the developed country Parties and by other countries and sources. The

mechanism shall operate within a democratic and transparent system of governance.

2. Pursuant to the objectives of this Convention, the Conference of the Parties shall at its first meeting determine the policy, strategy and programme priorities, as well as detailed criteria and guidelines for eligibility for access to and utilization of the financial resources including monitoring and evaluation on a regular basis of such utilization. The Conference of the Parties shall decide on the arrangements to give effect to paragraph 1 above after consultation with the institutional structure entrusted with the operation of the financial mechanism.

3. The Conference of the Parties shall review the effectiveness of the mechanism established under this Article, including the criteria and guidelines referred to in paragraph 2 above not less than two years after the entry into force of this Convention and thereafter on a regular basis. Based on such review. it shall take appropriate action to improve the effectiveness of the mechanism if necessary.

4. The Contracting Parties shall consider strengthening existing financial institutions to provide financial resources for the conservation and sustainable use of biological diversity.

Article 22: Relationship with Other International Conventions

1. The provisions of the present convention shall not affect the rights and obligation of any Contracting Party deriving from any existing international agreement, except where the exercise of those right and obligations would cause serious damage or a threat to biological diversity.

2. Contracting Parties shall implement this Convention with respect to the marine environment consistently with the rights and obligations of States under the law of the sea.

Article 23: Conference of the Parties

1. A Conference of the Parties is hereby established. The first meeting of the Conference of the Parties shall be convened by the Executive Director of the United Nations Environment Programme not later than one year after the entry into force of this Convention. Thereafter, ordinary meetings of the conference of the Parties shall be held at regular intervals to be determined by the Conference at its first meeting.

2. Extraordinary meeting of the Conference of the Parties shall be held at such other times as may be deemed necessary by the conference, or at the written request of any party, provided that, within six months of the request being communicated to them by the Secretariat, it is supported by at least one third of the parties.

3. The Conference of the Parties shall by consensus agree upon and adopt rules of procedure for itself and for any subsidiary it may establish, as well as financial rules governing the funding of the Secretariat. At each ordinary meeting, it shall adopt a budget for the financial period until the next ordinary meeting.

4. The Conference of the Parties shall keep under review the implementation of the present Convention, and for this purpose, shall:

   (a) Establish the form and the intervals for transmitting the information to be submitted in accordance with Article 26 and consider such information as well as reports submitted by any subsidiary body.

   (b) Review scientific, technical and technological advice on biological diversity provided in accordance with Article 25;

   (c) Consider and adopt, as required, protocols in accordance with article 28;

   (d) Consider and adopt, as required, in accordance with articles 29 and 30 amendments to this Convention and its annexes;

   (e) Consider amendments to any protocol, as well as to any annexes thereto, and if so decided, recommend their adoption to the Parties to the protocol concerned;

   (f) Consider and adopt, as required, in accordance with Article 30, additional annexes to this Convention;

   (g) Establish such subsidiary bodies, particularly to provide scientific and technical advice, as are deemed necessary for the implementation of this convention;

   (h) Contact, through the Secretariat, the executive bodies of conventions dealing with matters covered by this Convention with a view to establishing appropriate forms of cooperation with them; and

   (i) Consider and undertake any additional action that may be required for the achievement of the purposes of that convention in the light of experience gained in its operation.

5. The United Nations, its specialized agencies and the International Atomic Energy Agency, as well as any State not Party to this Convention, may be represented as observers at meetings of the conference of the Parties. Any other body or agency, whether governmental or non-governmental, qualified in fields relating to conservation and sustainable use of biological diversity which has informed the Secretariat of its wish to be represented as an observer at a meeting of the Conference of the Parties, may be admitted unless at least one third of the Parties present object. The admission and participation of observers shall be subject to the rules of procedure adopted by the conference of the Parties.

Article 24: Secretariat

1. A secretariat is hereby established. Its functions shall be:

   (a) To arrange for and service meetings of the conference of the Parties provided for in Article 23;

   (b) To perform the functions assigned to it by any protocol;

   (c) To Prepare reports on the execution of its functions under this Convention and present them to the conference of the Parties;

(d) To coordinate with other relevant international bodies, and in particular to enter into such administrative and contractual arrangements as may be required for the effective discharge of its functions; and

(e) To perform such other functions as may be determined by the conference of the Parties.

2. At its first ordinary meeting, the Conference of the Parties shall designate the secretariat from amongst those existing competent intentional organizations which have signified their willingness to carry out the secretariat functions under this Convention.

Article 25: Subsidiary Body on Scientific, Technical and Technological Advice

1. A subsidiary body for the provision of scientific, technical and technological advice is hereby established to provide the conference of the Parties and, as appropriate, its other subsidiary bodies with timely advice relating to the implementation of this convention. This body shall be open to participation by all Parties and shall be multidisciplinary. It shall comprise government representatives competent in the relevant field of expertise. It shall report regularly to the Conference of the Parties on all aspects of its work.

2. Under the authority of and in accordance with guidelines laid down by the Conference of the parties, and upon its request, this body shall:

(a) Provide scientific and technical assessments of the status of biological diversity;

(b) Prepare scientific and technical assessments of the effects of types of measures taken in accordance with the provision of this Convention;

(c) Identify innovative, efficient and state-of-the-art technologies and know-how relating to the conservation and sustainable use of biological diversity and advise on the ways and means of promoting development and/or transferring such technologies;

(d) Provide advice on scientific programmes and international cooperation in research and development related to conservation and sustainable use of biological diversity; and

(e) Respond to scientific, technical, technological and methodological questions that the Conference of the Parties and its subsidiary bodies may put to the body.

3. The functions, terms of reference, organization and operation of this body may be further elaborated by the Conference of the Parties.

Article 26: Reports

Each Contracting Party shall, at interval to be determined by the Conference of the Parties, present to the Conference of the Parities, reports on measures which it has taken for the implementation of the provisions of this Convention and their effectiveness in meeting the objectives of this Convention.

Article 27: Settlement of Disputes

1. In the event of a dispute between Contracting Parties concerning the interpretation or application of this Convention, the parties concerned shall seek solution by negotiation.
2. If the Parties concerned cannot reach agreement by negotiation, they may jointly seek the good offices of, or request mediation by, a third party.
3. When ratifying, accepting, approving or acceding to this Convention, or at any time thereafter a State or regional economic integration organization may declare in writing to the Depository that for a dispute not resolved in accordance with paragraph 1 or paragraph 2 above, it accepts one or both of the following means of dispute settlement as compulsory:
   (a) Arbitration in accordance with the procedure laid down in Part 1 of Annex II;
   (b) submission of the dispute to the International Court of Justice.
4. If the parties of the dispute have not, in accordance with paragraph 3 above, accepted the same or any procedure, the dispute shall be submitted to conciliation in accordance with part 2 of Annex II unless the Parties otherwise agree.
5. The provisions of this Article shall apply with respect to any protocol except as otherwise provided in the protocol concerned.

Article 28: Adoption of Protocols

1. The Contracting Parties shall cooperate in the formulation and adoption of protocols to this Convention.
2. Protocols shall be adopted at a meeting of the Conference of the Parties.
3. The text of any proposed protocol shall be communicated to the Contracting Parties by the Secretariat at least six months before such a meeting.

Article 29: Amendment of the Convention or Protocols

1. Amendments to this Convention may be proposed by any Contracting Party. Amendments to any protocol may be proposed by any Party to that protocol.
2. Amendments to this Convention shall be adopted at a meeting of the Conference of the Parties. Amendments to any protocol shall be adopted at a meeting of the Parties to the Protocol in question. The text of any proposed amendment to this Convention or to any protocol, except as may otherwise be provided in such protocol, shall be communicated to the Parties to the instrument in question by the secretariat at least six months before the meeting at which it is proposed for

adoption. The secretariat shall also communicate proposed amendments to the signatories to this Convention for information.

3. The Parties shall make every effort to reach agreement on any proposed amendment to this Convention or to any protocol by consensus. If all efforts at consensus have been exhausted, and no agreement reached, the amendment shall as a last resort be adopted by a two-third majority vote of the Parties to the instrument in question present and voting at the meeting, and shall be submitted by the Depository to all Parties for ratification, acceptance or approval.

4. Ratification, acceptance or approval of amendments shall be notified to the Depository in writing. Amendments adopted in accordance with paragraph 3 above shall enter into force among Parties having accepted them on the ninetieth day after the deposit of instruments of ratification, acceptance or approval by at least two thirds of the Contracting Parties to this Convention or of the Parties to the protocol concerned, except as may otherwise be provided in such protocol. Thereafter the amendments shall enter into force for any other Party on the ninetieth day after that party deposits its instrument of ratification, acceptance or approval of the amendments.

5. For the purposes of this Article, "Parties present and voting" means Parties present and casting an affirmative or negative vote.

Article 30: Adoption and Amendment of Annexes

1. The annexes to this Convention or to any protocol shall form an integral part of this Convention or of such protocol, as the case may be, and unless expressly provided otherwise, a reference to this Convention or its protocols constitutes at the same time a reference to any annexes thereto. Such annexes shall be restricted to procedural, scientific, technical and administrative matters.

2. Except as may be otherwise provided in any protocol with respect to its annexes, the following procedure shall apply to the proposal, adoption and entry into force of additional annexes to this Convention or of annexes to any protocol:

   (a) Annexes to this Convention or to any protocol shall be proposed and adopted according to the procedure laid down in Article 29;

   (b) Any Party that is unable to approve an additional annex to this Convention or an annex to any protocol to which it is Party shall so notify the Depository, in writing, within one year from the date of the communication of the adoption by the Depository. The Depository shall without delay notify all Parties of any such notification received. A Party may at any time withdraw a previous declaration of objection and the annexes shall thereupon enter into force for that Party subject to subparagraph (c) below;

   (c) On the expiry of one year from the date of the communication of the adoption by the Depository, the annex shall enter into force for all Parties to this Convention or to any protocol concerned which have not submitted a notification in accordance with the provisions of subparagraph (b) above.

3. The proposal, adoption and entry into force of amendments to annexes to this Convention or to any protocol shall be subject to the same procedure as for the proposal, adoption and entry into force of annexes to the Convention or annexes to any protocol.

4. If an additional annex or an amendment to an annex is related to an amendment to this Convention or to any protocol, the additional annex or amendment shall not enter into force until such time as the amendment to this Convention or to the protocol concerned enters into force.

Article 31: Right to Vote

1. Except as provided for in paragraph 2 below, each Contracting Party to this Convention or to any protocol shall have one vote.

2. Regional economic integration organisations, in matters within their competence, shall exercise their right to vote with a number of votes equal to the number of their member States which are Contracting Parties to the Convention or the relevant protocol. Such organisations shall not exercise their right to vote if their member States exercise theirs, and vice versa.

Article 32: Relationship Between this Convention and Its Protocols

1. A State or a regional economic integration organization may not become a Party to a Protocol unless it is, or becomes at the same time, a Contracting Party to this Convention.

2. Decisions under any protocol shall be taken only by the Parties to the protocol concerned. Any Contracting Party that has not ratified, accepted or approved a protocol may participate as an observer in any meeting of the parties to that protocol.

Article 33: Signature

This Convention shall be open for signature at Rio de Janeiro by all States and any regional economic integration organization from 5 or June 1992.

Article 34: Ratification, Acceptance or Approval

1. This Convention and any protocol shall be subject to ratification, acceptance or approval by States and by regional economic integration organizations. Instruments of ratification, acceptance or approval shall be deposited with the Depository.

2. Any organization referred to in paragraph 1 above which becomes a Contracting Party to this Convention or any protocol without any of its member States being a Contracting Party shall be bound by all the obligations under the Convention or the protocol, as the case may be. In the case of such organizations, one or more of whose member States is a Contraction Party to the Convention or relevant protocol, the organization and its member States shall decide on their respective responsibilities for the performance of their respective responsibilities

for the performance of their obligations under the Convention or protocol, as the case may be. In such cases, the organization and the member States shall not be entitled to exercise rights under the Convention or relevant protocol concurrently.

3. In their instruments of ratification, acceptance or approval, the organizations referred to in paragraph 1 above shall declare the extent of their competence with respect to the matters governed by the Convention or the relevant protocol. These organizations shall also inform the Depository of any relevant modification in the extent of their competence.

Article 35: Accession

1. This Convention and any protocol shall be open for accession by States and by regional economic integration organizations from the date on which the Convention or the protocol concerned is closed for signature. The instruments of accession shall be deposited with the Depository.

2. In their instruments of accession, the organizations referred to in paragraph 1 above shall declare the extent of their competence with respect to the matters governed by the Convention or the relevant protocol. These organizations shall also inform the Depository of any relevant modification in the extent of their competence.

3. The provisions of Article 34, paragraph 2, shall apply to regional economic integration organizations which accede to this Convention or any protocol.

Article 36: Entry into Force

1. This Convention shall enter into force on the ninetieth day after the date of deposit of the thirtieth instrument of ratification, acceptance, approval or accession.

2. Any protocol shall enter into force on the ninetieth day after the date of deposit of the number of instruments of ratification, acceptance, approval or accession, specified in that protocol, has been deposited.

3. For each Contracting Party which ratifies, accepts approves this Convention or accedes thereto after the deposit of the thirtieth instrument of ratification, acceptance, approval or accession, it shall enter into force on the ninetieth day after the date of deposit by such Contracting Party of its instrument of ratification, acceptance, approval or accession.

4. Any protocol, except as otherwise provided in such protocol, shall enter into force for a Contracting Party that ratifies, accepts or approves that protocol or accedes thereto after its entry into force pursuant to paragraph 2 above, on the ninetieth day after the date on which that Contracting Party deposits its instrument of ratification, acceptance, approval or accession, or on the date on which the Convention enters into force for that Contracting Party, whichever shall be the later.

5. For the purposes of paragraphs 1 and 2 above, any instrument deposited by a regional economic integration organization shall not be counted as additional to those deposited by member States of such organization.

Article 37: Reservations

No reservations may be made to this Convention.

Article 38: Withdrawals

1. At any time after two years from the date on which the present Convention has entered into force for a Contracting Party, that Contracting Party may withdraw from the Convention by giving written notification to the Depository.
2. Any such withdrawal shall take place upon expiry of one year after the date of its receipt by the Depository, or on such later date as may be specified in the notification of the withdrawal.
3. Any Contracting Party which withdraws from this Convention shall be considered as also having withdrawn from any protocol to which it is Party.

Article 39: Financial Interim Arrangements

Provided that it has been fully restructured in accordance with the requirements of Article 21, the Global Environment Facility of the United Nations Development Programme, the United Nations Environment Programme and the International bank for Reconstruction and Development shall be the institutional structure referred to in Article 21 on an interim basis, for the period between the entry into force of this Convention and the first meeting of the Conference of the Parties or until the Conference of the Parties decides which institutional structure will be designated in accordance with Article 21.

Article 40: Secretariat Interim Arrangements

The secretariat to be provided by the Executive Director of the United Nations Environment Programme shall be the secretariat referred to in Article 24, paragraph 2, on an interim basis for the period between the entry into force of this Convention and the first meeting of the Conference of the Parties.

Article 41: Depository

The Secretary-General of the United Nations shall assume the functions of Depository of this Convention and any protocols.

Article 42: Authentic Tests

The original of this Convention, of which the Arabic, Chinese, English, French, Russian and Spanish tests are equally authentic shall be deposited with the Secretary-General of the United Nations.

IN WITNESS WHEREOF the undersigned, being duly authorized to that effect, have signed this Convention.

Done at Rio de Janeiro on this fifth day of June, one thousand nine hundred and ninety-two.

ANNEXURE I

Identification and Monitoring

1. Ecosystems and habitats: containing high density, large numbers of endemic or threatened species, or wilderness; required by migratory species; of social, economic, cultural or scientific importance; or, which are representative, unique or associated with key evolutionary or other biological processes;
2. Species and communities which are: threatened; wild relatives of domesticated or cultivated species; of medicinal, agricultural or other economic value; or social, scientific or cultural importance; or importance for research into the conservation and sustainable use of biological diversity, such as indicator species; and
3. Described genomes and genes of social, scientific or economic importance.

ANNEXURE II

Part 1- Arbitration

Article 1

The claimant party shall notify the secretariat that the parties are referring a dispute to arbitration pursuant to Article 27. The notification shall state the subject-matter of arbitration and include, in particular, the articles of the Convention or the protocol, the interpretation or application of which are at issue. If the parties do not agree on the subject matter of the dispute before the President of the tribunal is designated, the arbitral tribunal shall determine the subject matter. The secretariat shall forward the information thus received to all Contracting Parties to this Convention or to the protocol concerned.

Article 2

1. In disputes between two parties, the arbitral tribunal shall consist of three members. Each of the parties to the dispute shall appoint an arbitrator and the two arbitrators so appointed shall designate by common agreement the third arbitrator who shall be the President of the tribunal. The latter shall not be a national of one of the parties to the dispute, nor have his or her usual place of residence in the territory of one of these parties, nor be employed by any of them, nor have dealt with the case in any other capacity.
2. In disputes between more than two parties, parties in the same interest shall appoint one arbitrator jointly by agreement.
3. Any vacancy shall be filled in the manner prescribed for the initial appointment.

Article 3

1. If the President of the arbitral tribunal has not been designated within two months of the appointment of the second arbitrator, the Secretary-General of the United Nations shall, at the request of a party, designate the President within a further two-month period.

2. If one of the parties to the dispute does not appoint an arbitrator within two months of receipt of the request, the other party may inform the Secretary-General who shall make the designation within a further two-month period.

Article 4

The arbitral tribunal shall render its decisions in accordance with the provisions of this Convention, any protocols concerned, and international law.

Article 5

Unless the parties to the dispute otherwise agree, the arbitral tribunal shall determine its own rules of procedure.

Article 6

The arbitral tribunal may, at the request of one of the parties, recommend essential interim measures of protection.

Article 7

The parties to the dispute shall facilitate the work of the arbitral tribunal and, in particular, using all means at their disposal, shall:

(a) Provide it with all relevant documents, information and facilities; and

(b) Enable it, when necessary, to call witnesses or experts and receive their evidence.

Article 8

The parties and the arbitrators are under an obligation to protect the confidentiality of any information they receive in confidence during the proceedings of the arbitral tribunal.

Article 9

Unless the arbitral tribunal determines otherwise because of the particular circumstances of the case, the costs of the tribunal shall be borne by the parties to the dispute in equal shares. The tribunal shall keep a record of all its costs, and shall furnish a final statement thereof to the parties.

Article 10

Any Contracting Party that has an interest of a legal nature in the subject-matter of the dispute which may be affected by the decision in the case, may intervene in the proceedings with the consent of the tribunal.

Article 11

The tribunal may hear and determine counterclaims arising directly out of the subject-matter of the dispute.

Article 12

Decisions both on procedure and substance of the arbitral tribunal shall be taken by a majority vote of its members.

Article 13

If one of the parties to the dispute does not appear before the arbitral tribunal or fails to defend its case, the other party may request the tribunal to continue the proceedings and to make its award. Absence of a party or a failure of a party to defend its case shall not constitute a bar to the proceedings. Before rendering its final decision, the arbitral tribunal must satisfy itself that the claim is well founded in fact and law.

Article 14

The tribunal shall render its final decision within five months of the date on which it is fully constituted unless it finds it necessary to extend the time-limit for a period which should not exceed five more months.

Article 15

The final decision of the arbitral tribunal shall be confined to the subject-matter of the dispute and shall state the reasons on which it is based. It shall contain the names of the members who have participated and the date of the final decision. Any member of the tribunal may attach a separate or dissenting opinion to the final decision.

Article 16

The award shall be binding on the parties to the dispute. It shall be without appeal unless the parties to the dispute have agreed in advance to an appellate procedure.

Article 17

Any controversy which may arise between the parties to the dispute as regards the interpretation or manner of implementation of the final decision may be submitted by either party for decision to the arbitral tribunal which rendered it.

Part 2- Conciliation

Article 1

A conciliation commission shall be created upon the request of one of the parties to the dispute. The commission shall, unless the parties otherwise agree, be composed of five members, two appointed by each Party concerned and a President chosen jointly by those members.

Article 2

In disputes between more than two parties, parties in the same interest shall appoint their members of the commission jointly by agreement. Where two or more parties have separate interests or there is a disagreement as to whether they are of the same interest, they shall appoint their members separately.

Article 3

If any appointments by the parties are not made within two months of the date of the request to conciliation commission, the Secretary-General of the United Nations shall, if asked to do so by the party that made the request, make those appointments within a further two-month period.

Article 4

If a President of the conciliation commission has not been chosen within two months of the last of the members of the commission being appointed, the Secretary-General of the United Nations shall, if asked to do so by a party, designate a President within a further two-month period.

Article 5

The conciliation commission shall take its decisions by majority vote of its members. It shall, unless the parties to the dispute otherwise agree, determine its own procedure. It shall render a proposal for resolution of the dispute, which the parties shall consider in good faith.

Article 6

A disagreement as to whether the conciliation commission has competence shall be decided by the commission.

Appendix - 4

(CITES) Convention on International Trade in Endangered Species

Many of the endangered plants and animals occur naturally in Third World countries and are exported to the Western Industrial countries. To help the former to cope with this problem, the IUCN lobbied for international trade controls and brought about the first Convention on International Trade in Endangered Species of World Flora and Fauna (CITES) in 1973. The trade agreement was enforced in 1975 after 10 nations had ratified the document. By 1992 the original list of ten signatories had swelled to 120 as more countries including India joined the plant conservation effort.

Three lists known as the Appendices mandate the treatment necessary to import or export the endangered species:

Appendix 1

It contains critically threatened species or genera. The CITES does not allow issue of permits for international trade unless under very exceptional circumstances. Transport of any specimen from this list if necessary it requires two permits- one to export the plant from the country and another to import it into the recipient country. Specimens can be moved only if removal will not jeopardize the survival of the species. The Appendix includes all lemurs and apes, many South American monkeys, the great whales, the cheetah, leopards, the Asian elephant, rhino, many birds of prey, cranes, pheasants and parrots, all sea turtles, some crocodiles and lizards, giant salamanders and some mussels.

Appendix 2

It lists species, whole genera, and even entire families of plants that are not as seriously troubled as the plants on Appendix 1. Substantial trade of the plants on Appendix 2 could result in the plant's being driven toward extinction. This list also includes plants which are not threatened but which resemble plants that are. For some groups, such as orchids or cacti, only an expert could distinguish the endangered species. Blanket protection is, therefore, given to several entire families. Import and export permits are required for all wild-collected plants listed in Appendix 2. Many tens of thousands of species, including all cacti and orchids, primates, cats, otters, whales, dolphins, tortoise and crocodile not mentioned in Appendix 1 come under this list. It also includes the African elephant, fur seals, the black stork birds of paradise, the coelacanth, some snails, birdwing butterflies and black corals.

Appendix 3

It lists those plants which are rare in one country but fairly common in others, and allows countries the option of restricting trade in those species.

CITES AND REGULATION OF TRADE IN ENDANGERED SPECIES

CITES plays a major role in controlling the international wildlife trade. One hundred and twenty countries including India are party to the Convention which aims not to eliminate the trade but to encourage rational and sustainable utilisation of living resources for human development.

International trade in wildlife and its products, including ivory and skins is now worth more than US $4-5 billion a year. Illegal trade in wildlife is probably the world's second largest illegitimate business (only narcotics are worth more), and one of the major foreign exchange earners in the international market. In the early seventies the size of he trade grew to such unprecedented proportions that it caused the depletion of many species from their natural habitat. This led to the formation of CITES in 1975. On the occasion of the 20th anniversary of CITES, Mr. Izgrev Topkov, secretary general of the organization says, "the animal and plant world would have been in a much worse state today if CITES did not exist". There are several instances in which CITES has been successful in controlling the extinction of species, e.g. the African leopard, Nile crocodile, the spectacled caiman in many South American countries, the vicuna in the Andean belt.

CITES role is to provide nations with a legal framework to fight the illegal trade of endangered species, which is among the seven largest problems of commercial fraud. The convention has also been responsible for improvement in the conditions of transport of live animals. Training programmes have been developed and cooperation between CITES, the Customs Cooperation Council (CCC) and INTERPOL is increasing and becoming more efficient everyday.

The approval of historic legislation calling upon nations doing tiger trade to enact laws to ban it, control stockpiles, preserve tiger habitats, and establish a regional network stemming the trade was a major development, mooted by several Asian nations. According to a recent CITES study, of the 81 countries with high levels of endangered species, only 15 have adequate legislations to support CITES. Mr. Claude Martin, the WWF director general said, the CITES cannot work to protect species in trade without the political will and backing of its members. The Convention is the deciding ground for future strategies- sustainable use or total protection. It is suggested that local communities need to be more involved in the management of biological resources and to benefit from their sustenance. If this point is missed or weakened, it is bounded to loose support of the majority of people who inhabit the poor and developing countries.

CITES decision at November, 1994 convention held in Florida, to allow to sell live white rhinos, is however, unfortunate as the decision was based on fudged figures. The aerial survey had figured only 1,210 rhinos as against 6,300 claimed by authorities of S. Africa. Although, South Africa is not only the largest reservoir of African rhinos, but also a powerful smuggling headquarters of horns, ivory, drugs, weapons and diamonds. The rhino population is at an extremely low level in Asia, mainly India, and continues to decrease in Africa as rampant poaching continues. The Rhino on Appendix I of the CITES includes all species that are threatened with extinction. It is believed that cover provided by CITES for trade of live rhino would definitely hasten the rhino's extinction.

India is a signatory to the CITES. Some restrictions on collection of plants and their products from wild populations and their export have been in force in India even before the ratification of the Convention. Carving and sale of African ivory has been banned in India from April 2, 1992, keeping in view the decision of the CITES to extend the ban on international trade in ivory of the African elephant.

Another organization, called TRAFFIC (Trade Record Analysis of Flora and Fauna in Commerce), monitors international trade of species. The organization was established in 1976 as a worldwide trade monitoring programme, jointly by WWF and IUCN. TRAFFIC-India was instituted in 1991 as part of TRAFFIC-International to be based in WWF-India headquarters, New Delhi. This division monitors trade in wildlife and its derivatives in the Indian context.

Appendix - 5

Some Interesting World Wide Web Sites

Sequence database searching

Database searching can be carried out to search for sequences similar to the sequence that interests you. The databases can also be utilized to find out the sequence of specific genes which are utilized for vector construction. Any database sequence and its associated documentation can be retrieved and written into a personal file or a personal data library. The National Center for Biotechnology Information (NCBI) in the NIH (National Institute of Health web (www) and others can be accessed through the Genetics Computer Group (GCG). Programs like the BLAST (Basic Local Alignment Search Tool), Entrez and NLM PubMed (MEDLINE) can be used to search databases maintained at the NCBI over the internet.), Bethesda, Maryland, USA maintain several databases. There are several programs that can be used to search these databases and some of these can be searched through the world wide ograms like FastA, TFastA and several others can be used to access the GCG database.

Available databases

Several databases are maintained throughout the world: GenBank, PIR-Protein, SWISS-PROT, EMBL, GenEMBL (specifying both the GenBank and EMBL) etc. GenBank is the NIH genetic sequence database, which is an annotated collection of all publicly available DNA sequences. GenBank is part of the International Nucleotide Sequence Database Collaboration, which is comprised of the DNA DataBank of Japan (DDBJ), the European Molecular Biology Laboratory (EMBL), and GenBank at NCBI. These three organizations exchange data on a daily basis.

Most journals require submission of sequence information to a database prior to publication so that an accession number may appear in the paper. The methods of submission of sequences vary with different databases. The databases offer services like PubMed MEDLINE, the Entrez search system, BLAST sequence similarity searching and other E-mail servers. In addition to the DNA sequence databases, there are protein sequence databases like the SWISS-PROT and networks which can be used for the prediction of secondary structure of proteins like the Protein Design Group, EMBL, Heidelberg, Europe. HEADLINE can be used to search for research literature on any topic.

General NCBI information can be obtained over the internet (www) at info@ncbi.nlm.nih.gov and the NCBI home page comes up. The Entrez can be clicked on if the need is to find the sequence of a particular gene. Entrez needs information about the name of the gene etc to start the search. The sequence information can then be downloaded as needed. On the other hand if a search needs to be made to find out any similar sequences to a particular protein, the BLAST program is utilized. In this program, the choice can be made to locate protein-protein similarity by specifying blastp and swissprot or Gen EMBL. The similarity sequences in nucleotide databases can also be specified by choosing the options. The information can be downloaded and utilized. The databases of fasta or tfasta can be searched for similarity sequences through the GCG programs. The GCG program is available only on the Wisconsin package. This runs on UNIX and one can logon to the system over the network by using a terminal emulation software eg.: EM340.

Interesting WWW sites

There are several interesting **World Wide Web** sites that can be explored by a scientific worker in the areas of Molecular Biology and Biotechnology. Some of these are given below.

1. http:// webcrawler.cs.washington.edu/webcrawler/webquery.html
   [*For general information.*]
2. http:// www.public.iastate.edu/~pedro/research_tools.html
   [*Molecular biology info.*]
3. http:// macserver.molbio.gla.ac.ukbest.gdb.org/best.html
   [*List of researchers in Molecular Biology.*]
4. http:// www.atcg.com/atcg
   [*Info on restriction enzymes.*]
5. http:// alpha.genebee.msu.su:80/supplier

   or
6. http:// web.frontier.net/MEDMarket/indexes/indexmfr.html
   [*Biotech companies and suppliers of reagents and chemicals.*]
7. http:// www.atcc.org/atcc.html
   [*American type culture collection of plasmids and strains.*]
8. http:// cgsc.biology.yale.edu/top.html
   [*Database of E. coli strains & genes.*]
9. http:// Yahoo. Com. Science search.
10. http://www.biotech.nwu.edu/nsf/weil/html
11. http://www.biotech.nwu.edu/nsf/callahan.html.
12. http://www.comunidadandina.org.
13. http://www.biodiv.org/chm/conv/art19.htm.
14. http://www.telegraph.co.uk.
15. http://www.bio.indiana.edu/people/terminator.html.
16. FAO (1999). Committee on Agriculture – Biotechnology, Rome
    http://www.fao.org/unfao/bodies/coag/coag15/x0074e.htm.

17. GRAIN (1995). Genetic Action Resources International.
http://www.grain.org/publications/reports/special.htm.

18. GRAIN (1998).
http://www.grain.org/publications/gtbc/issue3.htm.

19. Alarm over 'Frankenstein' foods.
http://www.telegraph.co.uk.

20. Ortiz, R. (1998). Critical role of plant biotechnology
Electronic Journal of Biotechnology http://www.ejb.org/content/vol1/issue3/full/7/

21. Bolivian farmers demand researchers drop patent on Andean food crop.
http://www.rafi.org/genotype/970618quin.html.

22. Biotechnology and ethics: a blueprint for the future.
http://www.biotech.nwu.edu/nsf/sheldon.html.

23. **Canadian Pest Management Regulatory Agency**
http://www.hc-sc.gc.ca/pmra-arla/qcont-e.html

24. **Health Canada Online**
http://www.hc-sc.gc.ca/english/

25. http://infosrv1.ctd.ornl.gov/TechResources/Human_Genome/publicat/primer/prim1.html

26. **Marker-Assisted Selection: Fast Track to New Crop Varieties Ag-west Biotech Inc.**
http://www.agwest.sk.ca/infsept98.html

27. **Natural Gene Transfers: How Nature Engineers Plants Ag-west Biotech Inc.**
http://www.agwest.sk.ca/genetrans.html

28. **Office of Biotechnology, Canadian Food Inspection Agency**
http://www.cfia-acia.agr.ca/english/ppc/biotech/bsco.html

29. **Primer on Molecular Genetics (Human Genome Management Information System, U.S. Dep't. of Energy)**

30. **The 1998-9 Citizens' Guide to Biotechnology (Biotechnology Industry Organization, Washington DC)**
http://www.bio.org/whatis/guidecit.html

31. **Transgenic Plants Ag-west Biotech Inc.**
http://www.agwest.sk.ca/transgenic_plants.html

32. Organic Farming Research Foundation
mark@ofrf.org, http://www.ofrf.org/

INTERNET ADDRESSES RELEVANT TO GENOME RESEARCH

Description	Address
Site relevant to the Human Genome Project	
CEPH (YAC mapping)	http://www.cephb.fr/bio/ceph-genethon-map.html
dBEST (EST sequences)	http://www.ncbi.nih.gov/dbEST/lindex.html
Genethon (SSLP maps)	http://www.genethon.fr/
Radiation hybrid database	http://www.ebi.ac.uk/RHdb
Whitehead Institute (physical mapping)	http.//www.genome.mit.edu
Sites relevant to other eukaryotic genome projects	
Microbial Eukaryotes	
Candido albicans	http://glamdring.ucsd.edu/others/dsmith/dictydb.html
Dictyostelium discoideum	http://glamdring.ucsd.edu/others/dsmith/dictyab.html
Plasmodium falciparum	http://ben.vub.ac.be/malaria/mad.html
	http://www.wehi.edu.au/biology/malaria/who.html
	http://parasite.arf.efl.edu/malaria.html
Saccharomyces cerevisiae	http://genome-www.stanford.edu/Saccharomyces/
	http://www.mips.biochem.mpg.de/
	http://www.sanger.ac.uk/Projects/
Schizosaccharomyces pombe	http://www.mips.biochem.mpg.de/
Invertebrates	
Bombyx mori	http://www.ab.a.u-tokyo.ac.jp/sericulture/shimada.html
Caenorhabditis	http://www.sanger.ac.uk/Projects/C_elegans/
	http://moulon.inra.fr/acedb/acedb.html
	http://www.ddbj.nig.ac.jp/htmls/c-elegans/html/CE_INDEX.html
Drosophila melanogaster	http://flybase.bio.indiana.edu/
Mosquito	http://klab.agsci.colostate.edu/
Vertebrates	
Chicken	http://www.ri.bbsrc.ac.uk/chickmap/chickgbase/manager.html
Cow	http://locus.jouy.inra.fr/cgi-bin/bovmap/lintro2.pl
Dog	http://mendel.berkeley.edu/dog.html
Mekada fish	http://nioll.bio.nagoya-u.ac.jp:8000/
Mouse	http://www.genome.wi.mit.edu/cgi-bin/mouse
	http://www.genethon.fr

Description	Address
	http://www.informatics.jax.org/
Pig	http://www.ri.bbsrc.as.uk/pigmap/pigbase/pigbase.html
Pufferfish	http://fugu.hgmp.mrc.ac.uk
Rat	http://ratmap.gen.gu.se/
Sheep	http://dirk.invermay.cri.nz/
Zebrafish	http://zfish.uoregon.edu/
Plants	
Arabidopsis thalliang	http://lenti.med.umn.edu/arabidopsis/arab/arab_top_page.html
	http://cbil.humgen.upenn.edu/-atgc/ATGCUPhtml
	http://www.arabidopsis.com/bb/b950928z.html
	http://nasc.nott.Ac.uk/JIC-contigs

Studies in Biotechnology Series **No. 1**

TRANSGENIC PLANTS

R. Ranjan

1. Introduction
2. Identification and Isolation of Genes
3. Gene Transfer Methods
4. Foreign Gene Expression in Plant
5. Engineered Resistance Against Fungal Pathogens
6. Engineered Resistance Against Virus Diseases
7. Engineered Resistance Against Bacterial Pathogens
8. Engineered Resistance Against Pest
9. Engineered Resistance Against Herbicides
10. Improvement of *Rhizobium* Inoculants by Genetic Engineering
11. Plant Genetic Engineering on Foods and Nutrition
12. Safety and Public Acceptance of Transgenic Plants
13. References

Studies in Biotechnology Series **No. 2**

HAIRY ROOT CULTURE AND SECONDARY METABOLITES PRODUCTION

Usha Mukundan / Himanshu G. Dawda / Seema Ratnaparkhi

1. Crown Gall and Hairy Root Syndrome
2. Establishment and Growth Characteristics of Hairy Roots
3. Secondary Metabolite Production by Hairy Root Cultures
4. Scaling up of Hairy Root Cultures and Future Prospects
5. References

Studies in Biotechnology Series **No. 3**

SYNTHETIC SEEDS FOR COMMERCIAL CROP PRODUCTION

U. Kumar

1. Discovery and Future Perspectives of Synthetic Seeds
2. Somatic Embryogenesis: Principles, concepts and Applications
3. Plant Tissue Culture Principles and Methodology
4. Production of Synthetic Seeds
5. Economics of Tissue Culture and Synthetic Seed
6. References

Studies in Biotechnology Series **No. 4**

GENETIC ENGINEERING AND ITS APPLICATIONS

P. Joshi

1. Introduction
2. Gene Organization and Expression
3. Enzymes in Genetic Engineering
4. Gene Cloning Vectors
5. Gene Isolation, Identification and Synthesis
6. Cloning of Specific Gene
7. Specific Gene Transfer
8. Expression of Induced Genes
9. Applications of Genetic Engineering
10. Perspectives
11. References

Studies in Biotechnology Series **No. 5**

TRANSGENIC PLANTS: PROMISE OR DANGER

B. L. Kakralya / Ishita Ahuja

Studies in Biotechnology Series **No. 7**

FISH BIOTECHNOLOGY

M. M. Ranga / Q. J. Shammi

1. Fish, Fisheries and Biotechnology
2. Animal Cell and Tissue Culture
3. Fish Genetics and Development of Transgenic Fishes
4. Enzymes in Genetic Engineering (Nucleic Acid Enzymology)
5. Fish Genomics: Trends and Prospects
6. Gene Cloning Vectors
7. Techniques for Genetic Engineering
8. Gene Cloning
9. Cryopreservation in Fishes
10. Techniques and Genetic Requirements for Transgenic Fish Production
11. Applications of Transgenic Fishes and Biotechnology
12. Biotechnology in Health Management for Aquaculture
13. Glossary
14. Selected Bibliography